Advances in Experimental Medicine and Biology

Volume 1446

Advances in Experimental Medicine and Biology provides a platform for scientific contributions in the main disciplines of the biomedicine and the life sciences. This series publishes thematic volumes on contemporary research in the areas of microbiology, immunology, neurosciences, biochemistry, biomedical engineering, genetics, physiology, and cancer research. Covering emerging topics and techniques in basic and clinical science, it brings together clinicians and researchers from various fields.

Advances in Experimental Medicine and Biology has been publishing exceptional works in the field for over 40 years, and is indexed in SCOPUS, Medline (PubMed), EMBASE, Reaxys, EMBiology, the Chemical Abstracts Service (CAS), and Pathway Studio.

2022 CiteScore: 6.2

Guoyao Wu
Editor

Nutrition and Metabolism of Dogs and Cats

Springer

Editor
Guoyao Wu
Department of Animal Science
Texas A&M University
College Station, TX, USA

ISSN 0065-2598 ISSN 2214-8019 (electronic)
Advances in Experimental Medicine and Biology
ISBN 978-3-031-54191-9 ISBN 978-3-031-54192-6 (eBook)
https://doi.org/10.1007/978-3-031-54192-6

This Springer imprint is published by the registered company Springer Nature Switzerland AG
The registered company address is: Gewerbestrasse 11, 6330 Cham, Switzerland

Contents

Chapter 1
Recent Advances in the Nutrition and Metabolism of Dogs and Cats

Guoyao Wu

Abstract Domestic dogs (facultative carnivores) and cats (obligate carnivores) have been human companions for at least 12,000 and 9000 years, respectively. These animal species have a relatively short digestive tract but a large stomach volume and share many common features of physiological processes, intestinal microbes, and nutrient metabolism. The taste buds of the canine and feline tongues can distinguish sour, umami, bitter, and salty substances. Dogs, but not cats, possess sweet receptors. α-Amylase activity is either absent or very low in canine and feline saliva, and is present at low or substantial levels in the pancreatic secretions of cats or dogs, respectively. Thus, unlike cats, dogs have adapted to high-starch rations while also consuming animal-sourced foods. At metabolic levels, both dogs and cats synthesize de novo vitamin C and many amino acids (AAs, such as Ala, Asn, Asp, Glu, Gln, Gly, Pro, and Ser) but have a very limited ability to form vitamin D_3. Compared with dogs, cats have higher requirements for AAs, some B-complex vitamins, and choline; greater rates of gluconeogenesis; a higher capacity to tolerate AA imbalances and antagonism; a more limited ability to synthesize arginine and taurine from glutamine/proline and cysteine, respectively; and a very limited ability to generate polyunsaturated fatty acids (PUFAs) from respective substrates. Unlike dogs, cats cannot convert either β-carotene into vitamin A or tryptophan into niacin. Dogs can thrive on one large meal daily and select high-fat over low-fat diets, whereas cats eat more frequently during light and dark periods and select high-protein over low-protein diets. There are increasing concerns over the health of skin, hair, bone, and joints (specialized connective tissues containing large amounts of collagen and/or keratin); sarcopenia (age-related losses of skeletal-muscle mass and function); and cognitive function in dogs and cats. Sufficient intakes of proteinogenic AAs and taurine along with vitamins, minerals, and PUFAs are crucial for the normal structures of the skin, hair, bone, and joints, while mitigating sarcopenia and cognitive dysfunction. Although pet owners may have different perceptions about the feeding and management practice of their dogs and cats, the health and well-being of the companion animals critically depend on safe, balanced, and nutritive foods. The new knowledge covered in this volume of *Adv Exp Med Biol* is essential to guide

G. Wu (✉)
Department of Animal Science, Texas A&M University, College Station, TX 77843, USA
e-mail: g-wu@tamu.edu

G. Wu (ed.), *Nutrition and Metabolism of Dogs and Cats*, Advances in Experimental Medicine and Biology 1446, https://doi.org/10.1007/978-3-031-54192-6_1

the formulation of pet foods to improve the growth, development, brain function, reproduction, lactation, and health of the companion animals.

Keywords Analysis · Brain · Connective tissues · Digestion · Feeding · Health · Metabolism · Nutrition

Abbreviations

AA	Amino acid
EAA	Nutritionally essential amino acid
GIT	Gastrointestinal tract
HPLC	High-performance liquid chromatography
ME	Metabolizable energy
NEAA	Nutritionally nonessential amino acid
NRC	National Research Council
OPA	*o*-Phthaldialdehyde
PUFAs	Polyunsaturated fatty acids

1.1 Introduction

Dogs and cats are two different mammalian species of the same order (i.e., Carnivora) in accordance with taxonomy. The domestic dog (*Canis familiaris*) is a domesticated descendant of the gray wolf (an obligate carnivore; Bergström et al. 2020). Dogs and wolves share 99.9% of their DNAs and differ in 29 genes, including 10 genes related to starch digestion and fat metabolism (Axelsson et al. 2013). The domestic cat (*Felis catus*) is the only domesticated species in the family Felidae (Clutton-Brock 1999). Dogs and cats have been human companions for at least 12,000 and 9000 years, respectively (Perri 2016; Serpell 2015) and have also been used as animal models in biomedical research to discover cures for human diseases. In recent years, the numbers of dogs and cats, along with the pet food market, have increased substantially in many countries, reaching 510 and 400 million globally, respectively, in 2020 (Li and Wu 2023).

There are approximately 450 breeds of dogs (Ostrander et al. 2019) and 60 breeds of cats (Driscoll et al. 2009) with different shapes, sizes, and colors worldwide. Puberty or sexual maturity occurs in dogs and cats at 6–12 and about six months, respectively, whereas the life span of dogs and cats is 10–16 and 12–18 years, respectively, depending on size, breed, and nutritional status. Based on their anatomical, metabolic, and natural feeding characteristics, dogs (facultative carnivores) and cats (obligate carnivores) are classified as carnivores in animal nutrition (McDonald et al. 2011) and veterinary medicine (Dyce et al. 1996) textbooks. The National Research

Council (NRC 2006) recognizes that the dog is a carnivore anatomically but has many metabolic characteristics of omnivores. Due to differences in the interpretation and understanding of biology, authors of scientific papers have described the dogs as either a carnivore (e.g., Jardim-Messeder et al. 2017; Yang 2014) or an omnivore (Levillain et al. 2008; MacDonald et al. 1984). This is purely a matter of definition and does not dispute the fact that the dog has evolved to effectively use high-starch rations (e.g., 30% starch, on a dry matter basis; Félix et al. 2012). However, an animal that can digest large amounts of starch-rich foods and consume animal-sourced foods as part of its diet may not necessarily be classified as an omnivore in biology. For example, the rainbow trout, which can use high-starch rations (e.g., 36% starch, on a dry matter basis) is a carnivore, but not an omnivore, in taxonomy (Li et al. 2022).

Dogs and cats share many common features of nutrition and metabolism, but dogs differ from cats in some characteristics. The formulation of current commercial foods for dogs and cats is based on available findings from scientific research, but no product is ideal for all animals due to differences in their genetics, behaviors, physiological conditions, ambient environments, and diseased states (Larsen and Oberbauer 2023; Zoran 2023). In addition, most homemade pet foods, which are increasingly popular among owners of companion animals, are imbalanced in nutrients and may even contain toxic substances (Larsen and Oberbauer 2023). Articles of select topics are highlighted in this volume of *Advances in Experimental Medicine and Biology* (Connolly and Wu 2024; He et al. 2024; Ing and Steiner 2024; Li and Wu 2024a, b; Lopez et al. 2024; Naughton et al. 2024; Singh et al. 2024; Stockman 2024; Wu 2024b). As for other mammals (Wu 2022a), adequate knowledge about this fundamental knowledge is essential to commercially manufacture and home-make nutritive pet foods and to feed and care for dogs and cats for ensuring their optimal growth, development, reproduction, lactation, and health.

1.2 Gastrointestinal Tract, Nutrition, and Health of Dogs and Cats

Both dogs and cats have a relatively shorter digestive tract and a large stomach volume but relatively longer canine teeth and a tighter digitation of molars than omnivorous mammals such as humans and pigs (He et al. 2024). There are taste buds in the tongues of dogs and cats for the selection of food based on its flavor and texture. Canine and feline taste buds express transmembrane receptors to distinguish sour (PKD2L1), umami (mGluR4, Tas1R1 and Tas1R3), bitter (Tas2R), and salty (ENaC) substances. Unlike humans, dogs and cats are not highly sensitive to salt intake. Dogs, but not cats, possess sweet receptors (Tas1R2 and Tas1R3), so that dogs prefer sweet foods but cats are insensitive to the sweetness of dietary sugars (Pekel et al. 2020). Both dogs and cats are sensitive to the tastes of AAs (e.g., glutamate, aspartate, glycine, and arginine), other organic acids, and nucleotides.

Interestingly, there are epigenetic effects on taste responses to dietary nutrients in mammals including dogs and cats.

Both dogs and cats have no detectable or a very low activity of salivary α-amylase but dogs, unlike cats, possess a substantial activity of pancreatic α-amylase to digest dietary starch. However, both animals have active pancreatic secretions of lipases and other related enzymes (e.g., co-lipase and cholesterol ester hydrolase) for the intestinal digestion of dietary triacylglycerols (Wu 2018). Unlike many mammals, the vitamin B_{12} (cobalamin)-binding intrinsic factor is produced in dogs primarily by pancreatic ductal cells and to a lesser extent the gastric mucosa but in cats exclusively by the pancreatic tissue.

Microbes are present in the gastrointestinal tract of both dogs and cats to degrade some of dietary AAs, while fermenting dietary carbohydrates (particularly fibers), AAs, and lipids to short-chain fatty acids. For example, the gastric fluid of dogs (Buddington and Malo 2003) and cats (Adler et al. 2016) contains bacteria (~10^3/ml; primarily Gram positive and aerobic, lactic acid-producing, e.g., *Bifidobacterium*, *Lactobacillus*, and *Streptococcus*). In addition, there are bacteria (mainly *Streptococci* and *Lactobacilli*) in the fluid of the canine duodenum and jejunum (normally $\leq 10^4$/ml; Suchodolski 2011). For comparison, the duodenum fluid of cats fed a canned diet contains 2.2×10^5 to 1.6×10^8 bacteria, with anaerobic bacteria being 7.5×10^4 to 1.1×10^8, whereas the anaerobic bacteria of the feline duodenum are mainly *Bacteroides* spp, *Fusobacterium* spp, and *Eubacterium*, but its aerobic bacteria are mainly *Pasteurella* species (Johnston et al. 1993). Furthermore, bacterial counts in the colon of dogs are 10^9 to 10^{11}/ml, with the main bacteria in the hindgut being Fusobacteria (24–40%), Bacteroidetes (32–34%), and Firmicutes (15–28%) (Suchodolski 2011). The large intestine of the cat has the same predominating bacterial species and the same pH as those in dogs (Maskell and Johnson 1993; Suchodolski 2011). In both dogs and cats, AAs (e.g., aspartate, glutamate and glutamine) and short-chain fatty acids (acetate, propionate, and butyrate) are the major metabolic fuels for enterocytes and colonocytes, respectively, to maintain the integrity, health, and function of these cells.

Gastrointestinal disorders (e.g., vomiting, diarrhea, anorexia, and inflammation) are among common clinical signs in dogs and cats, resulting in malnutrition and weight loss (Ing and Steiner 2024). Some causes of chronic diarrhea are within the gastrointestinal tract while others are secondary to dysfunction outside the digestive system. Useful biomarkers of chronic intestinal disease (enteropathy) exist in serum (e.g., low concentrations of vitamin B_{12}, albumin, calcium, cholesterol, methylmalonic acid, folate, and α1-proteinase inhibitor) and feces (e.g., high concentrations of α1-proteinase inhibitor, calprotectin, and calgranulin C). After determination that the clinical signs are due to primary gastrointestinal disease and that there is no parasitism, specific diets (e.g., hydrolyzed protein diets, highly digestible and fiber-enhanced diets, and probiotics and/or prebiotics). For chronic enteropathies, the prognosis is generally good for symptom resolution or at least improvement. When dietary interventions are not effective, antibiotic, anti-inflammatory, or immunosuppressant therapy may be adopted. Pancreatitis is a common gastrointestinal disease in dogs and cats, and many patients with mild to moderate diseases can be successfully treated with a low-fat and AAs-adequate diet. A novel pharmaceutical, fuzapladib (Panoquell-CA1) appears to be promising for treating more severe forms of acute pancreatitis in dogs. Thus, nutrition and gastrointestinal health are closely linked.

1.3 Characteristics of Nutrient Metabolism in Dogs and Cats as the Biochemical Basis for Their Dietary Requirements for Nutrients

Dogs and cats have evolved differentially in some aspects of nutrition, metabolism, chemical sensing, and feeding behavior (Li and Wu 2024a). For example, dogs have adapted to omnivorous diets containing taurine-abundant meat and starch-rich plant ingredients, but domestic cats must consume animal-sourced foods for survival, growth, and development. In addition, commercial foods for cats normally contain small amounts of dietary fibers that are both highly fermentable (water-soluble) and less fermentable (water-insoluble) in the large intestine (Zoran 2023).

Both dogs and cats synthesize many AAs (such as Ala, Asn, Asp, Glu, Gln, Gly, Pro, and Ser) from other proteinogenic AAs; palmitate from acetyl-CoA; glucose from glucogenic AAs, lactate, glycerol, propionate, and odd-carbon number fatty acids; and vitamin C from glucose (Che et al. 2021; Li and Wu 2024a; Oberbauer and Larsen 2021). However, these two species of animals have a limited ability to form de novo (a) arginine due to low activities of key synthetic enzymes; and (b) vitamin D_3 due to low concentrations of 7-dehydrocholesterol and a high activity of 7-dehydrocholesterol Δ^7 reductase to rapidly convert 7-dehydrocholesterol into cholesterol in the skin (How et al. 1994). Compared with dogs, cats have greater endogenous nitrogen losses and higher dietary requirements for AAs (particularly arginine, taurine and tyrosine), B-complex vitamins (niacin, thiamin, folate and biotin), and choline; exhibit greater rates of gluconeogenesis; are less sensitive to AA imbalances and antagonism; are more capable of concentrating urine through renal reabsorption of water; and cannot tolerate high levels of dietary starch due to limited pancreatic α-amylase activity. In addition, most breeds of dogs can form sufficient taurine from cysteine, whereas young adult dogs have high abilities to synthesize arachidonic acid from linoleic acid; eicosapentaenoic acid and docosahexaenoic acid from α-linolenic acid; all-*trans*-retinol from β-carotene; and niacin from tryptophan. However, the activities of these synthetic pathways are either absent or limited in all age groups of cats due to (a) little or low activities of key enzymes (including pyrroline-5-carboxylate synthase, cysteine dioxygenase, Δ^6-desaturase, β-carotene dioxygenase, and quinolinate phosphoribosyltransferase) and (b) the rapid removal of metabolic intermediates by competing enzymes (e.g., in the case of niacin synthesis). For example, unlike dogs, cats cannot convert either β-carotene into vitamin A or L-tryptophan into niacin. Note that golden retrievers, American Cocker Spaniels, and a small population (1.3–2.5%) of the Newfoundlands may be genetically predisposed to taurine deficiency possibly due to gene mutations (Backus et al. 2003, 2006; Kaplan et al. 2018; Kittleson et al. 1997).

Dogs can thrive on one large meal daily, select high-fat over low-fat diets, and consume sweet substances. By contrast, cats eat more frequently during light and dark periods, select high-protein over low-protein diets, refuse dry food, enjoy a consistent diet, and cannot taste sweetness. Thus, cats are not a smaller size of dogs and should not be fed canine foods.

Animals (including dogs and cats) have dietary requirements for AAs, but not nitrogen or protein (Wu 2018). In cats, a deficiency of dietary taurine results in both blindness and cardiomyopathy (Zoran 2023). Of particular note, over the past decade, there has been active debate regarding a possible role for a deficiency of dietary taurine in the pathogenesis of dilated cardiomyopathy in dogs fed plant-based foods with no or a limited amount of animal-sourced ingredients (Mansilla et al. 2019; McCauley et al. 2020; Walker et al. 2022). This issue can be resolved through both dietary supplementation with taurine to affected dogs and the assessment of their cardiovascular function. Adequate amounts and proportions of dietary AAs are crucial for the growth and health of dogs and cats (Singh et al. 2024). Nitrogen balance, growth, and metabolic studies have been useful to assess the requirements of these animals for AAs. In recent years, direct or indirect indicator AA oxidation methods have been employed to identify the requirements of dogs for lysine, methionine, phenylalanine, threonine, and tryptophan. Such techniques are being adopted for feline studies. To ensure the physiological requirements of domestic cats and dogs are met, data on true digestibilities of dietary nutrients (including AAs) are needed and may be obtained through ileal cannulation. Based on ethical considerations, in vivo studies with mink (another carnivorous species) and in vitro digestion methods can provide insights into the digestibilities of food ingredients in dogs and cats. The digestible indispensable AA score (DIAAS) has been proposed to evaluate the quality of proteins in individual or mixed food ingredients. This approach is based on only the most limiting AA but ignores AA balances in diets and the dietary requirements of dogs and cats for all proteinogenic AAs (including AAs that are synthesized by animal cells) and taurine (Wu 2022b). In addition, adequate knowledge of AA chemistry, experience, and reliable methods are necessary for accurate analyses of AAs (particularly cysteine, methionine, proline, and tryptophan).

Aged adult dogs and cats have greater requirements for dietary AAs than young adults to mitigate insulin insensitivity, chronic inflammation, and sarcopenia (progressive loss of skeletal-muscle mass and function) (Li and Wu 2023; Stockman 2024). In addition, cognitive, digestive, cardiovascular, and immunological functions decline with aging in dogs and cats, just like other mammals such as humans, possibly due to the accumulation of a protein called β-amyloid in the brain, excessive oxidative stress in tissues, abnormalities of intestinal structures and functions, and impaired adaptive/innate immunity (Stockman 2024; Wu 2024b). Thus, there is growing interest in nutritional solutions and strategies to prevent age-related skeletal and neurological dysfunction, while increasing longevity and improving the quality of life. Notably, enhancing the intake of animal-sourced protein can minimize muscle loss in senior dogs and cats. Moreover, dietary supplementation with ω3 polyunsaturated fatty acids (e.g., docosahexaenoic acid), antioxidant vitamins, arginine, and taurine can help alleviate inflammation, osteoarthritis, cognitive dysfunction, as well as symptoms of renal disease, cancer, and gastrointestinal disorders (Pan et al. 2013, 2018a). Interestingly, consumption of medium chain triglycerides (substrates for the hepatic production of ketone bodies as metabolic fuels for the brain) can improve cognitive function in aged dogs (Pan et al. 2018b). Likewise, dietary supplementation with 0.12% tryptophan reduces territorial aggression, barking and staring behaviors in dogs, as well as vocalization, agonistic behavior, exploring, scratching, and agonistic interactions in cats (DeNapoli et al. 2000). Furthermore, oral administration of tyrosine can prevent stress-induced cardiac, behavioral, and cognitive abnormalities in animals, while improving their mood.

1.4 Nutrition and Health of Connective Tissues in Dogs and Cats

The skin and hair are two specialized connective tissues with important biological and social significance. In mammals, the hair is mostly made up of protein and the skin contains protein as the most abundant dry matter. The keratinization of the skin and the normal growth of hair may account for 25–30% of the daily protein requirements by adult cats and dogs (Connolly and Wu 2024). Furthermore, many AAs are degraded in both the skin and hair follicles. Thus, in adult dogs and cats, up to 35% of daily protein intake is used to maintain their skin and hair. Keratin is the major protein in both the skin's epidermis and the hair. This protein is rich in cysteine, serine, and glycine. Filaggrin, which aids on the aggregation and compaction of keratin intermediate filaments, is a physiologically important protein in the epidermis of the skin and hair. Filaggrin is produced by the granular cells of the skin's epidermis and by the inner root sheath cells of the hair follicles and contains high amounts of serine, glutamate, glutamine, glycine, arginine, and histidine. Collagen and elastin (highly abundant in glycine and proline plus 4-hydroxyproline) are the major proteins in both the dermis and hypodermis. Taurine and 4-hydroxyproline are abundant free AAs in the skin of dogs and cats, and free 4-hydroxyproline is also abundant in their hair. The epidermis of the skin synthesizes melanin (the pigment in the skin and hair) from tyrosine and produces *trans*-urocanate from histidine.

Bone is another connective tissue that provides the skeleton of animals and accounts for 12–15% of the body weight. The bone is composed of bone matrix and bone marrow and is required for locomotion and the protection of internal organs, as well as the maintenance of calcium and phosphorous homeostasis in the body. Chemically, there are 30% organic matter, 60% inorganic matter, and 10% water in the bone, with collagen and elastin representing 90% and 1% of the organic component in the bone matrix, respectively (Lopez et al. 2024). On the other hand, joints consist of cartilage, tendons, ligaments, synovial joint capsules and membranes, and a synovial joint cavity filled with synovial fluid. Joints are responsible for resisting mechanical stress/trauma, aiding in locomotion, and supporting the overall musculoskeletal system. Notably, collagen accounts for about 70% of dry matter in the cartilage. Sufficient intakes of dietary nutrients (e.g., AAs, vitamin C, and minerals) are essential for maintaining healthy bones and joints, while preventing disorders of bone (e.g., *fractures and n*utritional osteopathies*)* and joints (e.g., fractures and inflammation) that cause pain and reduce the quality of life (Connolly and Wu 2024).

Cysteine, glutamate, glutamine, glycine, proline, serine, and tyrosine have traditionally been considered as NEAAs in canine and feline nutrition (NRC 2006), but these AAs are physiologically essential for maintaining the structure, integrity, and health of the skin and hair (Connolly and Wu 2024), as well as the bone and joints (Lopez et al. 2024). In addition, taurine has also been considered as an NEAA for most breeds of dogs but it is nutritionally essential in cats, as indicated previously. Animal-sourced foodstuffs (e.g., feather meal and poultry by-product meal) are excellent sources of all these functional AAs in the diets of dogs and cats to maintain the structure, function, and health of the specialized connective tissues (Li et al. 2021; Li and Wu 2022). Note that plant-sourced foods contain relatively low amounts of cysteine, methionine, glycine, proline, and serine, and lack taurine. This knowledge must be used to formulate pet foods.

1.5 Brain Development and Cognitive Function

The brain is the central commander of all physical activities and the expression of emotions in mammals (e.g., dogs and cats). Its development and cognitive health critically depend on the neural network, including neurons, glial cells (namely, non-neuronal cells), and neurotransmitters (communicators between neurons). The latter include L-glutamate, L-aspartate, glycine, γ-aminobutyrate, D-aspartate, D-serine, NO, CO, H_2S, dopamine, norepinephrine, epinephrine, and serotonin (Wu 2024b). In addition, some non-neurotransmitter metabolites of proteinogenic AAs, such as taurine, creatine, and carnosine, also play important roles in brain development, cognitive health, behavior, and mood of dogs and cats. Much evidence shows that cats require dietary ω3 (α-linolenic acid, eicosapentaenoic acid, and docosahexaenoic acid) and ω6 (linoleic acid and arachidonic acid) PUFAs for the development of the central nervous system. As an essential component of membranes of neurons and glial cells, cholesterol is also crucial for cognitive development and function. Thus, inadequate protein or lipid nutrition will lead to mood disorders. Some AAs (e.g., arginine, glycine, methionine, serine, taurine, tryptophan, and tyrosine) can help to alleviate behavioral and mood disorders (e.g., depression, anxiety and aggression). As abundant providers of all these functional AAs and lipids, animal-sourced foods (e.g., liver, intestinal mucosa, and meat) play important roles in brain development, cognitive function, and mood of dogs and cats. This may explain, in part, why dogs and cats prefer to eat visceral organs of their prey. Adequate provision of all nutrients to the companion animals in all phases of the life cycle (pregnancy, lactation, postnatal growth, and adulthood) is essential for optimizing their neurological health, while preventing cognitive dysfunction and abnormal behavior.

1.6 Feeding Frequencies and Management of Dogs and Cats

Dogs can rapidly consume a large meal that is sufficient to meet daily nutrient requirements, but they usually eat 4 to 8 meals/day and drink water during the light period (NRC 2006). In contrast, cats do not have clear-cut circadian rhythms, although some cats are nocturnal (e.g., resting during the daytime and going out or walking at night) and their existing rhythms can be altered by minor disturbances such as human noises (NRC 2006). In addition, cats generally eat frequent, small meals and drink water throughout the day and night, and can voluntarily eat 12–20 meals/day. Furthermore, sufficient consumption of water plays an important role in maintaining whole-body health, while preventing urinary and kidney diseases.

Dietary intakes of energy and other nutrients (particularly excessive calcium, phosphorus, and magnesium) by both dogs and cats must be controlled to prevent obesity and renal disease. Their appetite is regulated by (a) signals produced in the hypothalamus and peripheral organs (e.g., the digestive tract and white adipose tissue), as well as (b) the concentrations of glucose, fatty acids, AAs, and their metabolites in the blood (Li and Wu 2024a). Dogs prefer a high-fat diet over a low-fat diet and select diets containing 25–30% of energy as protein, and can eat powdered, dry, semi-moist, or canned diets. In contrast, cats are picky eaters, preferring moist and warm meat with good palatability factors (e.g., high animal-sourced fats, high protein, and high taurine) as food. Indeed, cats refuse foods that are low in animal fat regardless of dietary protein content (Zoran 2023). Excess consumption of high palatable foods can increase risks for intestinal dysfunction and whole-body insulin resistance in animals. Management of dog and cat feeding is increasingly important (Naughton et al. 2024), as more and more dogs and cats have become obese over the past decade because of overeating and a lack of exercise. Both dogs and cats prefer to eat meat over starch-rich foods. Sufficient provision of arginine along with taurine in well-balanced diets may be able to prevent the development of obesity and metabolic syndrome (including endothelial dysfunction) in dogs and cats, as reported for rats (Jobgen and Wu 2022a, b; Jobgen et al. 2003).

Naughton et al. (2024) reported living conditions (e.g., indoors and outdoors, living space available), fundamental care (e.g., feeding and exercise), and owners' perceptions of dog's preferences (e.g., living conditions) in the urban and rural areas of Northern Ireland. Interestingly, more female dogs were spayed in the city locations than on the farms. In addition, most respondents in the study of Naughton et al. (2024), regardless of their household location, declared that they did not monitor their animal's body weight or body condition. To feed their dogs, the respondents from the city locations mainly followed veterinary advice or instructions on food labels, whereas the respondents from the farm locations mostly relied on their visual inspection of dog conditions. The respondents from the city locations believed that their dogs are happier with living indoors, whereas the respondents from farm locations believed that dogs can get all the needed exercise if kept outdoors and are happier with living outdoors. Thus, there are differences in basic care and management of dogs by their owners between the city and farm locations.

1.7 Safety of Foods for Dogs and Cats

Both commercial and homemade petfoods must be cooked before they are fed to dogs and cats to prevent food-borne infections by microbial pathogens and improve nutrient digestibilities. In addition, pet foods contaminated with toxic substances can make dogs or cats sick and can even result in mortality. Thus, analysis of food is vital for the safe guarding of its quality. Gizzerosine [2-amino-9-(4-imidazolyl)-7-azanonanoic acid] is a toxic AA that might be formed from histamine and lysine at high temperatures during ingredient (e.g., fishmeal and meat and bone meal) processing and transportation, as well as product transportation and storage. Li and Wu (2024b) reported the development of a simple, rapid, sensitive, specific, and automated method for the analysis of gizzerosine in foodstuffs by high-performance liquid chromatography (HPLC). This technique involves precolumn derivatization with *o*-phthaldialdehyde (OPA) in the presence of *N*-acetylcysteine (instead of the usual 2-mercaptoethanol or ethanethiol reagent). OPA reacted with gizzerosine in an autosampler at room temperatures, and their derivative was directly injected into the HPLC column. The highly fluorescent gizzerosine-OPA derivative was well separated from the OPA-derivatives of all natural AAs known to be present in physiological fluids (e.g., plasma), proteins, and foodstuffs and was detected at an excitation wavelength of 340 nm and an emission wavelength of 450 nm. The detection limit for gizzerosine was at least 6 pmol/ml in an assay solution (HPLC vial) or at least 0.09 pmol per injection into the HPLC column. When gizzerosine was extracted from foodstuffs, its detection limit was at least 875 pmol/g foodstuff or at least 0.21 mg/kg foodstuff. The total time for chromatographic separation (including column regeneration) was only 20 min per sample. This routine HPLC method is useful for the analysis of gizzerosine in foodstuffs and animal tissues.

1.8 Conclusion

Optimal growth, development, and health of dogs and cats depend on adequate intakes of dietary nutrients (including AAs, carbohydrates, lipids, minerals, vitamins, and water). Dogs (facultative carnivores) and cats (obligate carnivores) have a relatively short digestive tract and share many common features of nutrition and metabolism (e.g., dietary requirements for proteinogenic AAs, as well as abilities to synthesize some AAs, glucose, and vitamin C in a tissue-specific manner). However, these two carnivores differ from each other in some aspects, including the secretion of pancreatic α-amylase, starch digestion, basal rates of whole-body AA oxidation, and the syntheses of arginine, taurine, PUFAs, vitamin A, and niacin from respective precursors. There is increasing concern over the health of the skin, hair, bone, and joints. These specialized connective tissues consist of proteins (e.g., collagen and keratin) as the major organic matters. Collagen and keratin are rich in glycine, proline, and 4-hydroxyproline, and keratin also contains high amounts of cysteine, glutamate, glutamine, serine, and tyrosine. All these AAs have been traditionally classified as nutritionally nonessential AAs, but along with other nutrients play important roles in maintaining the structure, integrity and health of all the connective tissues. Although pet owners may have different perceptions about how their dogs or cats should be fed or managed, adequate knowledge about nutrition and metabolism is crucial for the health and well-being of the companion animals. Finally, identifying toxic substances in pet foods through accurate analysis (e.g., the rapid and sensitive determination of gizzerosine by HPLC) is critical to ensure food safety for all dogs and cats worldwide. We hope that readers will benefit from reading papers in this special issue of *Advances in Experimental Medicine and Biology.*

Acknowledgements The author thanks Dr. Gonzalo Cordova (Editor of Springer Nature) for the opportunity and support of editing this special issue, as well as all authors and reviewers of its articles for their tireless efforts to make this publication possible. Appreciation is also extended to Mr. Karthikeyan Krishnan, Mr. Vishnu Muthuswamy, and their teams for their excellent work on coordinating manuscript submission and the production of all the articles. The author's work was supported by Texas A&M AgriLife Research (H-8200).

Conflict of Interest The author declares no conflict of interest.

References

Adler CJ, Malik R, Browne GV, Norris JM (2016) Diet may influence the oral microbiome composition in cats. Microbiome 4:23

Axelsson E, Ratnakumar A, Arendt M-L et al (2013) The genomic signature of dog domestication reveals adaptation to a starch-rich diet. Nature 495:360–364

Backus RC, Cohen G, Pion PD, Good KL, Rogers QR, Fascetti AJ (2003) Taurine deficiency in Newfoundlands fed commercially available complete and balanced diets. J Am Vet Med Assoc 223:1130–1136

Backus RC, Ko KS, Fascetti AJ, Kittleson MD, MacDonald KA, Maggs DJ, Berg JR, Rogers QR (2006) Low plasma taurine concentration in Newfoundland dogs is associated with low plasma methionine and cyst(e)ine concentrations and low taurine synthesis. J Nutr 136:2525–2533

Bergström A, Frantz L, Schmidt R et al (2020) Origins and genetic legacy of prehistoric dogs. Science 370:557–564

Buddington RK, Malo C (2003) Postnatal development of nutrient transport in the intestine of dogs. Am J Vet Res 64:635–645

Che DS, Nyingwa PS, Ralinala KM, Maswanganye GMT, Wu G (2021) Amino acids in the nutrition, metabolism, and health of domestic cats. Adv Exp Med Biol 1285:217–231
Clutton-Brock J (1999) Cats. In: A natural history of domesticated mammals, 2nd edn. Cambridge University Press, Cambridge, England, pp 133–140
Connolly ED, Wu G (2024) Functions and metabolism of amino acids in the hair and skin of dogs and cats. Adv Exp Med Biol 1446:135–154. https://doi.org/10.1007/978-3-031-54192-6_6
DeNapoli JS, Dodman NH, Shuster L, Rand WM, Gross KL (2000) Effect of dietary protein content and tryptophan supplementation on dominance aggression, territorial aggression, and hyperactivity in dogs. J Am Vet Med Assoc 217:504–508
Driscoll CA, Clutton-Brock J, Kitchener AC, O'Brien SJ (2009) The taming of the cat. Sci Am 300:68–75
Dyce KM, Sack WO, Wensing CJG (1996) Textbook of veterinary anatomy. W.B. Saunders Company, Philadelphia
Félix AP, Gabeloni LR, Brito CBM, Oliveira SG, Silva AVF, Maiorka A (2012) Effect of β-mannanase on the digestibility of diets with different protein sources in dogs determined by different methodologies. J Anim Sci 90:3060–3067
He W, Connolly ED, Wu G (2024) Characteristics of the digestive tract of dogs and cats. Adv Exp Med Biol 1446:15–38. https://doi.org/10.1007/978-3-031-54192-6_2
How KL, Hazewinkel HAW, Mol JA (1994) Dietary vitamin D dependence of cat and dog due to inadequate cutaneous synthesis of vitamin D. Gen Comp Endocrinol 96:12–18
Ing NH, Steiner JM (2024) The use of diets in the diagnosis and treatment of common gastrointestinal diseases in dogs and cats. Adv Exp Med Biol 1446:39–53. https://doi.org/10.1007/978-3-031-54192-6_3
Jardim-Messeder D, Lambert K, Noctor S, Pestana FM, de Castro Leal ME, Bertelsen MF, Alagaili AN, Mohammad OB, Manger PR, Herculano-Houzel S (2017) Dogs have the most neurons, though not the largest brain: Trade-off between body mass and number of neurons in the cerebral cortex of large carnivoran species. Front Neuroanat 11:118
Jobgen WS, Wu G (2022a) L-Arginine increases AMPK phosphorylation and the oxidation of energy substrates in hepatocytes, skeletal muscle cells, and adipocytes. Amino Acids 54:1553–1568
Jobgen WS, Wu G (2022b) Dietary L-arginine supplementation increases the hepatic expression of AMP-activated protein kinase in rats. Amino Acids 54:1569–1584
Jobgen WS, Lee M-J, Fried SK, Wu G (2023) L-Arginine supplementation regulates energy-substrate metabolism in skeletal muscle and adipose tissue of diet-induced obese rats. Exp Biol Med 248:209–216
Johnston K, Lamport A, Batt RM (1993) An unexpected bacterial flora in the proximal small intestine of normal cats. Vet Rec 132:362–363
Kaplan JL, Stern JA, Fascetti AJ, Larsen JA, Skolnik H, Peddle GD, Kienle RD, Waxman A, Cocchiaro M, Gunther-Harrington CT, Klose T, LaFauci K, Lefbom B, Lamy MM, Malakoff R, Nishimura S, Oldach M, Rosenthal S, Stauthammer C, O'Sullivan L, Visser LC, Williams R, Ontiveros E (2018) Taurine deficiency and dilated cardiomyopathy in golden retrievers fed commercial diets. PLoS ONE 13:e0209112
Kittleson MD, Keene B, Pion PD, Loyer CG (1997) Results of the multicenter spaniel trial (MUST): taurine- and carnitine-responsive dilated cardiomyopathy in American cocker spaniels with decreased plasma taurine concentration. J Vet Intern Med 11:204–211
Larsen JA, Oberbauer AM (2023) Dog nutrition. In: Phillips CJC (ed) Encyclopedia of animal nutrition, 2nd edition. CABI, Wallingford, pp 176–179
Levillain O, Rabier D, Duclos B, Gaudreau P, Vinay P (2008) L-arginine metabolism in dog kidney and isolated nephron segments. Metabolism 57:9–23
Li P, He WL, Wu G (2021) Composition of amino acids in foodstuffs for humans and animals. Adv Exp Med Biol 1332:189–210
Li P, Wu G (2022) Functional molecules of intestinal mucosal products in animal nutrition and health. Adv Exp Med Biol 1354:263–277

Li P, Wu G (2023) Amino acid nutrition and metabolism in domestic cats and dogs. J Anim Sci Biotechnol 14:19

Li P, Wu G (2024a) Characteristics of nutrition and metabolism in dogs and cats. Adv Exp Med Biol 1446:55–98. https://doi.org/10.1007/978-3-031-54192-6_4

Li P, Wu G (2024b) Analysis of gizzerosine in foodstuffs by HPLC involving pre-column derivatization with *o*-phthaldialdehyde. Adv Exp Med Biol 1446:237–250. https://doi.org/10.1007/978-3-031-54192-6_11

Li XY, Han T, Zheng SX, Wu G (2022) Hepatic glucose metabolism and its disorders in fish. Adv Exp Med Biol 1354:207–236

Lopez AN, Bazer FW, Wu G (2024) Functions and metabolism of amino acids in bones and joints of cats and dogs. Adv Exp Med Biol 1446:155–175. https://doi.org/10.1007/978-3-031-54192-6_7

Mansilla WD, Marinangeli CPF, Ekenstedt KJ, Larsen JA, Aldrich G, Columbus DA, Weber L, Abood SK, Shoveller AK (2019) The association between pulse ingredients and canine dilated cardiomyopathy: addressing the knowledge gaps before establishing causation. J Anim Sci 97:983–997

Maskell IE, Johnson JV (1993) Digestion and absorption. In: Burger I (ed) The Waltham book of companion animal nutrition. Pergamon Press, Oxford, UK, pp 25–44

MacDonald ML, Rogers QR, Morris JG (1984) Nutrition of the domestic cat, a mammalian carnivore. Annu Rev Nutr 4:521–562

McCauley SR, Clark SD, Quest BW, Streeter RM, Oxford EM (2020) Review of canine dilated cardiomyopathy in the wake of diet-associated concerns. J Anim Sci 98:skaa155

McDonald P, Edwards RA, Greenhalgh JFD, Morgan CA, Sinclair LA (2011) Animal nutrition, 7th edn. Prentice Hall, New York

National Research Council (NRC) (2006) Nutrient requirements of dogs and cats. National Academies Press, Washingto DC

Naughton V, Grzelak T, Naughton PJ (2024) Association between household location (urban vs. rural) and fundamental care provided to domestic dogs (*Canis familiaris*) in Northern Ireland. Adv Exp Med Biol 1446:217–236. https://doi.org/10.1007/978-3-031-54192-6_10

Oberbauer AM, Larsen JA (2021) Amino acids in dog nutrition and health. Adv Exp Med Biol 1285:199–216

Ostrander EA, Wang G-D, Larson G, Vonholdt BM, Davis BW, Jagannathan V, Hitte C, Wayne RK, Zhang Y-P (2019) Dog10K: An international sequencing effort to advance studies of canine domestication, phenotypes, and health. Natl Sci Rev 6:810–824

Pan Y, Araujo JA, Burrows J, de Rivera C, Gore A, Bhatnagar S, Milgram NW (2013) Cognitive enhancement in middle-aged and old cats with dietary supplementation with a nutrient blend containing fish oil, B vitamins, antioxidants and arginine. Br J Nutr. 110:40–49

Pan Y, Kennedy AD, Jönsson TJ, Milgram NW (2018a) Cognitive enhancement in old dogs from dietary supplementation with a nutrient blend containing arginine, antioxidants, B vitamins and fish oil. Br J Nutr 119:349–358

Pan Y, Landsberg G, Mougeot I, Kelly S, Xu H, Bhatnagar S, Gardner CL, Milgram NW (2018b) Efficacy of a therapeutic diet on dogs with signs of cognitive dysfunction syndrome (CDS): a prospective double blinded placebo controlled clinical study. Front Nutr 5:127

Pekel AY, Mülazımoğlu SB, Acar N (2020) Taste preferences and diet palatability in cats. J Appl Anim Res 48:281–292

Perri A (2016) A wolf in dog's clothing: Initial dog domestication and Pleistocene wolf variation. J Archaeol Sci 68:1–4

Serpell JA (2015) History of companion animals and the companion animal sector. In: Sandøe P, Corr S, Palmer C (eds) Companion animal ethics. Wiley, New York, pp 8–23

Singh P, Banton S, Bosch G, Hendriks WH, Shoveller AK (2024) Beyond the bowl: understanding amino acid requirements and digestibility to improve protein quality metrics for dog and cat foods. Adv Exp Med Biol 1446:99–134. https://doi.org/10.1007/978-3-031-54192-6_5

Stockman J (2024) Nutrition and aging in dogs and cats. Adv Exp Med Biol 1446:203–215. https://doi.org/10.1007/978-3-031-54192-6_9

Suchodolski JS (2011) Intestinal microbiota of dogs and cats: a bigger world than we thought. Vet Clin North Am Small Anim Pract 41:261–272

Walker AL, DeFrancesco TC, Bonagura JD, Keene BW, Meurs KM, Tou SP, Kurtz K, Aona B, Barron L, McManamey A, Robertson J, Adin DB (2022) Association of diet with clinical outcomes in dogs with dilated cardiomyopathy and congestive heart failure. J Vet Cardiol 40:99–109

Wu G (2018) Principles of animal nutrition. CRC Press, Boca Raton, Florida

Wu G (2022a) Nutrition and metabolism: foundations for animal growth, development, reproduction, and health. Adv Exp Med Biol 1354:1–24

Wu G (2022b) Amino acids: biochemistry and nutrition, 2nd edn. CRC Press, Boca Raton, Florida

Wu G (2024) Roles of nutrients in the brain development, cognitive function, and mood of dogs and cats. Adv Exp Med Biol 1446:177–202. https://doi.org/10.1007/978-3-031-54192-6_8

Yang B (2014) Transport characteristics of urea transporter-B. Subcell Biochem 73:127–135

Zoran DL (2023) Cat (Felis catus), essential nutrients. In: Phillips CJC (ed) Encyclopedia of animal nutrition, 2nd edition. CABI, Wallingford, pp 100–102

Chapter 2
Characteristics of the Digestive Tract of Dogs and Cats

Wenliang He, Erin D. Connolly, and Guoyao Wu

Abstract As for other mammals, the digestive system of dogs (facultative carnivores) and cats (obligate carnivores) includes the mouth, teeth, tongue, pharynx, esophagus, stomach, small intestine, large intestine, and accessory digestive organs (salivary glands, pancreas, liver, and gallbladder). These carnivores have a relatively shorter digestive tract but longer canine teeth, a tighter digitation of molars, and a greater stomach volume than omnivorous mammals such as humans and pigs. Both dogs and cats have no detectable or a very low activity of salivary α-amylase but dogs, unlike cats, possess a relatively high activity of pancreatic α-amylase. Thus, cats select low-starch foods but dogs can consume high-starch diets. In contrast to many mammals, the vitamin B_{12} (cobalamin)-binding intrinsic factor for the digestion and absorption of vitamin B_{12} is produced in: (a) dogs primarily by pancreatic ductal cells and to a lesser extent the gastric mucosa; and (b) cats exclusively by the pancreatic tissue. Amino acids (glutamate, glutamine, and aspartate) are the main metabolic fuels in enterocytes of the foregut. The primary function of the small intestine is to digest and absorb dietary nutrients, and its secondary function is to regulate the entry of dietary nutrients into the blood circulation, separate the external from the internal milieu, and perform immune surveillance. The major function of the large intestine is to ferment undigested food (particularly fiber and protein) and to absorb water, short-chain fatty acids (serving as major metabolic fuels for epithelial cells of the large intestine), as well as vitamins. The fermentation products, water, sloughed cells, digestive secretions, and microbes form feces and then pass into the rectum for excretion via the anal canal. The microflora influences colonic absorption and cell metabolism, as well as feces quality. The digestive tract is essential for the health, survival, growth, and development of dogs and cats.

Keywords Dogs · Cats · Digestive tract · Feeding · Health · Nutrition

W. He · E. D. Connolly · G. Wu (✉)
Department of Animal Science, Texas A&M University, College Station, TX 77843, USA
e-mail: g-wu@tamu.edu

G. Wu (ed.), *Nutrition and Metabolism of Dogs and Cats*, Advances in Experimental Medicine and Biology 1446, https://doi.org/10.1007/978-3-031-54192-6_2

Abbreviations

AA	Amino acid
BW	Body weight
CP Crude	Protein
DM	Dry matter
IF	Intrinsic factor
GIT	Gastrointestinal tract
SCFAs	Short-chain fatty acids
SGLT-1	Sodium-glucose linked transporter-1
TAGs	Triacylglycerols

2.1 Introduction

Based on their anatomical, metabolic, and natural feeding characteristics, dogs and cats are classified as carnivores in classic animal nutrition and veterinary medicine textbooks (Dyce et al. 1996; McDonald et al. 2011). The digestive system of these animals includes the mouth, teeth, tongue, pharynx, esophagus, stomach, small intestine, large intestine, and accessory digestive organs (salivary glands, pancreas, liver, and gallbladder). Both dogs (facultative carnivores) and cats (obligate carnivores) are endotherms, which use nutrient metabolism and heat dissipation mechanisms to maintain a stable internal temperature (Table 2.1). These two animal species have a relatively shorter digestive tract, a more rapid rate for foods to pass through the gastrointestinal tract (GIT), and a shorter time of food retention in the GIT, when compared with other nonruminant mammals such as humans and pigs (Morris 2002; Wu 2018). Most aspects of nutrient digestion and absorption, including the underlying processes and molecular mechanisms, are generally similar among mammals with different mature body weights (BWs), including dogs and cats (Case et al. 2011; Che et al. 2021; Oberbauer and Larsen 2021; Wu 2018). The digestive system of cats shares some similarities to, but also differs from, dogs in many aspects (Buff et al. 2014; Legrand-Defretin 1994; Zoran 2010). The following sections describe the salient characteristics of the digestive tracts of dogs and cats with an emphasis on the mouth, stomach, small intestine, and large intestine, so as to better understand their natural feeding behaviors and adaptation to various food sources with different digestibilities and nutritive values.

Table 2.1 Thermoneutral temperature zones as well as plasma glucose and amino acid concentrations in dogs, cats, and pigs[a]

Animal	Stage of life	Thermo neutral temp. (°C)	Body temp. (°C)	Glucose Conc, in plasma in the fasting state (mM)	Plasma concentrations of amino acids in the fed state (μM)[e]							
					Ala	Arg	Cit	Gln	Glu	Gly	Leu	Met
Dogs[b]	Neonatal	30	36	3.5–5.0	315 ± 46	278 ± 80	88 ± 39	564 ± 174	112 ± 82	283 ± 85	212 ± 39	52 ± 33
	Adult	20–25 –30–35	37.5–39.2	3.9–5.5	332 ± 162	104 ± 36	38 ± 12	658 ± 213	38 ± 12	171 ± 42	135 ± 48	52 ± 16
Cats[c]	Neonatal	34	37	3.5–5.0	–	–	–	–	–	–	–	–
	Adult	30–36	36.7–38.9	3.9–5.5	462 ± 160	95 ± 38	18 ± 6	664 ± 134	73 ± 38	398 ± 279	146 ± 49	64 ± 28
Pigs[d]	Neonatal	32–34	38.5–39.0	4.4–5.5	1049 ± 478	163 ± 63	89 ± 25	469 ± 120	156 ± 63	824 ± 237	186 ± 51	100 ± 51
	Adult	20–26	37	5.0–5.5	352 ± 53	165 ± 19	60 ± 7.5	491 ± 59	92 ± 15	688 ± 81	196 ± 18	47 ± 5.6

[a]Adapted from Kliegman et al. (1983); Lecheta et al. (2020); Levy et al. (2015); Pedraz and Mortola (1991); and Sagawa et al. (2009). The thermoneutral temperature zone is defined as the range of ambient temperatures within which the animal can maintain its normal body temperature through the control of sensible heat loss. Values are the means ± SD. [b]Facultative (optional) carnivores with the normal gestation length of 63 days (that can vary by breed and individual). The live-born litter size of dogs is usually 5–6 puppies and ranges from 1 to 15, depending on breeds, as well as maternal age, health, diet, and management

[c]Obligate carnivores with the normal gestation length of 65 days (that can vary by breed and individual). The live-born litter size of cats is usually 3–5 kittens and ranges from 1 to 12, depending on breeds, as well as maternal age, health, diet, and management

[d]The body temperature and thermoneutral zone of neonatal and adult pigs are adapted from Mount (1959) and Wu (2018), respectively

[e] Blazer-Yost and Jezyk (1979) for neonatal (< 2 weeks of age) and adult dogs (> 5 years of age), Heinze et al. (2009) for adult cats, Flynn and Wu (1996) for neonatal pigs (4 days of age), and Wu G (n = 8; unpublished observations) for adult pigs (20 months of age)

Conc. = concentrations; temp. = temperature; "–", data are not available

2.2 The Mouth of Dogs and Cats

An adult dog has 42 teeth within its mouth, including 12 incisors, 4 canines, 16 premolars, and 10 molars, to chew and crush solid food (Lee et al. 2020). An adult cat has the same number of incisor and canine teeth as those in dogs but only 10 premolars, and 4 molars, totaling 30 teeth (Lee et al. 2020). Compared with dogs, cats have relatively longer canine teeth, as well as shorter incisors and molars and, therefore, can effectively hunt their prey (e.g., rats and mice) and ingest their whole tissues (Pekel et al. 2020). Furthermore, in cats, as opposed to dogs, teeth do not play a part in hunting except to dispatch the prey. However, cats use sharp claws, which remain sharp by being retracted when not in use, to capture their prey, and then use their long canine teeth to sever the prey's spinal cord with a bite to the back of the neck. The cat's canines have root endings close to the tooth surface, allowing it to "feel" for spaces between vertebrae so that the spinal cord can be severed quickly and with less power if a crushing bite is used (Lombardero et al. 2021). Notably, like other mammalian carnivores but unlike omnivores and herbivores, the jaws of dogs and cats only allow for up and down motion, and these carnivores cannot chew side to side. Therefore, dogs and cats have a lower ability to grind food than most of other mammals.

The four pairs of common major salivary glands in dogs (Gil et al. 2018) and cats (Harrison et al. 2001) around the jaw, under the tongue, and near the throat are the parotid, mandibular, sublingual, and zygomatic glands. In addition, the cat has another major pair of the molar salivary gland that is located superficially near the angle of the mouth. These glands are located on either side of the mouth. Hundreds of minor salivary glands exist within the wall of the oral cavity and oropharynx; they are named based on anatomical location (e.g., labial, lingual, buccal, and palatine) and have no connective tissue capsules. The salivary glands secrete saliva to mix with food, lubricate the passage of the food to the stomach during swallowing, moisten and protect the oral mucous membrane, maintain tooth integrity, aid in nutrient digestion, and facilitate heat dissipation through the tongue (evaporative cooling) called "panting." Saliva also has the roles of buffering, the clearance of debris, and antimicrobial activity (via lysozyme).

Salivation is stimulated by the pleasant texture, smell, and amount [particularly dry matter (DM)] of food and increased dramatically in response to a rise in the ambient temperature. The saliva of dogs and cats contains lingual lipase [hydrolyzing triacylglycerols (TAGs)], proteases (hydrolyzing protein), and lysozyme (killing bacteria) (Nelson et al. 1977). The enzymatic activity of lysozyme in canine and feline saliva is greater than that in the human saliva (Tenovuo et al. 2000). There were reports that cats have either no (Larmas and Scheinin 1972) or a limited (McGeachin and Akin 1979) activity of α-amylase in their saliva. Similarly, there is no detectable activity (Chauncey et al. 1963) or protein (Pasha et al. 2018) of α-amylase in the saliva of dogs such as Labrador retrievers and Beagles. Interestingly, using the Western blot analysis and a validated spectrophotometric method, Contreras-Aguilar et al. (2017) reported the presence of α-amylase protein and a low enzymatic activity of α-amylase in the saliva of male beagles. This enzyme hydrolyzes starch and glycogen into disaccharides. There may be considerable inter-individual variations and differences in salivary α-amylase expression between breeds of animals including cats and dogs.

Taste buds in the tongue help dogs and cats select food based on its flavor and texture. During evolution, the dog has adapted to omnivorous diets containing taurine-rich meat and starch-abundant plant ingredients (Axelsson et al. 2013; Baker and Czarnecki-Maulden 1991). Taste buds of the canine (Kitchell 1978) and feline (Beauchamp et al. 1977) tongues express receptors (transmembrane proteins) to distinguish sour (PKD2L1), umami (mGluR4, Tas1R1 and Tas1R3), bitter (Tas2R), and salty (ENaC) substances. Unlike humans, dogs and cats do not develop very sensitive salt receptors (Ladd and Raisz 1949). Interestingly, dogs, but not cats, possess sweet receptors (Tas1R2 and Tas1R3; Li et al. 2005, 2006). Thus, dogs and cats are not highly sensitive to dietary salt in terms of chemical sensing. In addition, dogs prefer sweet foods, whereas cats are insensitive to dietary sugar (Bradshaw 2006). Of particular note, dietary salts can substantially enhance canine taste responses to sugars (Kumazawa and Kurihara 1990a), and there is synergism between dietary monosodium glutamate and nucleotides in canine taste nerve responses (Kumazawa and Kurihara 1990b). In contrast to salts and sweet sugars, both dogs (Ferrell 1984; Ugawa and Kurihara 1993) and cats (Bradshaw et al. 1996) are sensitive to the tastes of AAs (e.g., glutamate, aspartate, glycine, and arginine), other organic acids, and nucleotides. Interestingly, dietary salts had an enhancing effect on the taste of amino acids (AAs) in dogs but not in rats (Ugawa and Kurihara 1993). Physiological roles of AAs other than glutamate and aspartate for the selection of foods remain largely unknown (Wu 2020a,b). There are epigenetic effects on taste responses to dietary nutrients in mammals including dogs and cats (Kuo 1967).

2.3 The Stomach of Dogs and Cats

The stomach of dogs consists of distinct proximal and distal regions, whereas cats have a uniform gastric mucosa (Emås and Grossman 1967). The proximal region of the canine stomach and the feline stomach expands during food storage so that the stomach of both dogs and cats can accommodate a large meal at one time. This also allows the partially digested food to remain in the stomach for a prolonged period of time [~4–8 h in dogs (Bourreau et al. 2004) and ~3–6 h in cats (Goggin et al. 1999)], compared to ~0.5 h in humans and pigs. Bourreau et al. (2004) reported that, when adult dogs consumed a meal that met 50% of the daily energy requirements, large breeds of dogs (e.g., ~60 kg BW) retained food in the stomach for a longer period and had a lower rate of gastric emptying, compared with small breeds of dogs (e.g., 3.5 kg BW). Similarly, there is a strong inverse relationship between BW and gastric emptying time in adult humans weighing 59–93 kg (Lavigne et al. 1978). In dogs, gastric emptying is slowed by dietary fat (Lin et al. 1990; Meyer et al. 1994), increased by gastrin (a peptide hormone that stimulates the secretion of HCl by the parietal cells of the stomach and aids in gastric motility) (Jin et al. 1994), and influenced by meal size (Van Kruiningen et al. 1987).

The volume of the stomach of dogs with all BW ranges is ~90 ml/kg BW (Bartges 2003), whereas the volume of the feline stomach is reduced with age, being ~100, 70, 60, and 45 ml/kg BW at the BW of 0.5–0.99, 1.0–1.49, 1.5–3.99, and 4–6 kg, respectively (Bartges 2003). Like other carnivores, the stomach of dogs and cats is relatively simple, and its volume represents 60–70% of the total capacity of the digestive system (Bartges 2003), which is much greater than the value for nonruminant monogastric mammals (e.g., approximately 10% in adult pigs; Lærke and Hedemann 2012). The gastric fluid of dogs (Buddington 2003) and cats (Adler et al. 2016) contains bacteria (~10^3/ml; primarily Gram positive, aerobic, lactic acid-producing, e.g., *Bifidobacterium*, *Lactobacillus*, and *Streptococcus*), which may help to digest dietary fiber. Compared with adult dogs, adult cats have higher gastric fluid pH (Table 2.2). The stomach of dogs and cats secretes gastric lipase (partly hydrolyzing triacylglycerols (TAGs) into fatty acids and glycerol), pepsin (which begins digestion of proteins into polypeptides), hydrochloric acid, and some mucus which lubricates and protects the stomach wall (Emås and Grossman 1967; Nelson et al. 1977). In both dogs and cats, as in other mammals and in birds, gastric secretion is stimulated by protein intake and the volume of the meal (Wu 2018). In dogs, gastric lipase is inactive at pH 1.5 and 7.0 and is reduced substantially at pH 6. At pH 4, gastric lipase is 13 times more active with long-chain than short-chain TAGs (Carrière et al. 1991). The canine gastric lipase has a low activity toward long-chain TAGs but effectively hydrolyzes medium-chain TAGs. In dogs with pancreatic insufficiency, their ability to digest medium-chain TAGs and long-chain TAGs is reduced by about 25% and 95%, respectively (Rutz et al. 2004). In cats, gastric lipase is primarily localized to the epithelial cells of the gastric mucosa before weaning (Knospe and Plendl 1997) but is present mainly in gastric fluid after weaning (Pansu et al. 1987).

In most mammals, the gastric vitamin B_{12} (cobalamin)-binding intrinsic factor (IF) and haptocorrin (also known as the R-protein or transcobalamin I; another vitamin B_{12}-binding protein) for the digestion and absorption of vitamin B_{12} are released by gastric mucosal cells (Vaillant et al. 1990). Interestingly, in dogs, pancreatic ductal cells are the primary source of IF and, to a much lesser extent, the gastric mucosa. By contrast, in cats, there is no production of IF or haptocorrin by the gastric mucosa, and the IF in the lumen of the small intestine is derived exclusively from the pancreatic tissue (Siani et al. 2023).

2.4 The Small Intestine of Dogs and Cats

Like other carnivores (Herring et al. 2021; Li et al. 2021b), both dogs and cats have a relatively short GIT (NRC 2006). Larger breeds of dogs (> 25 kg BW as adults) have a proportionately smaller GIT, comprising 3–4% of BW, compared to 6–7% of BW in smaller breeds (e.g., medium breeds of dogs with 10–25 kg BW as adults; small breeds of dogs with <10 kg BW as adults; Meyer et al. 1993). Compared with adult

Table 2.2 The digestive tract of adult dogs, cats, and pigs[a]

Animal	Saliva pH	GIT wt (% of BW)	GIT filling (% of BW)	Stomach fluid pH	Gastric half-EMPT (min)	SI length	LI length	Liver wt (% of BW)	Luminal fluid pH			SI volume (% of GIT)	Retention time of food in whole GIT (h)
									Duo	Jej	Ile		
Dog	7.34–7.8	6–7[b] 3–4[c]	3–4[b] 1O[c]	2.05[f] 1.08–1.26[g]	72–240	3.9 m per 0.75 m BL[e]	0.5 m per 0.75 m BL[e]	6.82[d] 2.3–3.9[e]	5.7	6.4	6.6	25	23
Cat	7.5	5–6[d] 2–3[e]	3[e]	2.5	22–25[f] 449[g]	1.7 m per 0.5 m BL[e]	0.4 m per 0.5 m BL[e]	4.07[d] 2.46[e]	6.2	6.7	7.0	15	13
Pig	6.0–7.0	2.9[d] 2.3[e]	10[d] 20[e]	2–3[d] 2[e]	160–240[e]	10 m[d] 19 m[e] per 1 m BL	4 m[d] 5 m[e] per 1 m BL	2.7[d] 1.5[e]	6.1	6.6	6.7	51[d] 40[e]	43

[a]Adapted from Kararli (1995), Meyer et al. (1993), and NRC (2006) for dogs; Brosey et al. (2000) and Latimer (1967) for cats; and Gregory et al. (1990), Laerke and Hedemann MS (2012), Merchant et al. (2011), Patterson et al. (2008), and Widdowson (1985) for pigs
[b]Small breed of dogs
[c]Large breed of dogs
[d]Neonatal
[e]Adult
[f]Fasted state
[g]Fed state
[h]Pigs with a normal birth weight
BL = body length; BW = body weight; Duo = duodenum; EMPT = emptying; GIT = gastrointestinal tract; h = hour; Ile = ileum; Jej = jejunum; m = meter; SI = small intestine; wt = weight

dogs, adult cats have higher luminal fluid pH in the small intestine, a shorter small intestine, and a smaller volume of the small intestine (Table 2.2). In both dogs and cats, the movement of chyme (partially digested food combined with gastric juices) from the stomach to the small intestine stimulates the secretion of pancreatic juice into the duodenum, whereas the release of cholecystokinin from the small intestine is enhanced by partially digested food, fat, and Phe-containing peptides (Backus et al. 1997; Maskell and Johnson 1993; Stephens et al. 1975). The endogenous secretions of nitrogen [primarily crude protein (CP)] are greater in cats than in dogs (Hendriks et al. 1996, 2002). The primary function of the small intestine is to digest and absorb dietary nutrients, and its secondary function is to regulate the entry of dietary nutrients into the blood circulation, separate the external from the internal milieu, and perform immune surveillance.

In dogs and cats, as in other mammals, protein-derived small peptides (di- and tri-peptides) and free AAs are absorbed into the enterocytes via peptide transporter-1 and specific AA transporters, respectively (Wu 2022). Once inside these cells, the small peptides are hydrolyzed to free AAs, which are either utilized locally or transported through the basolateral membrane into the lamina propria of the intestinal mucosa before entering the portal vein. Starch- and glycogen-derived glucose is absorbed into canine and feline enterocytes mainly via the sodium-glucose linked transporter-1 (SGLT-1) and exits the cells via glucose transporter-2 (Batchelor et al. 2011). Fructose is absorbed into canine (Buddington and Malo 2003) and feline (Buddington and Diamond 1992) enterocytes by glucose transporter-5. The small intestine of dogs has a high capacity to absorb both AAs and glucose (Laganière et al. 1984), whereas the small intestine of cats also has a high capacity to absorb AAs but has a much lower ability to absorb glucose than dogs (Batchelor et al. 2011).

As noted previously, there is little or no salivary α-amylase activity in both cats and dogs. Kienzle (1993a) reported the presence of a low activity of pancreatic α-amylase activity in adult cats. Thus, the pancreatic enzyme present in the feline small intestine can digest a small amount of glycogen in meat but cannot digest a large amount of cooked or uncooked starch (Kienzle 1993b). Nonetheless, cats can tolerate a low proportion of starch-rich ingredients (e.g., grains and potatoes; providing < 10% of starch in diets, on a DM basis) used for pet food processing (Morris et al. 1977). In contrast, dogs have a much greater ability to digest dietary starch (e.g., 40% starch, on a DM basis) than cats and wolves due to higher activities of enzymes for hydrolyzing starch and disaccharides (McDonald et al. 1984). This is consistent with the greater expression of the pancreatic α-amylase (catalyzing the breakdown of starch and glycogen to oligosaccharides maltose and maltriose) and disaccharidases (hydrolyzing disaccharides to monosaccharides). For example, compared with wolves, mRNA levels for the pancreatic α-amylase gene (*AMY2B*) and serum α-amylase activity in dogs are 28- and 4.7-fold greater, respectively (Axelsson et al. 2013). In addition, mRNA levels for the intestinal maltase-glucoamylase gene (*MGAM*) and maltase-glucoamylase activity in dogs are 12- and 2-fold greater than those in wolves, respectively (Axelsson et al. 2013). Likewise, there are 3.1- and 1.6-fold greater lactase, 4.4- and 2.9-fold greater sucrase, and 4.6- and 3.1-fold greater maltase activities in the proximal and middle regions of the small intestine, respectively, in dogs than in cats (Batchelor et al. 2011). Furthermore, SGLT1 (a brush border transport protein responsible for the majority of glucose uptake by enterocytes) is expressed in the small intestine of dogs but is absent in 18 out of 19 wolves. For comparison, SGLT1 is expressed in feline enterocytes; however, mRNA and protein abundances for SGLT1 in the proximal, middle, and distal segments of the small intestine are about twofold greater in dogs than in cats (Batchelor et al. 2011). These findings provide a biochemical basis to explain why domestic dogs are capable of using grain-based rations with high-starch content (Li and Wu 2024).

In contrast to protein and starch/glycogen, the digestion of TAGs in dogs requires bile salts (the conjugates of bile acid with taurine; i.e., taurocholic acid, taurodeoxycholic acid, and taurochenodeoxycholic acid) (Knoebel and Ryan 1963). Bile salts are produced by the liver, stored in the gallbladder, and secreted into the duodenum through a bile duct in response to feeding. The bile salt mixes with TAGs and phospholipids to form micelles to facilitate the action of pancreatic lipase and other lipases (e.g., colipase and carboxyl ester lipase). Absorbed long- and medium-chain TAGs enter intestinal lymph vessels and the portal vein, respectively (Wu 2018), with the similarities and differences being summarized in Table 2.3. In both dogs and cats, the absorption of vitamin B_{12} takes place primarily in the ileum.

The number of bacteria (mainly *Streptococci* and *Lactobacilli*) in the fluid of the canine duodenum and jejunum rarely exceeds 10^4/ml (Suchodolski 2011). In the duodenum of healthy dogs, Firmicutes is the most abundant microbial species (46.4%), followed by Proteobacteria (26.6%), Bacteroidetes (11.2%), Spirochaetes (10.3%), Fusobacteria (3.6%), and Actinobacteria (1%) (Xenoulis et al. 2008). In the jejunum of healthy adult dogs, Proteobacteria is the most abundant microbial species (46%), followed by Firmicutes (15%), Actinobacteria (11.2%), Spirochaetes (14.2%), Bacteroidetes (6.2%), and Fusobacteria (5.4%) (Suchodolski et al. 2009). The number of bacteria (mainly *E. coli* and anaerobic bacteria) in the canine ileum is ~10^6/ml (NRC 2006). These data extend the report of (Johnston et al. 1993) that, in the canine duodenum, anaerobic bacteria are mainly *Bacteroides* spp, *Fusobacterium* spp, *Eubacterium*, whereas aerobic bacteria are mainly *Pasteurella* species. In dogs, the abundance of bacteria increases along the length of the GIT (namely ileum > jejunum > duodenum), and the microbiota in the small intestine inhibits the colonization of pathogenic microorganisms. Compared with dogs, the feline small intestine contains much more bacteria. For example, the duodenum fluid of cats fed a canned diet contains 2.2×10^5 to 1.6×10^8 bacteria, with anaerobic bacteria being 7.5×10^4 to 1.1×10^8 (Johnston et al. 1993). In addition, the anaerobic bacteria of the feline duodenum are mainly *Bacteroides* spp, *Fusobacterium* spp, *Eubacterium*, whereas aerobic bacteria are mainly *Pasteurella* species (Johnston et al. 1993).

Aspartate, glutamate, and glutamine are the major sources of energy for the small intestine of dogs (Pinkus and Windmueller 1977; Weber et al. 1977; Weber and Veach 1979; Windmueller 1982; Windmueller and Spaeth 1974) and cats (Windmueller 1982), as reported for carnivorous fish (Jia et al. 2017; Li et al. 2020, 2022, 2021b) and noncarnivorous mammals including humans, cattle, pigs and rats (Gilbreath et al. 2021; He and Wu 2022; Wu 2009; Wu and Rhoads 2024; Zhang et al. 2021). About 95% of dietary aspartate and glutamate as well as 70% of dietary glutamine are extracted by the small intestine in first-pass metabolism (Wu 2022). The small intestine of dogs and cats (Pinkus and Windmueller 1977; Weber and Veach 1979; Windmueller and Spaeth 1974) takes up about 30 and 20% of glutamine, respectively, but no aspartate or glutamate, from the arterial blood. In addition, the mucosa and luminal microbes of the small intestine degrade about 40% of dietary arginine, 40% of dietary proline, and 35–40% of dietary branched-chain AAs (primarily in the mucosa), as well as 5–30% of other proteinogenic AAs in diets (either exclusively or almost entirely in microbes) (Wu 2022; Yu et al. 1990, 1996). Data on the composition of all these AAs in plant- and animal-sourced foodstuffs are now available in the literature (Hou et al. 2019; Li and Wu 2020, 2023a, b; Li et al. 2021a). In dogs, enterocytes appear to oxidize glutamine and butyrate at similar rates, which are greater than those for glucose and propionate, in the presence of a mixture of the substrates (5 mM each) (Beaulieu et al. 2002). To date, little is known about the relative contribution of glucose as well as short- and long-chain fatty acids to ATP production in feline enterocytes.

Table 2.3 Digestion and absorption of long- and medium-chain triacylglycerols in the small intestine of dogs and cats[a]

Variable	Long-chain TAGs	Medium-chain TAGs
Paracellular absorption in the stomach and SI	No	Some
Hydrolysis by gastric lipase in the stomach	Limited	High
Requirement for bile salt for micellarization in the SI	Yes	No
Hydrolysis by pancreatic lipase in the SI	High	Very high
Solubility of fatty acids in water	Insoluble	Very high
Absorption of digestion products from the intestinal lumen into the intestinal epithelium	Mainly via diffusion but some via FA transporters (e.g., CD36)	Mainly via transmembrane transporters
Re-esterification and chylomicron formation in Ent	Yes	No
Primary transport of digestion products from the intestinal epithelium into blood	Lymphatic vessels	Portal vein

[a]Adapted from Knoebel and Ryan (1963), and Rutz et al. (2004) for dogs; and Bergstrom et al. (1954), Salas et al. (2014), and Trevizan et al. (2010) for cats

CD36 cluster of differentiation 36 (also known as fatty acid translocase; transporting long-chain fatty acids in an energy- independent manner); *Ent* enterocytes; *FA* fatty acid; *SI* small intestine; *TAGs* triacylglycerols

2.5 The Large Intestine of Dogs and Cats

Compared with omnivores (e.g., pigs), dogs and cats have a relatively short hindgut, including an unsacculated colon and a small cecum (Table 2.2), with pH values in the luminal fluid being 5.5–7.5 depending on diets. Cats have a less developed cecum than dogs (Griffin 2019). The large intestine of dogs contains much more bacteria than the small intestine. For example, bacterial counts in the colon of dogs are 10^9 to 10^{11}/ml, which help to digest fiber and other nutrients (Suchodolski 2011). The main bacteria in their hindgut are Fusobacteria (24–40%), Bacteroidetes (32–34%), and Firmicutes (15–28%), whereas Proteobacteria (5–6%) and Actinobacteria (0.8–1.4%) are quantitatively minor microbial species (Maskell and Johnson 1993; Suchodolski et al. 2008). For comparison, the main bacteria in the hindgut of healthy pigs are Firmicutes (78%), followed by Bacteroidetes (14%), Actinobacteria (1.6%), and Proteobacteria (0.2%) (Hoogeveen et al. 2020). Meng et al. (2020) reported that the large intestine of 28-day-old pigs weaned at 21 days of age contained only 0.01% Fusobacteria (Meng et al. 2020). Thus, the composition of bacteria in the hindgut differs greatly between dogs and omnivores. The large intestine of cats has the same predominating bacterial species and the same pH as dogs (Maskell and Johnson 1993; Suchodolski 2011).

The main stimuli for motility in the large intestine of mammals including dogs and cats are distention by its contents and the amounts of the undigested material entering the colon. In particular, colonic contents stimulate gut contractions to affect the time of their transit into the rectum. Previous studies showed little effect of canine BW on orocecal transit time (Weber et al. 2002, 2003), fecal microflora profile (Simpson et al. 2002), or the absorption of nutrients by the small intestine (Weber et al. 2002). In dogs and cats, as in other mammals, the major functions of the large intestine are to ferment food that escapes the small intestine to produce various metabolites [including short-chain fatty acids (SCFAs), L-lactate, CO_2, H_2S, ammonia, amines, phenols, indoles, and microbial proteins]; synthesize and absorb certain vitamins (e.g., B-complex vitamins and vitamin K); absorb water, electrolytes, and SCFAs; regulate intestinal immune cells; produce lactic acid to provide a hostile environment against pathogens; and form and store feces for excretion (Azzouz and Sharma 2023; Kerns 2005; Wernimont et al. 2020). The feces consist of unabsorbed nutrients (including water and AAs), sloughed cells, digestive secretions, microbial metabolites, and microbes and eventually pass into the rectum for excretion via the anal canal. The microflora influences colonic absorption and cell metabolism, as well as fecal quality.

SCFAs (including acetate, propionate, and butyrate) are highly abundant in the large intestine of mammals including dogs and cats. For example, the content of acetate, propionate, and butyrate in the feces (35% DM) of healthy adult dogs fed a 35%-CP diet was 280, 93, and 50 mmol/kg DM, respectively (Beloshapka et al. 2012), whereas the content of acetate, propionate, and butyrate in the feces (35% DM) of healthy adult cats was 130, 42, and 24 mmol/kg feces, respectively (Summers et al. 2020). As for other mammals such as rats and pigs (Wu 2018), SCFAs are the major sources of energy for colonocytes of the large intestine in dogs and cats, thereby powering these cells in metabolism (McManus et al. 2002; Moreno et al. 2022; Rondeau et al. 2003). Results of studies involving dogs have shown that their colonocytes appear to oxidize butyrate at highest rates than glucose > glutamine > propionate, in the presence of a mixture of the substrates (5 mM each) (Beaulieu et al. 2002). Thus, inclusion of appropriate levels of fermentable fibers in canine and feline diets is beneficial for the health of their hindguts (NRC 2006; Rochus et al. 2014). In support of this view, dietary supplementation with fructans (highly fermentable substrates in the large intestine) decreases pH, while increasing concentrations of SCFAs and the abundance of *Bifidobacterium* spp. and *Lactobacilli* spp. in the feces of healthy adult dogs (Flickinger et al. 2003; Propst et al. 2003; Swanson et al. 2002; Willard et al. 2000). Similar effects of dietary supplementation with Brewer's spent grain and carrot pomace (Eisenhauer et al. 2019) or dried apple pomace (De Brito et al. 2021) have been reported for dogs. Although plant-based foods are normally not fed to cats, inclusion of small amounts (e.g., 3-5%) of fibers (water-soluble and insoluble) in feline diets plays an important role in stimulating intestinal motility and maintaining GIT homeostasis (Moreno et al. 2022). An ideal proportion of soluble (e.g., fructans and galactans) to insoluble (e.g., cellulose and hemicellulose) fibers in diets is 1:3, with the total content of dietary fiber being < 5% of DM (Doebelin 2021; Gross et al. 2010). Thus, diets have great impacts on the gastrointestinal microbiome of cats and dogs.

2.6 Digestibilities of Nutrients in Dogs and Cats

Due to ethical concerns, there are no data on the true ileal digestibilities of dietary nutrients in dogs or cats, although measurements of apparent digestibility values based on the collection of fecal samples have been made (Li and Wu 2023a, b). In recent years, the cecectomized rooster model has been used to assess the digestibilities of nutrients in foodstuffs for dogs and cats (Oba et al. 2020; Reilly et al. 2021). However, it is unknown whether results from an avian model can be applied to mammals including dogs and cats. Many factors (e.g., heating, cooking, dietary intakes of fiber and bioactive substances, and ages of animals) that influence apparent nutrient digestibilities in dogs have similar effects in cats (Algya et al. 2018; Buffington et al. 1989; Harper and Turner 2000; Muir et al. 1996; Morris et al. 1977; Quigley et al. 2004; Rodríguez et al. 2016; Shields 1993; Silvio et al. 2000; Weber et al. 2003).

The small intestine of both dogs and cats has a high ability to digest proteins and lipids in cooked or uncooked diets (Carciofi et al. 2008; Li and Wu 2023a, b; Vhile et al. 2005), as well as cooked or extruded starch (95% of apparent digestibility; Murray et al. 1999). Thus, apparent total tract digestibilities of organic nutrients in properly processed foods are generally greater than 75% in dogs (Table 2.4). For example, the apparent digestibility of starch in the extruded diets containing 43–66% plant ingredients (cassava flour, corn grain, sorghum, brewer's rice, lentil, or pea) is

99% in the total digestive tract of neutered adult mixed-breed dogs (Carciofi et al. 2008). The apparent digestibilities of CP, lipids, organic matter, and gross energy in the cooked diets containing these ingredients are 80–89%, 88–93%, 79–88%, and 78–88%, respectively, in the total digestive tract of these animals (Carciofi et al. 2008). The apparent digestibilities of starch, total carbohydrates, lipids, and CP in the cooked diets containing both plant ingredients (primarily wheat and corn grains) and meat (lamb meal, poultry meal, or fish meal) are 99, 84, 97, and 84%, respectively, in the total digestive tract of adult dogs (Tjernsbekk et al. 2014). Likewise, cooking corn and wheat grains increases the apparent digestibility of their starch in the total GIT of cats by 11 and 4%, respectively. Cooking can also enhance the digestibilities of AAs in dogs (Algya et al. 2018) and starch in cats (Morris et al. 1977). However, overheating plant-sourced foodstuffs reduces the digestibilities of dietary protein and AAs (Li and Wu 2023a, b).

By contrast, uncooked or unextruded starch is poorly digested by the small intestine of dogs and cats (Murray et al. 1999). Despite relatively low activities of mammalian enzymes digesting carbohydrates, the apparent digestibility of starch in a meat-based diet containing 22% cornstarch is 94% in the total digestive tract of nearly mature cats, likely due to the extensive fermentation in the large intestine. This digestibility value is similar to that for starch in finely ground raw corn grain containing 12% starch and in coarsely ground raw wheat grain containing 12% (92.5%) (Table 2.5). Compared with raw coarse grinding, the fine grinding of corn and wheat grains increases the apparent digestibility of their starch in the total GIT of cats by 15 and 5%, respectively. Thus, the size of particles in diets also affects the digestibilities of their nutrients in companion animals.

Inclusion of 7.5% fiber in diets reduced the apparent ileal digestibilities of DM and organic matter in adult dogs but did not affect the apparent ileal digestibilities of AAs (Muir et al. 1996). Similar results were reported for adult cats fed fiber-supplemented diets (Morris et al. 1977). By contrast, dietary supplementation with *yucca schidigera* (60–120 ppm) enhances the digestibility and absorption of nutrients in these companion animals, thereby reducing the odor of pet stools (Beynen and Saris 2014; Dickson 2021; Lowe et al. 1997). Interestingly, the inclusion of 20% lactose in diets for cats can cause diarrhea and reduce apparent digestibility of dietary CP (Morris et al. 1977). Data on true ileal digestibilities of protein and other nutrients in dogs and cats are needed to precisely formulate their diets.

2.7 Conclusion

Dogs (facultative carnivores) and cats (obligate carnivores) have a relatively short digestive tract that is characteristic of carnivores and distinct from noncarnivorous mammals. Dogs, but not cats, have receptors for sweet substances in their tongue, and both animals are not highly sensitive to dietary salt in terms of chemical sensing. The volume of the stomach of dogs and cats accounts for 60–70% of the volume of their

Table 2.4 Digestibilities of nutrients in adult dogs fed diets containing different sources of protein ingredients with or without added β-mannanase[a]

Variable	Diets containing PBM or SBM as protein ingredients			
	Poultry by-product meal		Soybean meal	
	0.00% β-Mannanase	0.01% β-Mannanase	0.00% β-Mannanase	0.01% β-Mannanase
Content of ingredients in diets (%, as-fed basis)				
Poultry by-product meal (PBM)	37.50	37.50	–	–
Soybean meal (SBM)	–	–	30.00	30.00
β-Mannanase	–	0.01	–	0.01
Corn grain	50.21	50.21	44.01	44.0
Poultary fat	6.00	6.00	6.00	6.00
Liquid poultry liver hydrolysate	3.00	3.00	3.00	3.00
Powdered poultry liver hydrolysate	1.00	1.00	1.00	1.00
Hydrated aluminosilicate (Celite)	1.00	1.00	1.00	1.00
Vitamin-mineral supplement	0.50	0.50	0.50	0.50
Sodium chloride	0.50	0.50	0.50	0.50
Antioxidants[b]	0.05	0.05	0.05	0.05
Mold inhibitor (ammonium propionate)	0.24	0.24	0.24	0.24
Composition of nutrients in diets				
Calculated ME (kcal/g of dry matter)	4.11	4.09	3.95	3.94
Dry matter (%, as-fed basis)	92.50	91.98	92.75	91.87
Crude protein (%, dry matter basis)	28.11	28.15	28.63	28.71
Lipids (%, dry matter basis)	15.62	15.55	13.50	13.43
Nitrogen-free extract (%, dry matter basis)[c]	43.77	43.93	46.39	46.46
Crude fiber (%, dry matter basis)	3.59	3.54	4.46	4.41

(continued)

Table 2.4 (continued)

Variable	Diets containing PBM or SBM as protein ingredients			
	Poultry by-product meal		Soybean meal	
	0.00% β-Mannanase	0.01% β-Mannanase	0.00% β-Mannanase	0.01% β-Mannanase
Minerals (%, dry matter basis)	8.91	8.83	7.02	6.99
Acid-insoluble ash (%, dry matter basis)	1.44	1.41	1.16	1.16
Apparent digestibility of nutrients in the total GIT and determined ME				
Dry matter	79.9^{e}	79.7^{e}	78.8^{e}	82.5^{d}
Organic matter	86.1^{d}	85.9^{d}	82.5^{e}	85.5^{d}
Crude protein	81.5^{e}	82.3^{e}	82.5^{e}	85.9^{d}
Lipids	93.9^{d}	94.3^{d}	89.7^{e}	92.2^{d}
Gross energy	86.6^{d}	86.7^{d}	83.5^{e}	86.9^{d}
Determined ME (kcal/g of dry matter)	4.4^{d}	4.3^{d}	4.1^{e}	4.4^{d}

[a]Adapted from Felix et al. (2012). The mean body weight and age of the Beagle dogs (4 males and 4 females) were 8.6 kg and 2.2 years, respectively. This study involved a 5-day period of adaptation and a 5-day period of total fecal collection
[b]Butyrated hydroxyanisole and butyrated hydroxytoluene
[c]Consisting of sugars, starches, glycogen, fructan, pectins, hemicellulose (partial), and lignin (partial)
GIT gastrointestinal tract, ME metabolizable energy
[d,e]Within a row, means without sharing the same superscript letter differ ($P < 0.05$)
"–", data are not available

total digestive tract, which is much greater than the value for omnivores (e.g., approximately 10% in adult pigs). Dogs differ from cats in some aspects of the digestive anatomy. Compared with adult dogs, adult cats have higher gastric fluid pH, higher luminal fluid pH in the small intestine, a shorter gut (both small and large intestines), a lower volume of the small intestine, and a shorter retention time of food in the whole GIT. In both dogs and cats, the small intestine functions to digest and absorb dietary nutrients, regulate the entry of dietary nutrients into the blood circulation, separate the external from the internal milieu, and perform immune surveillance. These carnivores have a high ability to digest dietary protein and fat in the small intestine. Its enterocytes use glutamate, glutamine, and aspartate as the major metabolic fuels. Dogs, but not cats, can tolerate high percentages of cooked starch in diets due to a relatively high activity of pancreatic α-amylase. Their large intestine plays important roles in fermenting undigested food and endogenous secretions to SCFAs, as well as absorbing water, SCFAs and vitamins into the blood circulation. SCFAs are the

Table 2.5 Apparent digestibilities of nutrients in the total gastrointestinal tract of nearly mature, short-haired cats[a]

Experiment No.	Content of nutrients in diet (% of DM)						Apparent digestibility of nutrients (%)			
	Starch	Glucose	Fiber	CP	Lipids	Minerals	Starch	Fiber	CP	Lipids
1. Meat-based diet without AC	0.6	1.2	0	44.9	44.2	8.3	–	–	94.4	99.3
1. Raw corn starch	22.2	1.1	ND	34.6	33.8	6.1	94.2	ND	90.9	99.0
1. Cellulose	-	1.0	22.7	33.8	33.3	5.7	–	0	92.6	99.4
2. Corn grain[b]										
Raw coarse grinding[b]	12.2	0.04	ND	45.5	31.9	ND	79.4	ND	86.9	ND
Raw fine grinding	12.2	0.04	ND	45.5	31.9	ND	93.7	ND	88.0	ND
Coarse grinding and cooked[b, c]	12.2	0.04	ND	45.5	31.9	ND	88.1	ND	87.1	ND
2. Wheat grain[b]										
Raw coarse grinding[b]	12.1	0.02	ND	46.2	31.5	ND	92.5	ND	88.7	ND
Raw fine grinding	12.1	0.02	ND	46.2	31.5	ND	97.2	ND	89.5	ND
Coarse grinding and cooked[b, c]	12.1	0.02	ND	46.2	31.5	ND	96.2	ND	90.0	ND

[a]Adapted from Morris et al. (1977). All diets were based on meat (a mixture of beef and mutton)-based diet. The mean body weight of the cats was 3.4 kg
[b]15% of the air-dried corn or wheat grain was added to the meat-based diet
[c]The grain was moistened with water and then autoclaved at 100 kN/m^2 for 45 min
AC added carbohydrate, *CP* crude protein (N × 6.25), *DM* dry matter, *ND* not determined. "–", data are not available

primary sources of energy for colonocytes. The digestive tract is essential for the growth, whole-body homeostasis, health, and survival of dogs and cats.

Acknowledgements We thank Prof. Nancy Ing (DVM and PhD, Department of Animal Science at Texas A&M University) for critically reviewing and providing helpful comments on this manuscript. We are also grateful to Prof. Jürgen Zentek (DVM and PhD, Freie Universität, Berlin, Germany) for kindly sharing his published articles with us. This work was supported by Texas A&M AgriLife Research (H-8200).

Conflict of Interest The author declares no conflict of interest.

References

Adler CJ, Malik R, Browne GV, Norris JM (2016) Diet may influence the oral microbiome composition in cats. Microbiome 4:23

Algya KM, Cross TL, Leuck KN, Kastner ME, Baba T, Lye L, de Godoy NRC, Swanson KS (2018) Apparent total-tract macronutrient digestibility, serum chemistry, urinalysis, and fecal characteristics, metabolites and microbiota of adult dogs fed extruded, mildly cooked, and raw diets. J Anim Sci 96:3670–3683

Axelsson E, Ratnakumar A, Arendt M-L et al (2013) The genomic signature of dog domestication reveals adaptation to a starch-rich diet. Nature 495:360–364

Azzouz LL, Sharma S (2023) Physiology, large intestine. StatPearls. Treasure Island, FL

Backus RC, Howard KA, Rogers QR (1997) The potency of dietary amino acids in elevating plasma cholecystokinin immunoreactivity in cats is related to amino acid hydrophobicity. Regul Pept 72:31–40

Baker DH, Czarnecki-Maulden GL (1991) Comparative nutrition of cats and dogs. Annu Rev Nutr 11:239–263

Bartges JW (2003) Enteral nutrition. The 28th World Small Animal Veterinary Association World Congress Proceedings, October 24–27, 2003. Bangkot, Thailand. Retrieved from https://www.vin.com/apputil/content/defaultadv1.aspx?meta=Generic&pId=8768&id=3850156. Accessed on 10 Nov 2023

Batchelor DJ, Al-Rammahi M, Moran AW, Brand JG, Li X, Haskins M, German AJ, Shirazi-Beechey SP (2011) Sodium/glucose cotransporter-1, sweet receptor, and disaccharidase expression in the intestine of the domestic dog and cat: two species of different dietary habit. Am J Physiol 300:R67-75

Beauchamp GK, Maller O, Rogers JG Jr (1977) Flavor preferences in cats (*Felis catus* and *Panthera* sp). J Comp Physiol Psychol 91:1118–1127

Beaulieu AD, Drackley JK, Overton TR, Emmert LS (2002) Isolated canine and murine intestinal cells exhibit a different pattern of fuel utilization for oxidative metabolism. J Anim Sci 80:1223–1232

Beloshapka AN, Wolff AK, Swanson KS (2012) Effects of feeding polydextrose on faecal characteristics, microbiota and fermentative end products in healthy adult dogs. Br J Nutr 108:638–644

Bergstrom S, Borgstrom B, Carlsten A (1954) On the mechanism of intestinal fat absorption in the cat. Acta Physiol Scand 32:94–98

Beynen AC, Saris DHJ (2014) Beating odour with *yucca schidigera.* All About Feed. Retrieved from https://www.allaboutfeed.net/animal-feed/feed-additives/beating-odour-with-yucca-schidigera. Accessed on 10 Nov 2023

Blazer-Yost B, Jezyk PF (1979) Free amino acids in the plasma and urine of dogs from birth to senescence. Am J Vet Res 40:832–838

Bourreau J, Hernot D, Bailhache E, Weber M, Ferchaud V, Biourge V, Martin L, Dumon H, Nguyen P (2004) Gastric emptying rate is inversely related to body weight in dog breeds of different sizes. J Nutr 134:2039S-2041S

Bradshaw JWS (2006) The evolutionary basis for the feeding behavior of domestic dogs (*Canis familiaris*) and cats (*Felis catus*). J Nutr 136:1927S-1931S

Bradshaw JWS, Goodwin D, Legrand-Defretin V, Nott HMR (1996) Food selection by the domestic cat, an obligate carnivore. Comp Biochem Physiol A 114:205–209

Brosey BP, Hill RC, Scott KC (2000) Gastrointestinal *volatile* fatty acid concentrations and pH in cats. Am J Vet Res 61:359–361

Buddington RK (2003) Postnatal changes in bacterial populations in the gastrointestinal tract of dogs. Am J Vet Res 64:646–651

Buddington RK, Diamond J (1992) Ontogenetic development of nutrient transporters in cat intestine. Am J Physiol 263:G605-616

Buddington RK, Malo C (2003) Postnatal development of nutrient transport in the intestine of dogs. Am J Vet Res 64:635–645
Buff PR, Carter RA, Bauer JE, Kersey JH (2014) Natural pet food: a review of natural diets and their impact on canine and feline physiology. J Anim Sci 92:3781–3791
Buffington CA, Branam JE, Dunn GC (1989) Lack of effect of age on digestibility of protein, fat and dry matter in Beagle dogs. In: Burger IH, Rivers JPW (eds) Nutrition of the dog and cat. Cambridge University Press, Cambridge, UK, pp 397
Carrière F, Moreau H, Raphel V, Laugier R, Benicourt C, Junien JL, Verger R (1991) Purification and biochemical characterization of dog gastric lipase. Eur J Biochem 202:75–83
Carciofi AC, Takakura FS, de-Oliveira LD, Teshima E, Jeremias JT, Brunetto MA, Prada F (2008) Effects of six carbohydrate sources on dog diet digestibility and post-prandial glucose and insulin response. J Anim Physiol Anim Nutr 92:326–336
Case LP, Daristotle L, Hayek MG, Raasch MF (2011) Canine and feline nutrition, 3rd edn. Mosby, Maryland Heights, Missouri
Chauncey HH, Henrigues BL, Tanzer JM (1963) Comparative enzyme activity of saliva from the sheep, hog, dog, rabbit, rat, and human. Arch Oral Biol 8:615–627
Che DS, Nyingwa PS, Ralinala KM, Maswanganye GMT, Wu G (2021) Amino acids in the nutrition, metabolism, and health of domestic cats. Adv Exp Med Biol 1285:217–231
Contreras-Aguilar MD, Tecles F, Martínez-Subiela S, Escribano D, Bernal LJ, Cerón JJ (2017) Detection and measurement of alpha-amylase in canine saliva and changes after an experimentally induced sympathetic activation. BMC Vet Res 13:266
Gross KL, Jewell DE, Yamka RM, Schoenherr WD, Khoo C, Debraekeleer J, Friesen KG, Zicker SC (2010) Macronutrients. In: Thatcher CD, Remillard RL, Roudebush P (eds) Small animal clinical nutrition hand MS. Mark Morris Institute, Topeka, KS, pp 49–105
De Brito CBM, Menezes Souza CM, Bastos TS, Mesa D, Oliveira SG, Félix AP (2021) Effect of dietary inclusion of dried apple pomace on faecal butyrate concentration and modulation of gut microbiota in dogs. Arch Anim Nutr 75:48–63
Dickson E (2021) Improving pet food digestibility with yucca schidigera. Retrieved from https://www.alltech.com/blog/improving-pet-food-digestibility-yucca-schidigera. Accessed on 10 Nov 2023
Doebelin C (2021) Cats & fiber: feeding plant matter? Retrieved from https://thelittlecarnivore.com/en/blog/cats-and-fiber-feeding-plant-matter. Accessed on 10 Nov 2023
Dyce KM, Sack WO, Wensing CJG (1996) Textbook of veterinary anatomy. W.B. Saunders Company, Philadelphia
Eisenhauer L, Vahjen W, Dadi T, Kohn B, Zentek J (2019) Effects of Brewer's spent grain and carrot pomace on digestibility, fecal microbiota, and fecal and urinary metabolites in dogs fed low- or high-protein diets. J Anim Sci 97:4124–4133
Emås S, Grossman MI (1967) Comparison of gastric secretion in conscious dogs and cats. Gastroenterology 52:29–34
Félix AP, Gabeloni LR, Brito CBM, Oliveira SG, Silva AVF, Maiorka A (2012) Effect of β-mannanase on the digestibility of diets with different protein sources in dogs determined by different methodologies. J Anim Sci 90:3060–3067
Ferrell F (1984) Preference for sugars and nonnutritive sweeteners in young beagles. Neurosci Biobehav Rev 8:199–203
Flickinger EA, Schreijen EMWC, Patil AR, Hussein HS, Grieshop CM, Merchen NR, Fahey GC (2003) Nutrient digestibilities, microbial populations, and protein catabolites as affected by fructan supplementation of dog diets. J Anim Sci 81:2008–2018
Flynn NE, Wu G (1996) An important role for endogenous synthesis of arginine in maintaining arginine homeostasis in neonatal pigs. Am J Physiol 271:R1149-1155
Gilbreath KR, Bazer FW, Satterfield MC, Wu G (2021) Amino acid nutrition and reproductive performance in ruminants. Adv Exp Med Biol 1285:43–61
Gil F, Arencibia A, García V, Ramírez G, Vázquez JM (2018) Anatomic and magnetic resonance imaging features of the salivary glands in the dog. Anat Histol Embryol 47:551–559

Goggin JM, Hoskinson JJ, Kirk CA, Jewell D, Butine MD (1999) Comparison of gastric emptying times in healthy cats simultaneously evaluated with radiopaque markers and nuclear scintigraphy. Vet Radiol Ultrasound 40:89–95

Gregory PC, McFadyen M, Rayner DV (1990) Pattern of gastric emptying in the pig: relation to feeding. Br J Nutr 64:45–58

Griffin S (2019) Feline abdominal ultrasonography: what's normal? what's abnormal? the normal gastrointestinal tract. J Feline Med Surg 21:1039–1046

Harper EJ, Turner CL (2000) Age-related changes in apparent digestibility in growing kittens. Reprod Nutr Dev 40:249–260

Harrison JD, Fouad HM, Garrett JR (2001) Variation in the response to ductal obstruction of feline submandibular and sublingual salivary glands and the importance of the innervation. J Oral Pathol Med 30:29–34

He WL, Wu G (2022) Oxidation of amino acids, glucose, and fatty acids as metabolic fuels in enterocytes of developing pigs. Amino Acids 54:1025–1039

Hendriks WH, Moughan PJ, Tarttelin MF (1996) Gut endogenous nitrogen and amino acid excretions in adult domestic cats fed a protein-free or an enzymatically hydrolyzed casein-based diet. J Nutr 126:955–962

Hendriks WH, Sritharan K, Hodgkinson SM (2002) Comparison of the endogenous ileal and faecal amino acid excretion in the dog (*Canis familiaris*) and the rat (*Rattus rattus*) determined under protein-free feeding and peptide alimentation. J Anim Physiol Anim Nutr 86:333–341

Herring CM, Bazer FW, Wu G (2021) Amino acid nutrition for optimum growth, development, reproduction, and health of zoo animals. Adv Exp Med Biol 1285:233–253

Hoogeveen AME, Moughan PJ, de Haas ES, Blatchford P, McNabb C, Montoya CA (2020) Ileal and hindgut fermentation in the growing pig fed a human-type diet. Br J Nutr 124:567–576

Hou YQ, He WL, Hu SD, Wu G (2019) Composition of polyamines and amino acids in plant-source foods for human consumption. Amino Acids 51:1153–1165

Jia SC, Li XY, Zheng SX, Wu G (2017) Amino acids are major energy substrates for tissues of hybrid striped bass and zebrafish. Amino Acids 49:2053–2063

Jin HO, Lee KY, Chang TM, Chey WY, Dubois A (1994) Secretin: a physiological regulator of gastric emptying and acid output in dogs. Am J Physiol 267:G702-708

Johnston K, Lamport A, Batt RM (1993) An unexpected bacterial flora in the proximal small intestine of normal cats. Vet Rec 132:362–363

Kararli TT (1995) Comparison of the gastrointestinal anatomy, physiology, and biochemistry of humans and commonly used laboratory animals. Biopharm Drug Dispos 16:351–380

Kerns N (2005) The canine digestion process. Whole Dog J. Retrieved from https://www.whole-dog-journal.com/health/digestion/the-canine-digestion-process. Accessed on 10 Nov 2023

Kienzle E (1993a) Carbohydrate metabolism of the cat. 1. activity of amylase in the gastrointestinal tract of the cat. J Anim Physiol Anim Nutr 69:92–101

Kienzle E (1993b) Carbohydrate metabolism of the cat. 2. digestion of starch. J Anim Physiol Anim Nutr 69:102–114

Kitchell RL (1978) Taste perception and discrimination by the dog. Adv Vet Sci Comp Med 22:287–314

Kliegman RM, Miettinen EL, Tserng KY (1983) Fetal and neonatal metabolism in the pup of a canine diabetic mother: fasting circulating fuels and neonatal glucose turnover. Diabetes 32:352–359

Knoebel LK, Ryan JM (1963) Digestion and mucosal absorption of fat in normal and bile-deficient dogs. Am J Physiol 204:509–514

Knospe C, Plendl J (1997) Histochemical demonstration of lipase activity in the gastric mucosa of the cat. Anat Histol Embryol 26:303–304

Kumazawa T, Kurihara K (1990a) Large enhancement of canine taste responses to sugars by salts. J Gen Physiol 95:1007–1018

Kumazawa T, Kurihara K (1990b) Large synergism between monosodium glutamate and 59-nucleotides in canine taste nerve responses. Am J Physiol 259:R420–R426

Kuo ZY (1967) The dynamics of behaviour development: an epigenetic view. Random House, New York
Ladd M, Raisz LG (1949) Response of the normal *dog* to dietary sodium chloride. Am J Physiol 159:149–152
Lærke HN, Hedemann MS (2012) The digestive system of the pig. In: Knudsen NJ, Kjeldsen HD, Jensen BB (eds) Nutritional physiology of pigs (Bach KE. Danish Pig Research Center, Copenhagen, Denmark, pp 1–27
Laganière S, Berteloot A, Maestracci D (1984) Digestive and absorptive functions along dog small intestine: comparative distributions in relation to biochemical and morphological parameters. Comp Biochem Physiol A 79:463–472
Larmas M, Scheinin A (1972) Studies on dog saliva. 3. observations on enzymes acting on ester bonds and glycosyl compounds. Acta Odontol Scand 30:357–370
Latimer HB (1967) Variability in body and organ weights in the newborn dog and cat compared with that in the adult. Anat Red 157:449–456
Lavigne ME, Wiley ZD, Meyer JH, Martin P, MacGregor IL (1978) Gastric emptying rates of solid food in relation to body size. Gastroenterology 74:1258–1260
Lecheta DR, Silva DKM, Santos GA, Cunha MS, Gaspar TT, Lopes BA, Deboleto SGC, Braz PH (2020) Effect of pre-anesthetic fasting on gastric emptying and plasma glucose in healthy dogs of different age groups. Brazil J Vet Res 40:289–292
Lee S, Lee K, Kim H, An J, Han J, Lee T, Jeong H, Cho Y (2020) Comparison of dental radiography and computed tomography: measurement of dentoalveolar structures in healthy, small-sized dogs and cats. J Vet Sci 21:e75
Legrand-Defretin V (1994) Differences between cats and dogs: a nutritional view. Proc Nutr Soc 53:15–24
Levy JK, Nutt KR, Tucker SJ (2015) Reference interval for rectal temperature in healthy confined adult cats. J Feline Med Surg 17:950–952
Li P, Wu G (2020) Composition of amino acids and related nitrogenous nutrients in feedstuffs for animal diets. Amino Acids 52:523–542
Li P, Wu G (2023) Composition of nutrients in rendered animal-sourced feedstuffs. In: Clive P (ed) Encyclopedia of animal nutrition, CABI, Wallingford, Oxon, UK
Li P, Wu G (2023b) Amino acid nutrition and metabolism in domestic cats and dogs. J Anim Sci Biotechnol 14:19
Li P, Wu G (2024a) Characteristics of nutrition and metabolism in dogs and cats. Adv Exp Med Biol 1446:55–98. https://doi.org/10.1007/978-3-031-54192-6_4
Li X, Li W, Wang H, Cao J, Maehashi K, Huang L, Bachmanov AA, Reed DR, Legrand-Defretin V, Beauchamp GK (2005) Pseudogenization of a sweet-receptor gene accounts for cats' indifference toward sugar. PLoS Genet 1:e27-35
Li X, Li W, Wang H, Bayley DL, Cao J, Reed DR, Bachmanov AA, Huang L, Legrand-Defretin V, Beauchamp GK, Brand JG (2006) Cats lack a sweet taste receptor. J Nutr 136(Suppl 7):1932S-1934S
Li XL, Zheng SX, Jia SC, Song F, Zhou CP, Wu G (2020) Oxidation of energy substrates in tissues of largemouth bass (*Micropterus salmoides*). Amino Acids 52:1017–1032
Li P, He WL, Wu G (2021a) Composition of amino acids in foodstuffs for humans and animals. Adv Exp Med Biol 1332:189–210
Li XY, Zheng SX, Wu G (2021b) Nutrition and functions of amino acids in fish. Adv Exp Med Biol 1285:133–168
Li XY, Han T, Zheng SX, Wu G (2022) Hepatic glucose metabolism and its disorders in fish. Adv Exp Med Biol 1354:207–236
Lin HC, Doty JE, Reedy TJ, Meyer JH (1990) Inhibition of gastric emptying by sodium oleate depends on length of intestine exposed by nutrient. Am J Physiol 259:G1031-1036
Lombardero M, Alonso-Peñarando D, Yllera MDM (2021) The cat mandible (I): anatomical basis to avoid Iatrogenic damage in veterinary clinical practice. Animals 11:405

Lowe JA, Kershaw SJ, Taylor AJ, Linforth RS (1997) The effect of *Yucca schidigera* extract on canine and feline faecal volatiles occurring concurrently with faecal aroma amelioration. Res Vet Sci 63:67–71
MacDonald ML, Rogers QR, Morris JG (1985) Aversion of the cat to dietary medium-chain triglycerides and caprylic acid. Physiol Behav 35:371–375
Maskell IE, Johnson JV (1993) Digestion and absorption. In: Burger I (ed) The waltham book of companion animal nutrition, Pergamon Press, Oxford, UK, pp 25–44
MacDonald ML, Rogers QR, Morris JG (1984) Nutrition of the domestic cat, a mammalian carnivore. Annu Rev Nutr 4:521–562
McDonald P, Edwards RA, Greenhalgh JFD, Morgan CA, Sinclair LA (2011) Animal nutrition, 7th edn. Prentice Hall, New York
McGeachin RL, Akin JR (1979) Amylase levels in the tissues and body fluids of the domestic cat (*Felis catus*). Comp Biochem Physiol B 63:437–439
McManus CM, Michel KE, Simon DM, Washabau RJ (2002) Effect of short-chain fatty acids on contraction of smooth muscle in the canine colon. Am J Vet Res 63:295–300
Meng Q, Luo Z, Cao C, Sun S, Ma Q, Li Z, Shi B, Shan A (2020) Weaning alters intestinal gene expression involved in nutrient metabolism by shaping gut microbiota in pigs. Front Microbiol 11:694
Merchant HA, McConnell EL, Liu F, Ramaswamy C, Kulkarni R, Basit AW, Murdan S (2011) Assessment of gastrointestinal pH, fluid and lymphoid tissue in the guinea pig, rabbit and pig, and implications for their use in drug development. Eur J Pharm Sci 42:3–10
Meyer H, Kienzle E, Zentek J (1993) Body size and relative weights of gastrointestinal tract and liver in dogs. J Vet Nutr 2:31–35
Meyer JH, Elashoff JD, Domeck M, Levy A, Jehn D, Hlinka M, Lake R, Graham LS, Gu YG (1994) Control of canine gastric emptying of fat by lipolytic products. Am J Physiol 266:G1017-1035
Moreno AA, Parker VJ, Winston JA, Rudinsky AJ (2022) Dietary fiber aids in the management of canine and feline gastrointestinal disease. J Am Vet Med Assoc 260:S33–S45
Morris JG (2002) Idiosyncratic nutrient requirements of *cats appear to be diet-*induced *evolutionary* adaptations. Nutr Res Rev 15:153–168
Morris JG, Trudell J, Pencovic T (1977) Carbohydrate digestion by the domestic *cat* (*Felis catus*). Br J Nutr 37:365–373
Muir HE, Murray SM, Fahey GC Jr, Merchen NR, Reinhart GA (1996) Nutrient digestion by ileal cannulated dogs as affected by dietary fibers with various fermentation characteristics. J Anim Sci 74:1641–1648
Murray SM, Fahey GC, Merchen NR, Sunvold GD, Reinhart GA (1999) Evaluation of selected high-starch flours as ingredients in canine diets. J Anim Sci 77:2180–2186
National Research Council (NRC 2006) Nutrient requirements of dogs and cats. National Academies Press, Washington DC
Nelson JH, Jensen RG, Pitas RE (1977) Pregastric esterase and other oral lipases—a review. J Dairy Sci 60:327–362
Oba PM, Utterback PL, Parsons CM, Swanson KS (2020) True nutrient and amino acid digestibility of dog foods made with human-grade ingredients using the precision-fed cecectomized rooster assay. Transl Anim Sci 4:442–451
Oberbauer AM, Larsen JA (2021) Amino acids in dog nutrition and health. Adv Exp Med Biol 1285:199–216
Pansu D, Vagne M, Reinberg A, Mechkouri M (1987) Circannual and circa-hemiannual rhythms of basal and stimulated gastric secretion in conscious cats. Chronobiol Int 4:59–67
Pasha S, Inui T, Chapple I, Harris S, Holcombe L, Grant MM (2018) The saliva proteome of dogs: Variations within and between breeds and between species. Proteomics 18:1700293
Patterson JK, Lei XG, Miller DD (2008) The pig as an experimental model for elucidating the mechanisms governing dietary influence on mineral absorption. Exp Biol Med 233:651–664
Pedraz C, Mortola JP (1991) CO_2 production, body temperature, and ventilation in hypoxic newborn cats and dogs before and after body warming. Pediatr Res 30:165–169

Pekel AY, Mülazımoğlu SB, Acar N (2020) Taste preferences and diet palatability in cats. J Appl Anim Res 48:281–292

Pinkus LM, Windmueller HG (1977) Phosphate-dependent glutaminase of small intestine: Localization and role in intestinal glutamine metabolism. Arch Biochem Biophys 182:506–517

Propst L, Flickinger EA, Bauer LL, Merchen NR, Fahey GC (2003) A dose–response experiment evaluating the effects of oligofructose and inulin on nutrient digestibility, stool quality, and fecal protein catabolites in healthy adult dogs. J Anim Sci 81:3057–3066

Quigley JD, Campbell JM, Polo J, Russell LE (2004) Effects of spray-dried animal plasma on intake and apparent digestibility in dogs. J Anim Sci 82:1685–1692

Reilly LM, von Schaumburg PC, Hoke JM, Davenport GM, Utterback PL, Parsons CM, de Godoy MRC (2021) Use of the precision-fed cecectomized rooster assay to determine standardized amino acid digestibility, true metabolizable energy content, and digestible indispensable amino acid scores of plant-based protein by-products used in canine and feline diets. Transl Anim Sci 5:1–9

Rochus K, Janssens GPJ, Hesta M (2014) Dietary fibre and the importance of the gut microbiota in feline nutrition: a review. Nutr Res Rev 27:295–307

Rodríguez C, Saborido N, Ródenas J, Polo J (2016) Effects of spray-dried animal plasma on food intake and apparent nutrient digestibility by cats when added to a wet pet food recipe. Anim Feed Sci Technol 216:243–250

Rondeau MP, Meltzer K, Michel KE, McManus CM, Washabau RJ, RJ, (2003) Short chain fatty acids stimulate feline colonic smooth muscle contraction. J Feline Med Surg 5:167–173

Rutz GM, Steiner JM, Bauer JE, Williams DA (2004) Effects of exchange of dietary medium chain triglycerides for long-chain triglycerides on serum biochemical variables and subjectively assessed well-being of dogs with exocrine pancreatic insufficiency. Am J Vet Res 65:1293–1302

Sagawa K, Li F, Liese R, Sutton SC (2009) Fed and fasted gastric pH and gastric residence time in conscious beagle dogs. J Pharm Sci 98:2494–2500

Salas A, Manuelian CL, Garganté M, Sanchez N, Fernández S, Compagnucci M, Cerón JJ, Jeusette I, Vilaseca L, Torre C (2014) Fat digestibility is reduced in old cats with subnormal cobalamin concentrations. J Nutr Sci 3:e62

Shields RG (1993) Digestibility and metabolizable energy measurement in dogs and cats. In: Proc Petfood Forum'93. Watt Publishing, Mt. Morris, IL, p 21–35

Siani G, Mercaldo B, Alterisio MC, Loria AD (2023) Vitamin B12 in cats: nutrition, metabolism, and disease. Animals 13:1474

Silvio J, Harmon DL, Gross KL, McLeod KR (2000) Influence of fiber fermentability on nutrient digestion in the dog. Nutrition 16:289–295

Simpson JM, Martineau B, Jones WE, Ballam JM, Mackie RI (2002) Characterization of fecal bacterial populations in canines: effects of age, breed and dietary fiber. Microb Ecol 44:186–197

Stephens JR, Woolson RF, Cooke AR (1975) Effects of essential and nonessential amino acids on gastric emptying in the dog. Gastroenterology 69:920–927

Suchodolski JS (2011) Intestinal microbiota of dogs and cats: a bigger world than we thought. Vet Clin North Am Small Anim Pract 41:261–272

Suchodolski JS, Camacho J, Steiner JM (2008) Analysis of bacterial diversity in the canine duodenum, jejunum, ileum, and colon by comparative 16S rRNA gene analysis. FEMS Microbiol Ecol 66:567–578

Suchodolski JS, Dowd SE, Westermarck E, Steiner JM, Wolcott RD, Spillmann T, Harmoinen JA (2009) The effect of the macrolide antibiotic tylosin on microbial diversity in the canine small intestine as demonstrated by massive parallel 16S rRNA gene sequencing. BMC Microbiol 9:210

Summers S, Quimby JM, Phillips RK, Stockman J, Isaiah A, Lidbury JA, Steiner JM, Suchodolski J (2020) Preliminary evaluation of fecal fatty acid concentrations in cats with chronic kidney disease and correlation with indoxyl sulfate and p-cresol sulfate. J Vet Intern Med 34:206–215

Swanson KS, Grieshop CM, Flickinger EA, Bauer LL, Chow J, Wolf BW, Garleb KA, Fahey GC (2002) Fructooligosaccharides and Lactobacillus acidophilus modify gut microbial populations,

total tract nutrient digestibilities and fecal protein catabolite concentrations in healthy adult dogs. J Nutr 132:3721–3731
Tenovuo J, Illukka T, Vähä-Vahe T (2000) Non-immunoglobulin defense factors in canine saliva and effects of a tooth gel containing antibacterial enzymes. J Vet Dent 17:136–140
Tjernsbekk MT, Tauson AH, Ahlstrøm O (2014) Ileal, colonic and total tract nutrient digestibility in dogs (*Canis familiaris*) compared with total tract digestibility in mink (*Neovison vison*). Arch Anim Nutr 68:245–261
Trevizan L, de Mello KA, Bigley KE, Anderson WH, Waldron MK, Bauer JE (2010) Effects of dietary medium-chain triglycerides on plasma lipids and lipoprotein distribution and food aversion in cats. Am J Vet Res 71:435–440
Ugawa T, Kurihara K (1993) Large enhancement of canine taste responses to amino acids by salts. Am J Physiol 264:R1071-1076
Vaillant C, Horadagoda NU, Batt RM (1990) Cellular localization of intrinsic factor in pancreas and stomach of the dog. Cell Tissue Res 260:117–122
Van Kruiningen HJ, Wojan LD, Stake PE, Lord PF (1987) The influence of diet and feeding frequency on gastric function in the dog. J Am Anim Hosp Assoc 23:146–153
Vhile SG, Skrede A, Ahlstrøm Ø, Hove K (2005) Comparative apparent total tract digestibility of major nutrients and amino acids in dogs (Canis familiaris), blue foxes (*Alopex lagopus*) and mink (*Mustela vison*). Anim Sci 81:141–148
Weber FL Jr, Maddrey WC, Walser M (1977) Amino acid metabolism of dog jejunum before and during absorption of keto analogues. Am J Physiol 232:E263-269
Weber FL Jr, Veach GL (1979) The importance of the small intestine in gut ammonium production in the fasting dog. Gastroenterology 77:235–240
Weber MP, Stambouli F, Martin LJ, Dumon HJ, Biourge VC, Nguyen PG (2002) Influence of age and body size on gastrointestinal transit time of radioopaque markers in healthy dogs. Am J Vet Res 63:677–682
Weber MP, Martin LJ, Biourge VC, Nguyen PG, Dumon HJ (2003) Influence of age and body size on orocecal transit time as assessed by use of the sulfasalzine method in healthy dogs. Am J Vet Res 64:1105–1109
Wernimont SM, Radosevich J, Jackson MI, Ephraim E, Badri DV, MacLeay JM, Jewell DE, Suchodolski JS (2020) The effects of nutrition on the gastrointestinal microbiome of cats and dogs: Impact on health and disease. Front Microbiol 11:1266
Widdowson EM (1985) Development of the digestive system: comparative animal studies. Am J Clin Nutr 41(Suppl):384–390
Willard MD, Simpson RB, Cohen ND, Clancy JS (2000) Effects of dietary fructooligosaccharide on selected bacterial populations in feces of dogs. Am J Vet Res 61:820–825
Windmueller HG (1982) Glutamine utilization by the small intestine. Adv Enzymol 53:201–237
Windmueller HG, Spaeth AE (1974) Uptake and metabolism of plasma glutamine by the small intestine. J Biol Chem 249:5070–5079
Wu G (2009) Amino acids: metabolism, functions, and nutrition. Amino Acids 37:1–17
Wu G (2018) Principles of Animal Nutrition. CRC Press, Boca Raton, Florida
Wu G (2020a) Metabolism and functions of amino acids in sense organs. Adv Exp Med Biol 1265:201–217
Wu G (2020b) Important roles of dietary taurine, creatine, carnosine, anserine and 4-hydroxyproline in human nutrition and health. Amino Acids 52:329–360
Wu G (2022) Amino Acids: Biochemistry and Nutrition, 2nd edn. CRC Press, Boca Raton, Florida
Wu G, Rhoads JM (2024) Arginine, citrulline, nitric oxide, and glutamine: In: Modern Nutrition in Health and Disease, 12th edition (Tucker KL, Ross AC, Jensen GL, Peterson KE, Touger-Decker R, Duggan CP eds). Jones & Bartlett Learning, LLC, Burlington, MA, pp 475-488
Xenoulis PG, Palculict B, Allenspach K, Steiner JM, Van House AM, Suchodolski JS (2008) Molecular-phylogenetic characterization of microbial communities imbalances in the small intestine of dogs with inflammatory bowel disease. FEMS Microbiol Ecol 66:579–589

Yu YM, Wagner DA, Tredget EE, Walaszewski JA, Burke JF, Young VR (1990) Quantitative role of splanchnic region in leucine metabolism: L-[1-^{13}C,^{15}N]leucine and substrate balance studies. Am J Physiol 259:E36-51
Yu YM, Burke JF, Tompkins RG, Martin R, Young VR (1996) Quantitative aspects of interorgan relationships among arginine and citrulline metabolism. Am J Physiol 271:E1098–E1109
Zhang Q, Hou YQ, Bazer FW, He WL, Posey EA, Wu G (2021) Amino acids in swine nutrition and production. Adv Exp Med Biol 1285:81–107
Zoran DL (2010) Obesity in dogs and cats: a metabolic and endocrine disorder. Vet Clin North Am Small Anim Pract 40:221–239

Chapter 3
The Use of Diets in the Diagnosis and Treatment of Common Gastrointestinal Diseases in Dogs and Cats

Nancy H. Ing and Joerg M. Steiner

Abstract The nutritional health of dogs and cats is important to pet owners around the world. Nutrition is inextricably linked to the health of the gastrointestinal system and vice versa. Gastrointestinal signs, such as vomiting, diarrhea, anorexia, or weight loss, are one of the most common reasons that dog and cat owners make non-routine appointments with veterinarians. Those patients are evaluated systematically to identify and/or rule out the causes of the symptoms. Some causes of chronic diarrhea are within the gastrointestinal tract while others are secondary to pathogenic factors outside the digestive system. Some useful biomarkers of chronic intestinal disease (enteropathy) exist in serum and feces. After determination that the clinical signs are due to primary gastrointestinal disease and that there is no parasitism, specific diets are used for at least two weeks. There are several types of diets for pets with chronic enteropathies. There are limited ingredient diets and hydrolyzed protein diets with reduced levels of allergens. There are also highly digestible and fiber-enhanced diets. Some diets contain probiotics and/or prebiotics. If symptoms do not improve and the patient is stable, a diet from a different class may be tried. For chronic enteropathies, the prognosis is generally good for symptom resolution or at least improvement. However, if interventions with novel diets do not ameliorate the symptoms of chronic enteropathy, then antibiotic, anti-inflammatory, or immunosuppressant therapy or further, more invasive diagnostics such as taking an intestinal biopsy, may be indicated. Pancreatitis is a common gastrointestinal disease in dogs and cats and patients may present with mild to severe disease. Many patients with mild to moderate disease can be successfully treated with early supportive care, including feeding a low-fat diet. A novel pharmaceutical, fuzapladib (Panoquell-CA1) looks very promising for treating more severe forms of acute pancreatitis in dogs. Maintenance on a low-fat diet may prevent pancreatitis in at-risk dogs. Future advances

N. H. Ing (✉)
Department of Animal Science, Texas A&M University, College Station, TX 77843, USA
e-mail: N-ing@tamu.edu

J. M. Steiner
Department of Small Animal Clinical Sciences, Texas A&M University, College Station, TX 77843, USA

G. Wu (ed.), *Nutrition and Metabolism of Dogs and Cats*, Advances in Experimental Medicine and Biology 1446, https://doi.org/10.1007/978-3-031-54192-6_3

in medicine will allow pet owners and veterinarians to use dietary management to maximize the health of their dogs and cats.

Keywords Allergen · Biomarker · Enteropathy · Intestinal dysbiosis · Microbiome · Pancreatitis · Limited antigen diet · Hydrolyzed protein diet · Easily digestible diet

Abbreviations

CAFR	Cutaneous adverse food reaction
FOS	Fructooligosaccharides
GI	Gastrointestinal
LFA-1	Leukocyte function-associated antigen 1 alpha
PERT	Pancreatic enzyme replacement therapy
SCFA	Short chain fatty acid

3.1 Introduction

Dogs and cats are the most popular pets in the USA, with 69% of households owning one or more dogs and 45% owning one or more cats in 2021 (Statistica.com). According to the American Pet Products Association, pet owners spent $125 billion on their pets in 2021, with the majority being spent for pet food. In addition, the popularity of owning dogs and cats as pets is growing in the USA and other developed countries around the world (Li and Wu 2023). Owning dogs and cats provides several benefits for pet owners, including unconditional love, social attention, and tactile interactions. Dog owners may have increased physical exercise from walking their dogs. Dog and cat owners have health benefits including decreased blood pressure, serum cholesterol and triglycerides, and sensitivity to mental stressors (Levine et al. 2013). Dog ownership reduces the risk of dying from all causes by 24% and even by 65% for owners with a prior coronary disease event (Kramer et al. 2019). People viewing a pet owner may judge them to be more mature, responsible, compassionate, and outgoing than non-pet owners (Gomez-Leal et al. 2021). Therefore, the health of dog and cat pets is very important to the physical and mental health of the owners (Saunders et al. 2017). Pet owners depend upon veterinarians and other professionals in the companion animal industry to provide expert advice and support for the health of their companion animals.

Nutrition is defined as "the process of providing or obtaining the food necessary for health and growth" (Oxford dictionary). Adequate nutrition is dependent upon numerous processes including food intake, digestion, absorption, metabolism, and excretion (Wu 2018). Some diseases in dogs and cats can reduce food intake, and in cats, finicky eating behaviors and stressors can reduce or abolish food intake, resulting in serious and even fatal diseases, e.g., hepatic lipidosis. Digestion is the physical and enzymatic breakdown of macronutrients (i.e., carbohydrates, fats, and proteins) into their simple components: sugars, fatty acids, and amino acids that can be absorbed via the gastrointestinal (GI) tract. Several segments of the GI tract are involved in digestion, absorption, and metabolism of nutrients, including the oral cavity, the stomach, and the small intestine. The liver and exocrine pancreas are also critical to digestion, absorption, and metabolism. The liver is the first organ to receive blood draining from the GI tract via the portal vein and is the central metabolic organ. The liver makes glycogen from blood glucose and stores it until the process is reversed to raise blood glucose to power the body. The liver also produces bile salts that are secreted into the small intestine to emulsify fats (Hou et al. 2020). The exocrine pancreas secretes digestive enzymes into the small intestine that break down macromolecules: proteins into amino acids, fat into fatty acids, and complex carbohydrates into simple sugars. Altogether, the health of the GI tract and supporting organs is absolutely essential for good nutrition for animals.

3.2 Nutrition and Gastrointestinal Health are Linked

Good nutrition is, in turn, required for the health of the GI tract. Recently gained knowledge highlights the importance of the microbiome of the GI tract to nutrition and health (Barko et al. 2018; Pilla and Suchodolski 2020). The microbiome includes bacteria (which most studies focus on), as well as fungi and viruses. These extend the metabolic and biosynthetic functions of the host. Utilizing micromolecular tools, we now know that the GI tract of dogs and cats contains 500 to 5000 different species of bacteria. The prevalence of bacteria is very low (usually only 1 species) in the stomach due to the acidic environment from gastric acid secretion. Bacterial numbers and diversity increase in the more caudal parts of the GI tract. While the small intestine has more bacterial species than the stomach, the large intestine has the most and the contents of the colon are largely the microbiome. The types of bacteria in the GI tract may be beneficial (e.g., *Lactobacillus* species) or pathogenic (e.g., some strains of *Escherichia. coli*), but they may also be potential pathogens (e.g., *Clostridium perfringens*). When the health of the GI tract is compromised, usually there is “intestinal dysbiosis,” which means that the microbiome is altered and the nutritional status of the patient is impaired.

3.3 Gastrointestinal Disease in Dogs and Cats is Common

Gastrointestinal clinical signs are one of the most common reasons that dog and cat owners take their pets to the veterinarian for diagnosis and treatment (O'Neill et al. 2014, 2021). GI symptoms of pets include vomiting, anorexia, and/or diarrhea, all of which can lead to weight loss and failure to thrive. Both diarrhea and vomiting are associated with malassimilation. Together, these problems become very concerning for pet owners. Chronic diarrhea is common in both dogs and cats. Causes of chronic gastrointestinal signs can be "primary," which include infectious, inflammatory, neoplastic, toxic, and mechanical diseases of the GI tract (Fig. 3.1). But there are also secondary causes of GI signs, including diseases of the exocrine pancreas (i.e., inflammation or lack of digestive enzymes), liver (i.e., inflammation, changes in blood flow, or lower levels of bile salts), and kidney (i.e., not removing toxins from the blood leading to gastrointestinal signs). Secondary GI diseases can also be due to hypoadrenocorticism as well as hypothyroidism in dogs and hyperthyroidism in cats. The challenge to the veterinarian is to identify the cause of the GI disease, if possible, and to create a viable treatment plan for each patient.

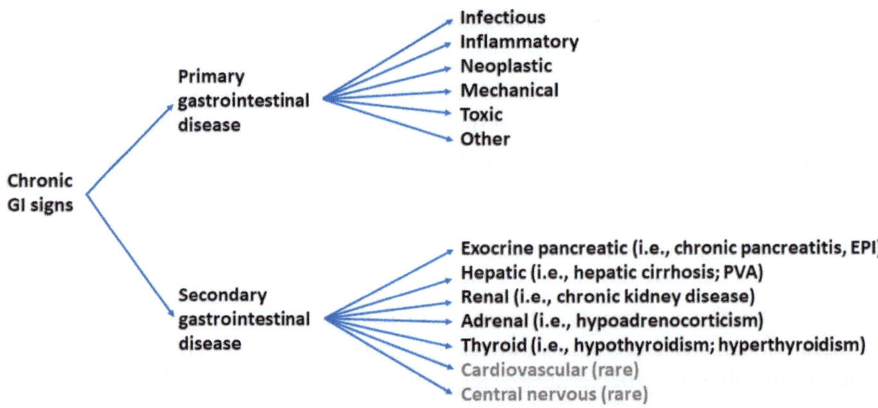

Fig. 3.1 Causes of chronic gastrointestinal signs in dogs and cats. This figure shows the many possible causes for dogs and cats with chronic gastrointestinal signs (GI signs), which can be broadly categorized into primary and secondary gastrointestinal diseases. Cardiovascular and central nervous system diseases are rare causes of chronic diarrhea and are shown in gray. EPI = exocrine pancreatic insufficiency; PVA = portosystemic vascular anomaly

3.4 Diagnosis of Dogs and Cats with Signs of Chronic Gastrointestinal Disease

Dogs and cats presenting to the veterinarian with GI signs, of which the chief complaint is often chronic diarrhea, require a systematic health evaluation. A complete history includes when the onset of symptoms occurred, an exact description of the problem, including a detailed description of the appearance of the feces, and the diets, treats, supplements, and medications being administered to the pet. A thorough physical examination is performed to identify any problems with the pet, including examination of the oral cavity and palpation of the abdomen to assess pain. The pet is also weighed, and the body weight, body condition score, and muscle condition score are compared to previously recorded measurements. Diagnostic tests include a complete blood cell count, blood chemistry panel, and urinalysis to systematically assess the animal, as well as the health of the kidney and liver. GI disease may cause protein loss, with lower serum albumin concentrations, as well as lower concentrations of serum calcium and cholesterol. Specific blood tests for thyroid hormones, cortisol, and pancreatic lipase (for pancreatitis) may be done. Fecal flotation is done to identify protozoal parasites as well as eggs of helminths. If present, the parasites are treated using specific anthelmintic agents for the parasites that were identified. If there is no evidence of an endoparasite, the patient is still treated with a broad-spectrum anthelminthic agent. Abdominal radiographs may be useful in identifying foreign objects in the GI tract, which usually cause acute clinical signs and thus are a rare cause of chronic signs of gastrointestinal disease. Despite this thorough assessment, a definitive cause of the chronic gastrointestinal signs may not be identified because there are so many different etiologies. Additional biomarkers may be useful to further assess the patient and guide decisions about treatment options.

3.5 Biomarkers of Chronic Gastrointestinal Disease

3.5.1 Functional Biomarkers

There are several functional biomarkers of GI health and disease in dogs and cats. Cobalamin (vitamin B_{12}) is an essential vitamin in dogs and cats and symptoms of cobalamin deficiency include diarrhea, anorexia, weight loss, immunodeficiency, and central and peripheral neuropathies (Heilmann and Steiner 2018). Dogs and cats can become deficient in cobalamin if they have chronic gastrointestinal disease (chronic enteropathy) as the absorption of cobalamin requires that it binds intrinsic factor in the duodenum (Kather et al. 2019). Since, in contrast to humans, where intrinsic factor is secreted by the gastric mucosa, intrinsic factor in dogs and cats is almost exclusively secreted by the pancreas, dogs and cats with exocrine pancreatic insufficiency also frequently have cobalamin deficiency. After it is bound to intrinsic factor, active absorption of cobalamin occurs in the ileum, which expresses the Cubam receptor for cobalamin. The Cubam receptor is composed of two gene products: amnionless (AMN) and cubulin (CUBN). Dogs with a mutation in one of these genes have been identified in giant schnauzers, border collie, beagles, and Australian shepherd breeds (Hanisch et al. 2018), similar to some people from Norway and Finland (Carmel et al. 2003). One of the dog breeds that is predisposed to having a low serum cobalamin concentration is the Chinese Shar Pei, although the specific mode of inheritance has not yet been identified (Bishop et al. 2012). However, in addition to these inherited causes of cobalamin deficiency, the most common cause of cobalamin deficiency in both dogs and cats is chronic enteropathy. Hypocobalaminemia is very common in the German shepherd dog, while golden retrievers have a lower risk of this disorder (Dandrieux et al. 2013). It is recommended that serum cobalamin levels are measured in all dogs and cats with gastrointestinal signs, especially if the symptoms are chronic (i.e., have persisted for more than 3 weeks). If cobalamin levels are low, supplementation should be initiated either orally or parenterally (Toresson et al. 2016; Hanisch et al. 2018).

Other functional markers of GI health in serum are methylmalonic acid and folate concentrations (Heilmann and Steiner 2018). The organic acid methylmalonic acid requires cobalamin to be converted to succinyl-CoA. Therefore, levels of methylmalonic acid increase in the blood as cellular cobalamin levels fall and in turn lead to a decrease in methylmalonyl-CoA mutase activity, which is required for metabolism of some amino acids and fats. Serum folate (vitamin B_9) is another essential dietary vitamin. Concentrations of folate relate to the health of the proximal small intestine, where this vitamin is absorbed. Chronic intestinal malabsorption (if longstanding and severe) can cause folate levels in the serum to drop, but it is not as sensitive a biomarker as serum cobalamin (Ullal et al. 2023). Hypofolatemia can be treated with oral supplementation with folate, but its effectiveness in treating chronic enteropathies has not been documented. The protein α1-proteinase inhibitor (α1-PI) is another functional marker of GI health. Its levels drop in serum in dogs and cats with protein-losing enteropathies. More importantly, fecal α1-PI concentrations increase in the feces in dogs and cats with increased gastrointestinal protein loss (Melgarejo et al. 1997). While most proteins lost into the GI tract are digested by proteases, α1-PI is resistant to proteolysis and can be measured in feces. Fecal alpha1-proteinase levels increase early in protein-losing enteropathies before clinical signs of hypoalbuminemia develop and identify the source of the problem as GI.

3.5.2 Biochemical Markers of Inflammation

Veterinary medicine has followed human medicine in characterizing several biochemical markers that relate to chronic inflammatory enteropathy in dogs. Studies have focused on serum C-reactive protein, serum perinuclear anti-neutrophilic cytoplasmic antibodies, serum 3-bromotyrosine, serum and urinary N-methylhistamine, fecal and serum calprotectin, fecal and serum calgranulin C, and serum soluble receptor for advanced glycation end products (Heilmann and Steiner 2018). Of the above, assessing concentrations of serum C-reactive protein and fecal calprotectin appear to be the most useful, especially for identifying objective improvement during a dietary trial. Both have been shown to correlate with the severity of the chronic inflammatory enteropathy in dogs. In addition, serum C-reactive protein and fecal calprotectin levels were useful in distinguishing partial responders in diet trials (which may need more time on the diet) from non-responders, which may require antibiotics, anti-inflammatory, or immunosuppressive treatments.

3.5.3 Assessment of the Intestinal Microbiome in Dogs and Cats with Enteropathies

Alterations of the intestinal and fecal microbiome have been identified in both dogs and cats with chronic enteropathies. The fecal dysbiosis index uses quantitative PCR on fecal DNA to compare the abundances of eight bacterial groups between healthy dogs and those with chronic enteropathy (AlShawaqfeh et al. 2017; Heilmann and Steiner 2018). The abundances of total bacteria are assessed together with those of *Blautia*, *Clostridium hiranosis*, *Escherichia coli*, *Faecalibacterium*, *Fusobacterium*, *Streptococcus*, and *Turicibacter*. The levels are used in an algorithm to generate a fecal dysbiosis index, which can be compared to that of a normal dog. A similar dysbiosis index has also been described for cats. These dysbiosis indices may be useful for identifying responses to dietary treatments (Rossi et al. 2020). The dysbiosis index discriminated between healthy dogs and those with chronic inflammatory enteropathy (AlShawaqfeh et al. 2017). In another study, 30 dogs with large bowel diarrhea that were fed a fiber-enhanced diet supplemented with probiotics for 30 days had their initial state of intestinal dysbiosis changed to one of normobiosis (Rossi et al. 2020).

3.6 Dietary Treatment Trials

While the causes of diarrhea in dogs and cats may be unclear, 50 to 60% of dogs and cats with chronic diarrhea respond to dietary treatment alone (Rudinsky et al. 2018). Treatment trials use a novel diet to feed the dog or cat for some weeks to assess any improvement in the symptoms. There is no one dietary treatment that works for every case. There is also no formula for which type of diet (described below) to try first. If no improvement is noted in a case of diarrhea within two weeks of feeding the novel diet, then the type of diet is changed. This is in contrast to dogs and cats with allergic skin disease due to dietary allergens, in which the trial duration is at least eight weeks (Olivery et al. 2015). There are several companies that make specialized diets to address chronic gastrointestinal disease, and the components of the diets vary widely. Many contain prebiotics, which are compounds that feed beneficial gut bacteria. The

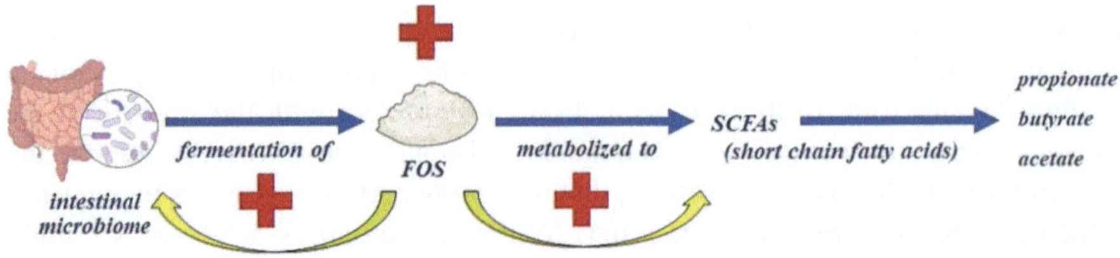

Fig. 3.2 Potential benefits of FOS. Fructooligosaccharides (FOS) are non-metabolizable poly-sugars that can be fermented by the intestinal microbiome to produce short chain fatty acids (SCFAs) that represent the major energy source for enterocytes. Adding FOS to the diet will enhance the proliferation of the beneficial bacteria of the intestinal microbiome and lead to the synthesis of additional SCFAs. This figure was created with BioRender.com

prebiotic fructooligosaccharides (FOS) are fermented by GI bacteria to produce short chain fatty acids (SCFAs; Fig. 3.2). Thus, the FOS feeds the bacteria and the SCFAs produced stimulate the proliferation of enterocytes (Venegas et al. 2019). This is in contrast to probiotics, which represent the beneficial bacteria themselves. While some pet food manufacturers spray their diets with bacterial spores, these diets containing probiotics have not been critically assessed and probiotics should instead be used as a separate nutraceutical.

There are several types of diets used in treatment trials for diarrhea in dogs and cats. The choice of therapeutic diet is based on assessment of clinical signs and a thorough diet history. Even so, multiple diets may need to be tried to better tailor the diet composition based on clinical response. Limited ingredient diets have one protein source and one carbohydrate source that the dog or cat has not eaten before. Both protein and even carbohydrate sources have some potential protein antigens that could trigger an immune reaction in the dog's GI tract. Thus, limiting both the protein as well as the carbohydrate sources is essential. These diets typically have the single protein and carbohydrate sources in the product name. Hydrolyzed protein diets have the proteins cleaved into small fragments to reduce their antigenicity. There is a difference between products in the degree of hydrolyzation and range of the sizes of the peptides, with the more hydrolyzed, smaller peptides being less allergenic. This is an excellent concept, but only a few studies have been done with these diets in dogs (Marks et al. 2002, Mandigers et al. 2010). The latter study used 26 dogs with chronic small bowel enteropathy and fed them either a hydrolyzed protein diet or a highly digestible diet for three years. They determined that more dogs remained in remission for diarrhea while eating the hydrolyzed protein diet than the dogs eating the highly digestible diet. Highly digestible diets are moderate in fat because fat is difficult to digest. These diets typically contain prebiotics such as FOS. These are not digested in the small bowel but reach the large bowel to serve as a source of fermentation by the beneficial bacteria there. The highly digestible diets can also contain other components, such as an altered n3/n6 fatty acid ratio or ginger. One of these highly digestible diets showed a good response in the aforementioned study but was not as effective at reducing diarrhea long term as was the hydrolyzed protein diet tested (Mandigers, et al. 2010). There are also fiber-enhanced diets that can be effective at controlling large bowel diarrhea. A recent study of feeding a diet containing a proprietary mixture of fibers containing antioxidants and polyphenols showed it greatly reduced chronic large bowel diarrhea in 31 dogs (Fritsch et al. 2022). The diet had 6% crude fiber on a dry matter basis. In addition to that, the diet's prebiotic fiber blend contained beet pulp, citrus pulp, pressed cranberries, flaxseed, and pecan shell flour with 59 mg/g insoluble fiber and 6.mg/g soluble fiber. In another study, a different fiber-enhanced diet fed along with a probiotic mixture successfully treated large bowel diarrhea in 30 dogs (Rossi et al. 2020). That diet was Intestinal Colitis Phase 1 from Forza10 and contained 21% dietary fiber including 8% raw fiber. Fiber-enhanced diets should be renamed to indicate what fiber types they contain: soluble vs. insoluble fibers and fermentable vs. non-fermentable fiber. One possible nomenclature could be a "soluble fiber-enhanced diet" in contrast to a "bulking fiber-enhanced diet," which would contain more insoluble fiber. The type of fiber influences the effects on the GI tract and digestion. It must also be mentioned that diets contain fiber sources, rather than a specific fiber. For example, psyllium is frequently used in dogs with chronic diarrhea, but psyllium is not a specific fiber, but rather a fiber source that contains both fermentable and non-fermentable fiber and may even contain substances with pharmaceutical activity.

As mentioned, the majority of dog and cat primary diarrheal diseases do respond to dietary interventions. However, if the diarrhea is unresponsive to sequential dietary treatment trials, the veterinarian may prescribe an antibiotic or even an anti-inflammatory and/or immunosuppressive drug. Ideally, this is preceded by an intestinal biopsy to characterize the inflammation and rule out, for example, a fungal infection. Typically, antibiotics reduce the diversity of the GI microbiome (Bottero et al. 2022), but many dogs and cats with chronic diarrhea will respond to antibiotic therapy (Westermarck et al. 2005; Kilpinen et al. 2011). The goal would be to use an antibiotic that is not associated with many side effects and does not have frequent other uses. The antibiotic most commonly used for this purpose is tylosin, a macrolide antibiotic that used to be used more frequently as a feed additive in poultry and is not used in humans in Western countries. Along with fecal transplantation, prebiotics, probiotics, and antibiotics, all target the microbiome of the GI tract. When there is GI disease, the diversity of the microbiome in the gut is decreased.

3.7 Allergens in Dog and Cat Diets

Skin is the main allergic-response organ of the dog. There are cases in which the diet of the dog causes cutaneous adverse food reactions (CAFRs), which can be severe. The reactions may cause itchy skin so that the animal scratches and removes patches of hair or chews its hair off everywhere it can reach. Compared to the number of times owners bring their dogs and cats in for any diagnosis, only 1 to 2% turn out to be CAFRs for dogs and only 0.2% for cats (Olivry and Mueller 2017). For dogs, the most common dietary allergens are (ranked from the top) beef, dairy products, chicken, wheat, and lamb (Mueller et al. 2016). Similarly, for cats, the most common dietary allergens that cause CAFR are beef, fish, chicken, wheat, corn, dairy products, and lamb. Importantly, the ingredients on the pet food label do not always fully correlate with the contents (Olivry and Mueller 2018). For example, if a pet food manufacturer does not completely clean their equipment between making different products, there will be carryover into subsequent foods. It is also noteworthy that even contamination of pet food with storage mites has been reported to cause CAFR in dogs and cats (Olivry and Mueller 2019). Diagnosis of CAFR is done by dietary restriction-provocation trials. The suspected allergens are removed from the diet for at least eight weeks and remission of skin symptoms is recorded. Then, the pet may have a provocation diet with individual suspected allergens to identify them specifically so the owner can avoid feeding them in the future.

3.8 Pancreatitis

Dogs and cats are susceptible to the inflammatory disease of the pancreas: pancreatitis. The causes of acute pancreatitis can often not be determined and the condition is labeled as idiopathic. However, some cases are related to eating unusual and/or high fat foods (Steiner 2022). Miniature schnauzers are much more susceptible to pancreatitis than other breeds of dogs. Yorkshire terriers, cocker spaniels, dachshunds, and poodles are also commonly affected by pancreatitis, while golden retrievers are not (Cridge et al. 2022a). The risk of having increased pancreatic lipase in serum (one diagnostic indicator of pancreatitis) is greater in Eskimo dogs, fox terriers, pointers, and greyhounds, while that risk is low in beagles, Labrador retrievers, bulldogs, and golden retrievers (Cridge et al. 2022b). In dogs, but not cats, hypertriglyceridemia (> 500 mg/dL) is a risk factor for developing pancreatitis. Some prescription drugs are thought to cause pancreatitis in people, and the same might also be true in dogs and cats. The pathology of pancreatitis involves decreased secretion of digestive enzymes into the small intestine and activation of the protease trypsin in situ, within pancreatic acinar cells. The accumulation of activated proteases in the tissue causes edema, bleeding, inflammation, and necrosis (Fig. 3.3; Cridge et al. 2022a). White blood cells are attracted to the inflamed pancreas and secrete cytokines. Along with non-esterified fatty acids from fat necrosis, the cytokines go into the circulation and can cause generalized disease including intravascular coagulation, kidney or lung failure, and multiorgan failure leading to death.

The symptoms of severe pancreatitis in dogs are anorexia, vomiting, weakness, abdominal pain, dehydration, and diarrhea (Steiner 2022). Symptoms of severe pancreatitis in cats are less well defined but include anorexia, lethargy, dehydration, and hypothermia. Milder pancreatitis cases in dogs and cats can have vague symptoms such as anorexia and diarrhea. In people, abdominal pain is a major symptom of pancreatitis and pet owners may overlook this symptom in their dogs and cats with pancreatitis. The diagnosis of pancreatitis is based on history, physical examination, and clinical diagnostics. In both dogs and cats, the most specific diagnostic test for pancreatitis is serum pancreatic lipase immunoreactivity using species-specific

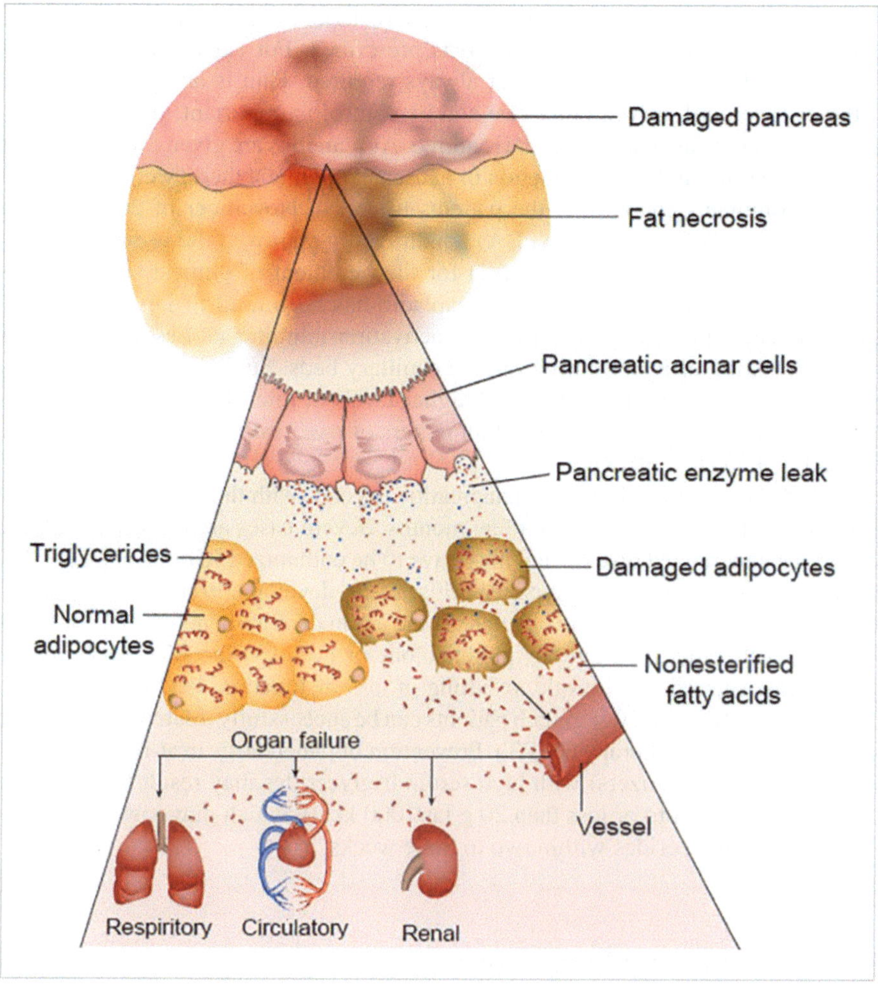

Fig. 3.3 Pathogenesis of pancreatitis. Pancreatic enzyme leakage and activation in situ leads to inflammation and fat necrosis, releasing non-esterified fatty acids that cause systemic inflammation and organ failure. This figure was reused with permission from Cridge et al. (2022a) J Vet Intern Med 36:847–64

reagents. For dogs, two commercial tests for pancreatic lipase in serum (Spec cPL™ and SNAP cPL™) show high sensitivity and specificity (McCord et al. 2012). Ultrasonography and fine needle aspiration of the pancreas may provide information about the cells present and whether the condition is acute or chronic (Cridge et al. 2020, 2021).

Treatment of pancreatitis in dogs and cats revolves around supportive care. The best results for acute pancreatitis occur with early support, including fluid therapy, pain medication, antiemetics, and nutritional support, most preferably by enteral nutrition (Steiner 2022). For dogs, a low-fat diet (i.e., a diet containing equal to or less than 20 g of fat per 1000 kcal) is prescribed. For cats, a moderately fat-restricted diet can be used but is not always required. Pain is assumed to be part of the disease until proven otherwise and its control is important. In severe cases, constant intravenous opioids may be required. Antiemetics are given even to animals that are not vomiting for the reduction of nausea. Antiemetic therapy combined with appetite stimulants encourages the patients to eat. In November of 2022, the US Food and Drug Administration (FDA) conditionally approved the use of the novel medicine, fuzapladib (Panoquell-CA1) for the treatment of clinical signs in dogs with acute pancreatitis. Since 2018, fuzapladib has been used in Japan with good results. Fuzapladib sodium is an LFA-1 activation inhibitor and as such prevents the extravasation of neutrophils from the capillary beds. Studies of experimentally induced pancreatitis in dogs identified higher levels of microRNAs including miR-375 in serum, which may be useful biomarkers of pancreatitis in the future (Rouse et al. 2017; Lee et al. 2018).

The prognosis of mild cases of pancreatitis is good, with the patients maintained on the lower fat diets for at least a few months. Severe cases are closely monitored for serious generalized disease and may have to be euthanized (Cridge et al. 2022a). If the dog or cat survives acute pancreatitis, it can develop chronic disease and be susceptible to more pancreatitis events. These animals may benefit from immunosuppressant therapy, for example, prednisolone or cyclosporine. The pancreas can also become scarred and fibrotic, resulting in loss of function, leading to exocrine pancreatic insufficiency. Often such patients can be successfully treated by pancreatic enzyme replacement therapy (PERT). Prevention of pancreatitis in at-risk dogs (e.g., older miniature schnauzers) with high serum triglycerides may result from feeding a low-fat diet (equal to or less than 20 g fat/1000 kcal), which may lower the levels of the serum triglycerides within two to three weeks.

3.9 Summary and Perspectives

Good nutrition is critically linked to the health of the GI tract and the whole animal, and vice versa. It is common for dogs and cats to get chronic signs of GI disease (vomiting, diarrhea, and weight loss) that require medical intervention by a veterinarian. This begins with a complete workup, including history, physical examination, blood work, and urinalysis to attempt to define the possible primary or secondary causes of the GI symptoms. For chronic enteropathies caused by problems within the GI tract, there are several serum and fecal biomarker assays that may provide information guiding the disease diagnosis and assessing response during dietary trials. In trials with a novel diet, if symptoms do not improve within two weeks, a different class of a novel diet is tried. For chronic enteropathies, the symptoms improve for most dogs and cats on one or another novel diet, which the pet is then maintained on. For patients with GI symptoms that do not improve, antibiotic, anti-inflammatory, or immunosuppressive treatments can be tried. Acute pancreatitis is a life-threatening disease of dogs and cats. Control of pain and nausea allowing oral alimentation with a low-fat diet is key to successful treatment of pancreatitis. A newly approved drug, fuzapladib, that inhibits neutrophil extravasation appears to be effective in treating acute pancreatitis in dogs. The future promises more advances in diagnosing and treating GI diseases to enhance the health and longevity of our beloved dogs and cats.

References

AlShawaqfeh MK, Wajid B, Minamoto Y, Markel M, Lidbury JA, Steiner JM, Serpedin E, Suchodolski JS (2017) A dysbiosis index to assess microbial changes in fecal samples of dogs with chronic inflammatory enteropathy. FEMS Microbiol Ecol 93:fix136

Barko PC, McMichael MA, Swanson KS, Williams DA (2018) The gastrointestinal microbiome: a review. J Vet Intern Med 32:9–25

Bishop MA, Xenoulis PG, Bergoff N, Gruztner N, Suchodolski JS, Steiner JM (2012) Partial characterization of cobalamin deficiency in Chinese Shar Peis. The Vet J 191:41–45

Bottero E, Ferriani R, Benvenuti E, Ruggiero P, Astorina S, Giraldi M, Bertoldi L, Benvenuto G, Sattin E, Gianella P, Suchodolski JS (2022) Clinical evaluation and microbiota analysis in 9 dogs with antibiotic-responsive enteropathy: a prospective comparison study. J Vet Intern Med 36:1220–1228

Carmel R, Green R, Rosenblatt DS, Watkins D (2003) Update on cobalamin, folate, and homocysteine. Am Soc Hematol Educ Program 2003:62–81

Cridge H, Sullivant AM, Wills RW, Lee AM (2020) Association between abdominal ultrasound findings, the specific canine pancreatic lipase assay, clinical severity indices, and clinical diagnosis in dogs with pancreatitis. J Vet Intern Med 34:636–643

Cridge H, Twedt DC, Marolf AJ, Sharkey LC, Steiner JM (2021) Advances in the diagnosis of acute pancreatitis in dogs. J Vet Intern Med 35:2572–2587

Cridge H, Scott N, Steiner JM (2022b) Risk factors and clinical presentation in dogs with increased serum pancreatic lipase concentrations—a descriptive analysis. Animals 12:1581

Cridge H, Lim SY, Algül H, Steiner JM (2022a) New insights into the etiology, risk factors, and pathogenesis of pancreatitis in dogs: potential impacts on clinical practice J Vet Intern Med 36:847–864

Dandrieux JR, Noble PJM, Halladay LJ, McLean L, German AJ (2013) Canine breed dispositions for marked hypocobalaminemia or decreased folate concentration assessed by a laboratory survey. J Small Anim Prac 54:143–148

Fritsch DA, Wernimont SM, Jackson MI, MacLeay JM, Gross KL (2022) A prospective multicenter study of a fiber-supplemented dietary intervention in dogs with chronic large bowel diarrhea. BMC Vet Res 18:244

Gomez-Leal R, Costa A, Megias-Robles A, Fernandez-Berrocal P, Faria L (2021) Relationship between emotional intelligence and empathy towards humans and animals. Peer J. https://doi.org/10.7717/peerj.11274

Hanisch F, Toresson L, Spillmann T (2018) Cobalamin deficiency in dogs and cats. Tierarztl Prax Ausg Kleintiere Heimtiere 46:309–314

Heilmann RM, Steiner JM (2018) Clinical utility of currently available biomarkers in inflammatory enteropathies of dogs. J Vet Intern Med 32:1495–1508

Hou YQ, Hu SD, Li XY, He WL, Wu G (2020) Amino acid metabolism in the liver: nutritional and physiological significance. Adv Exp Med Biol 1265:21–37

Kather S, Gruetzner N, Kook PH, Dengler F (2019) Review of cobalamin metabolism in dogs. J Vet Intern Med 34:13–28

Kilpinen S, Spillmann T, Syrja P, Skrzypczak T, Louhelainen M, Westermarck E (2011) Effect of tylosin on dogs with suspected tylosin-responsive diarrhea: a placebo-controlled, randomized, double-blinded, prospective clinical trial. Acta Vet Scand 53:26

Kramer CK, Mehmood SM, Suen RS (2019) Dog ownership and survival: a systematic review and meta-analysis. Circ Cardiovasc Qual Outcomes 12:e005554

Lee H-B, Park H-K, Choi H-J, Lee S, Lee S-J, Lee J-Y, Cho E-H, Han H-J, Seok J-H, Son W-C (2018) Evaluation of circulating microRNA biomarkers in the acute pancreatic injury dog model. Int J Mol Sci 19:3048

Levine GN, Allen K, Braun LT, Christian HE, Friedmann E, Taubert KA, Thomas SA, Wells DL, Lange RA (2013) Pet ownership and cardiovascular risk—a scientific statement from the American heart association. Circulation 127:2353–2363

Li P, Wu G (2023) Amino acid nutrition and metabolism in domestic cats and dogs. J Anim Sci Biotechnol 14:19

Mandigers PJJ, Biourge V, van den Ingh TSGAM, Ankringa N, German AJ (2010) A randomized, open-label, positively-controlled field trial of a hydrolyzed protein diet in dogs with chronic small bowel enteropathy. J Vet Intern Med 24:1350–1357

Marks SL, Laflamme DP, McAloose D (2002) Dietary trial using a commercial hypoallergenic diet containing hydrolyzed protein for dogs with inflammatory bowel disease. Vet Ther 3:109

McCord K, Morley PS, Armstrong J, Simpson K, Rishniw M, Forman MA, Biller D, Parnell N, Arnell K, Hill S, Avgeris S, Gittelman H, Moore M, Hitt M, Oswald G, Marks S, Burney D, Twedt D (2012) A multi-institutional study evaluating the diagnostic utility of the Spec cPL and SNAP(R) cPL in clinical acute pancreatitis in 84 dogs. J Vet Intern Med 26:888–896

Melgarejo T, Tamayo A, Williams DA (1997) Fecal alpha$_1$-protease inhibitor (α_1-PI) for the diagnosis of canine protein-losing enteropathy. J Vet Int Med. 11:115

Mueller RS, Olivry T, Prelaud P (2016) Critically appraised topic on adverse food reactions of companion animals (2): common food allergen sources in dogs and cats. BMC Vet Res 12:9

Olivry T, Mueller RS (2017) Critically appraised topic on adverse food reactions of companion animals (3): prevalence of cutaneous adverse food reactions in dogs and cats. BMC Vet Res 13:51

Olivry T, Mueller RS (2018) Critically appraised topic on adverse food reactions of companion animals (5): discrepancies between ingredients and labeling in commercial pet foods. BMC Vet Res 14:24

Olivry T, Mueller RS (2019) Critically appraised topic on adverse food reactions of companion animals (8): storage mites in commercial pet foods. BMC Vet Res 15:385

Olivry T, Mueller RS, Prelaud P (2015) Critically appraised topic on adverse food reactions of companion animals (1): duration of elimination diets. BMC Vet Res 11:225

O'Neill DG, Church DB, McGreevy PD, Thomson PC, Brodbelt DC (2014) Prevalence of disorders recorded in cats attending primary-care veterinary practices in England. Vet J 202:286–291

O'Neill DG, James H, Brodbelt DC, Church DB, Pegram C (2021) Prevalence of commonly diagnosed disorders in UK dogs under primary veterinary care: results and applications. BMC Vet Res 17:69

Pilla R, Suchodolski JS (2020) The role of the canine gut microbiome and metabolome in health and gastrointestinal disease. Front Vet Sci 6:498

Rossi G, Cerquetella M, Gavazza A, Galosi L, Berardi S, Mangiaterra S, Mari S, Suchodolski JS, Lidbury JA, Steiner JM, Pengo G (2020) Rapid resolution of large bowel diarrhea after the administration of a combination of a high-fiber diet and a probiotic mixture in 30 dogs. Vet Sci 7:21

Rouse R, Rosenzweig B, Shea K, Knapton A, Stewart S, Xu L, Chockalingam A, Zadrozny L, Thompson K (2017) MicroRNA biomarkers of pancreatic injury in a canine model. Exp Tox Path 69:33–43

Rudinsky AJ, Rowe JC, Parker VJ (2018) Nutritional management of chronic enteropathies in dogs and cats. J Am Vet Med Assoc 253:570–578

Saunders J, Parast L, Babey SH, Miles JV (2017) Exploring the differences between pet and non-pet owners: Implications for human-animal interaction research and policy. PLoS ONE. https://doi.org/10.1371/journal.pone.0179494

Steiner JM (2022) Pancreatitis in dogs and cats. In: The Merck veterinary manual. Whitehouse Station, NJ, Merck & Co., Inc.

Toresson L, Steiner JM, Suchodolski JS, Spillmann T (2016) Oral cobalamin supplementation in dogs with chronic enteropathies and hypocobalaminemia. J Vet Intern Med 30:101–107

Ullal TV, Marks SL, Huebner SN, Taylor SL, Shelley CD (2023) Association of folate concentrations with clinical signs and laboratory markers of chronic enteropathy in dogs. J Vet Intern Med 2023:1–10

Venegas DP, De la Fuente MK, Landskron G, Gonzalez MJ, Quera R, Dijkstra G, Harmsen HJM, Faber KN, Hermoso MA (2019) Short chain fatty acids (SCFAs)-mediated gut epithelial and immune regulation and its relevance for inflammatory bowel diseases. Front Immunol 10:277

Westermarck E, Skrzypczak T, Harmoinen J, Steiner JM, Ruaux CG, Williams DA, Eerola E, Sundback P, Rinkinen M (2005) Tylosin-responsive chronic diarrhea in dogs. J Vet Intern Med 19:177–186

Wu G (2018) Principles of animal nutrition. CRC Press, Boca Raton, Florida

Sa[illegible] (2021) Exploring the differences between pet and non-pet [illegible] information research and policy [illegible]
[illegible]
Stro[illegible] (20[illegible]) [illegible] in dogs and cats. In: The Merck [illegible] manual. [illegible] Station, NJ [illegible]
Th[illegible] (20[illegible]) [illegible]
[illegible]
[illegible] (202[illegible]) [illegible]
[illegible]
20[illegible]
[illegible]
[illegible]
[illegible]
[illegible]
[illegible]

Chapter 4
Characteristics of Nutrition and Metabolism in Dogs and Cats

Peng Li and Guoyao Wu

Abstract Domestic dogs and cats have evolved differentially in some aspects of nutrition, metabolism, chemical sensing, and feeding behavior. The dogs have adapted to omnivorous diets containing taurine-abundant meat and starch-rich plant ingredients. By contrast, domestic cats must consume animal-sourced foods for survival, growth, and development. Both dogs and cats synthesize vitamin C and many amino acids (AAs, such as alanine, asparagine, aspartate, glutamate, glutamine, glycine, proline, and serine), but have a limited ability to form de novo arginine and vitamin D_3. Compared with dogs, cats have greater endogenous nitrogen losses and higher dietary requirements for AAs (particularly arginine, taurine, and tyrosine), B-complex vitamins (niacin, thiamin, folate, and biotin), and choline; exhibit greater rates of gluconeogenesis; are less sensitive to AA imbalances and antagonism; are more capable of concentrating urine through renal reabsorption of water; and cannot tolerate high levels of dietary starch due to limited pancreatic α-amylase activity. In addition, dogs can form sufficient taurine from cysteine (for most breeds); arachidonic acid from linoleic acid; eicosapentaenoic acid and docosahexaenoic acid from α-linolenic acid; all-*trans*-retinol from β-carotene; and niacin from tryptophan. These synthetic pathways, however, are either absent or limited in all cats due to (a) no or low activities of key enzymes (including pyrroline-5-carboxylate synthase, cysteine dioxygenase, Δ^6-desaturase, β-carotene dioxygenase, and quinolinate phosphoribosyltransferase) and (b) diversion of intermediates to other metabolic pathways. Dogs can thrive on one large meal daily, select high-fat over low-fat diets, and consume sweet substances. By contrast, cats eat more frequently during light and dark periods, select high-protein over low-protein diets, refuse dry food, enjoy a consistent diet, and cannot taste sweetness. This knowledge guides the feeding and care of dogs and cats, as well as the manufacturing of their foods. As abundant sources of essential nutrients, animal-derived foodstuffs play important roles in optimizing the growth, development, and health of the companion animals.

P. Li
North American Renderers Association, Alexandria, VA 22314, USA

G. Wu (✉)
Department of Animal Science, Texas A&M University, College Station, TX 77843, USA
e-mail: g-wu@tamu.edu

G. Wu (ed.), *Nutrition and Metabolism of Dogs and Cats*, Advances in Experimental Medicine and Biology 1446, https://doi.org/10.1007/978-3-031-54192-6_4

Keywords Dogs · Cats · Nutrition · Metabolism · Feeding · Health · Animal-sourced foodstuffs

Abbreviations

AA	Amino acid
BCAA	Branched-chain amino acid
BW	Body weight
CLA	Conjugated linoleic acid
CP	Crude protein
DHA	Docosahexaenoic acid
DM	Dry matter
EPA	Eicosapentaenoic acid
EAA	Nutritionally essential amino acid
GIT	Gastrointestinal tract
ME	Metabolizable energy
NEAA	Nutritionally nonessential amino acid
NRC	National Research Council
PEPCK	Phosphoenolpyruvate carboxykinase
PUFAs	Polyunsaturated fatty acids
RQ	Respiratory quotient
SCFAs	Short-chain fatty acids
TAGs	Triacylglycerols

4.1 Introduction

The metabolism of most nutrients in domestic dogs and cats is similar to that in other mammals (Baker and Czarnecki-Maulden 1991). Thus, the qualitative dietary requirements of dogs and cats for most nutrients [e.g., amino acids (AAs) that are not formed de novo in animal cells] are similar to those for omnivores (e.g., humans and pigs). However, dogs and cats have a relatively short digestive tract (He et al. 2024) and have evolved to have some unique feeding behaviors and metabolic characteristics that are distinct from most of other nonruminant mammals such as pigs, rats, and humans (Legrand-Defretin 1994; Morris 2002). In addition, dogs differ from cats in some of these physiological and nutritional aspects (Che et al. 2021; Oberbauer and Larsen 2021). For example, the dog has adapted to omnivorous diets containing taurine-rich meat and starch-rich plant ingredients, but the cat must consume at least a portion of animal-sourced foods or the same essential nutrients from synthetic supplements for survival, growth, and development because of their limited ability to synthesize taurine (Baker and Czarnecki-Maulden 1991). In addition, there are peculiar differences in the syntheses and requirements of certain nutrients between dogs and cats, including taurine (essential for tissue integrity; Pion et al. 1987), arginine (essential for maintaining the urea cycle in an active state; Wu and Morris 1998), ω3 and ω6 polyunsaturated fatty acids (PUFAs; essential for cell structure and function; Bauer 2007), vitamin A (essential for retinal health; Case et al. 2011), and niacin (essential for nutrient metabolism; NRC 2006). Therefore, the concentrations of many AAs, fatty acids, and their metabolites in plasma differ between dogs and cats fed the same diet (Hall 2018b). Regardless of their sources of foods, adequate knowledge of nutrient metabolism and requirements by dogs and cats is crucial to ensure their optimal growth, development, and health. The major objective of this article is to integrate the metabolic characteristics of dogs and cats with their nutrition and feeding management.

4.2 Energy Metabolism

Energy is defined as the capacity to do work (Wu 2018). Energy metabolism is the sum of the metabolism of carbohydrates, lipids, and AAs in animals. Resting metabolic rate (measured after an overnight fast) of a dog and a cat, like other animals, is an estimate of its minimum energy requirement. In dogs living in a thermoneutral environment with a 12-h light period/a 12-h dark period, their heat production rate varies with a circadian rhythm, with the lowest rate in the morning (e.g., 124 kcal/kg $BW^{0.75}$/day in beagles), gradually increasing to a maximum in the evening (e.g., 195 kcal/kg $BW^{0.75}$/day in beagles), and then declining through the night (Besch and Woods 1977). Such changes in heat production are largely independent of food intake. The rate of heat production by dogs during the dark period is similar to that during the light period (Woods and Besch 1971). In contrast to dogs, no circadian rhythm in heat production was detected in cats (Riond et al. 2003). Furthermore, thermoneutral zones differ between dogs and cats, and even among different breeds of the same species or different sizes of the same breed (Hill and Scott 2004; Middleton et al. 2017). At an ambient temperature of 25 °C, dogs, but not cats, are within their thermoneutral zone.

In dogs and cats, as in other mammals, the maintenance energy requirement supports energy equilibrium over a long period of time and is calculated as $70.5 \times BW^{0.73}$ kcal (BW is body weight in kg; Brody et al. 1934), which can be simplified as $70 \times BW^{0.75}$ kcal (BW is body weight in kg; Kleiber 1961). Metabolizable energy (ME) is commonly used to express the requirements of dogs and cats for dietary energy. The energy requirements of adult dogs and cats may be related to their body surface areas [0.105 and 0.110 m^2/kg of $BW^{0.67}$, respectively (Hill and Scott 2004)]. NRC (2006) recommended that intact active adult dogs, normal adult lean cats, and obese cats be fed ~130 kcal ME/kg $BW^{0.75}$/day, 100 kcal ME/kg $BW^{0.67}$/day, and 130 kcal ME/kg $BW^{0.4}$/day, respectively. Through an extensive meta-analysis, Bermingham et al. (2010) indicated that the maintenance energy requirement of adult cats based on their BW alone may not be accurate and many factors (e.g., age, sex, neuter status, and composition) other than BW should be taken into consideration. Their proposed allometric equations for predicting the maintenance energy requirements of adult light, normal, and heavy cats were 53.7 kcal/kg $BW^{-1.061}$, 46.8 kcal/kg $BW^{-1.115}$, and 131.8 kcal/kg $BW^{-0.366}$, respectively, and were 58.4 kcal/kg lean $mass^{-1.140}$.

Chronic imbalances between energy intake and expenditure result in obesity in dogs and cats, as in other animals (Larsen and Oberbauer 2023; Zoran 2023). As reported for many other mammals (e.g., humans, rats, and sheep), brown adipose tissue (BAT) is present in neonatal dogs and nearly absent from adult dogs (Holloway et al. 1985). Likewise, there is more BAT in neonatal cats than in adult cats (Clark et al. 2013; Loncar and Afzelius 1989). Maintaining BAT in an active state plays an important role in preventing obesity in cats. A recent study has indicated the browning and beiging of adipose tissue in dogs (Lyer et al. 2019). Both shivering and non-shivering mechanisms are responsible for increasing heat production in dogs exposed to a cold environment. Epinephrine and norepinephrine stimulate lipolysis in both BAT and white adipose tissues to provide fatty acids and glycerol as metabolic fuels (Wu 2018).

4.3 Lipid Metabolism in Dogs and Cats

4.3.1 Overview of Lipid Metabolism

Lipids are hydrocarbon compounds (e.g., fats, long-chain fatty acids, and cholesterol) that are soluble in organic solvents and, except for short- and medium-chain fatty acids, are insoluble in water. Fats contribute to the palatability and texture of pet foods, and also have important metabolic functions in animals (Wu 2018). The carbon of a fatty acid can be counted from either the carboxyl group (the Δ nomenclature system) or the methyl group (the ω nomenclature system). In dogs and cats, as in other animals, no double bond can be introduced beyond the Δ^9 carbon, and therefore, a ω3, ω6, ω7, or ω9 fatty acid (with the first double bond appearing in the 3rd, 6th, 7th, and 9th carbon counted from the methyl group, respectively) remains in the same class of the fatty acid despite its desaturation and elongation (Baker and Czarnecki-Maulden 1991). Animal-sourced, but not plant-sourced foodstuffs contain cholesterol. Cholesterol is an essential component of cell membranes and plays an important role in the metabolism, growth, development, and reproduction of animals, but an excessive amount of this lipid in the blood increases risk for cardiovascular disorders (Wu 2018).

Because no double bond can be formed beyond the Δ^9 position in all animals, both α-linolenic acid ($C_{18:3}$, ω3; $\Delta^{9,12,15}$) and linoleic acid ($C_{18:2}$, ω6; $\Delta^{9,12}$) are not synthesized by dogs and cats (Table 4.1). These two fatty acids are integral components of the plasma membrane of cells (including erythrocytes and respiratory epithelial cells) for cell integrity, transport, and function (Bryan et al. 2001;

Lemaitre et al. 2009), thereby exerting enormous physiological functions in tissues. In addition, α-linolenic acid and linoleic acid are the precursors for the synthesis of eicosapentaenoic acid (EPA; $C_{20:5}$, ω3; $\Delta^{5,8,11,14,17}$) plus docosahexaenoic acid (DHA; $C_{22:6}$, ω3; $\Delta^{4,7,10,13,16,19}$) and arachidonic acid ($C_{20:4}$, ω6; $\Delta^{5,8,11,14}$), respectively (Wu 2018). EPA, DHA, and arachidonic acid are also essential components of cell membranes to affect cell integrity, signaling, and homeostasis. Thus, both linolenic acid and linoleic acid must be supplied in the diets for dogs and cats to ensure their survival, growth, development, and health (Table 4.2).

Beef tallow and butter contain low content of linoleic acid and α-linolenic acid. Therefore, when fed to dogs and cats, these two foodstuffs should be used along with other ingredients such as corn, soybean, and safflower oils that are rich in linoleic acid and α-linolenic acid (NRC 2006). As reported for humans (Eyjolfson et al. 2004), conjugated linoleic acid (CLA) improves metabolic profiles in obese or overweight dogs (Schoenherr and Jewell 1999) and cats (Bartges and Cook 1999). For example, dietary supplementation with 0.3% CLA to overweight dogs for 120 days decreased fat deposition in subcutaneous tissue and the mass of whole-body white fat mass but increased blood levels of high-density lipoproteins (Rivera et al. 2011). CLA-rich foods are ruminant meat, fats, and dairy products, with the predominant CLA isomer (>90%) being c9,t11 (Larsen et al. 2003).

4.3.2 Lipid Metabolism in Dogs

Like most other mammals (e.g., humans, pigs, and rats), dogs can introduce one or more double bonds to a long-chain fatty acid between its Δ^1 and Δ^9 carbons through the action of Δ-desaturases (enzymes in the smooth endoplasmic reticulum). Dogs have Δ^6 desaturase activity and, therefore, can readily convert linoleic acid into arachidonic acid (Dunbar and Bauer 2002). Likewise, growing and young adult dogs can actively form EPA and DHA from α-linolenic acid. However, the synthesis of EPA and DHA is insufficient to meet the needs of these dogs (NRC 2006). Furthermore, as reported for other animals (Bordoni et al. 1988), Δ^6 desaturase activity in the tissues of dogs may decline markedly with increasing age, thereby reducing EPA and DHA syntheses in older dogs. By contrast, dogs express a high activity of stearoyl-CoA desaturase (a Δ^9 desaturase in the smooth endoplasmic reticulum) to readily convert palmitic acid ($C_{16:0}$; a saturated long-chain fatty acid) into ω7 and ω9 unsaturated fatty acids (Bauer 2007). The desaturation and elongation of long-chain fatty acids to form mono- and polyunsaturated fatty acids in the smooth endoplasmic reticulum are shown in Fig. 4.1.

Dogs have a wide tolerance for dietary lipids (e.g., at least 40% fat in their diets) and can be maintained on diets containing 5–8% lipids on the DM basis (NRC 2006). A minimum content of 5% lipids in a canine diet is adequate if it provides sufficient essential fatty acids. To meet their requirement for energy, sled dogs are often fed a diet containing at least 50% fat during the racing season (Hill 1998; Loftus et al. 2014). During fasting, dogs mobilize fat in white adipose tissue and

Table 4.1 De novo formation of nutrients in dogs, cats, and pigs

Nutrients and energy	Formation de novo			Comments
	Dog	Cat	Pig	
Water	Yes	Yes	Yes	Catabolism of AAs, FAs, and glucose produces water; insufficient formation
Amino acids (AAs)	Yes	Yes	Yes	Dependent on breeds, sex, age, physical activity, environmental factors, and disease
Fatty acids (FAs)	Yes	Yes	Yes	Elongation; introduction of new double bonds between Δ^1 and Δ^9 carbons
Glycogen	Yes	Yes	Yes	Synthesis primarily in the liver and skeletal muscle of all animals
Glucose	Yes	Yes	Yes	Synthesis in all animals; high rate of constant synthesis from AAs in cats; Synthesis: Pigs < Dogs < Cats
Arginine	Ins	LM	AD, Ins	Species differences in the expression of arginine-synthetic enzymes in enterocytes of the small intestine; insufficient synthesis in dogs possibly due to low intestinal activities of P5CS and POX; low intestinal activities of P5CS and OAT in cats
Taurine	Yes; Ins for some breeds	LM	Yes	Most breeds of dogs synthesize sufficient taurine from L-cysteine in liver; limited synthesis from L-cysteine in cats due to low hepatic activities of CDO and CSD
Felinine	No	Yes	No	Synthesized from cysteine in cats; used for intra-species communication

(continued)

Table 4.1 (continued)

Nutrients and energy	Formation de novo			Comments
	Dog	Cat	Pig	
Phe plus Tyr	No	No	No	Carbon skeletons of Phe and Tyr are not formed de novo; Phe – > Tyr in animals
Palmitoleic acid (ω7)	Yes	Yes	Yes	$C_{16:1}$; active synthesis from palmitic acid ($C_{16:0}$) by Δ^9 desaturase in all animals
Oleic acid (ω9)	Yes	Yes	Yes	$C_{18:1}$; formed from palmitic acid ($C_{16:0}$) by elongation and then Δ^9 desaturase
α-Linolenic acid (ω3)	No	No	No	$C_{18:3}$; no double bond can be formed beyond the Δ^9 carbon in animal cells
Linoleic acid (ω6)	No	No	No	$C_{18:2}$; no double bond can be formed beyond the Δ^9 carbon in animal cells
Arachidonic acid (ω6)	Yes	LM	Yes	$C_{20:4}$; active synthesis from linoleic acid in dogs and pigs; very limited synthesis from linoleic acid in cats due to very low Δ^6 desaturase activity
EPA (ω3)	Yes; Ins during gestation	LM	Yes	$C_{20:5}$; active synthesis from α-linolenic acid in dogs and pigs; very limited synthesis from α-linolenic acid in cats due to very low Δ^6 desaturase activity
DHA (ω3)	Yes; Ins during gestation	LM	Yes	$C_{22:6}$; active synthesis from α-linolenic acid in dogs and pigs; very limited synthesis from α-linolenic acid in cats due to very low Δ^6 desaturase activity

(continued)

Table 4.1 (continued)

Nutrients and energy	Formation de novo			Comments
	Dog	Cat	Pig	
Vitamin A	Yes; Ins	No	Yes; Ins	Active synthesis from β-carotene in dogs and pigs; no synthesis in cats due to the absence of β-carotene dioxygenase in the small intestine and other tissues
Vitamin D_3[a]	LM	LM	Yes	Very limited synthesis from 7-dehydrocholesterol in dogs and cats even when exposed to sunlight; active synthesis in pigs when exposed to sunlight
Vitamin C	Yes	Yes	Yes	Active synthesis from glucose in dogs, cats, and pigs; required in heat stress
Niacin[b]	Yes; Ins	No	Yes; Ins	Active synthesis from L-tryptophan in dogs and pigs; no synthesis in cats

AAs, amino acids, AD, age-dependent (a high rate of citrulline synthesis from glutamine/glutamate and proline at birth; inadequate synthesis of citrulline at 7- to 21-days of age; and increased synthesis of citrulline after weaning); CDO, cysteine dioxygenase; CSD, cysteinesulfinate decarboxylase; DCM, dilated cardiomyopathy; DHA, docosahexaenoic acid; EPA, eicosapentaenoic acid; Ins, insufficient; 4-LKs and 5-LKs, series 4 and 5 leukotrienes, respectively; 2- and 3-PGs, series 2 and 3 prostaglandins, respectively; LM, very limited; OAT, ornithine aminotransferase; P5CS, pyrroline-5-carboxylate synthase; POX, proline oxidase

[a] In both dogs and cats, the skin has a low concentration of 7-dehydrocholesterol and a high activity of 7-dehydrocholesterol Δ^7 reductase. The latter rapidly converts 7-dehydrocholesterol into cholesterol, thereby limiting the formation of vitamin D_3

[b] The liver of cats has all the enzymes for converting tryptophan, but also has a relatively high activity of picolinate decarboxylase. This results in the diversion of 2-amino-3-carboxymuconate semialdehyde (an intermediate of the niacin-synthetic pathway) toward the formation of α-ketoadipate rather than the production of nicotinate mononucleotide

rapidly utilize ketone bodies as metabolic fuels (De Bruijne and van den Brom 1986; McCue 2010). This may explain, in part, why the concentrations of acetoacetate and β-hydroxybutyrate in the plasma of fasting dogs are low (< 0.15 mM) during brief and long-term fasting (De Bruijne and van den Brom 1986).

Table 4.2 Dietary requirements of nutrients by dogs, cats, and pigs

Nutrients and energy	Dietary requirement			Comments on metabolism, function, and dietary requirements
	Dog	Cat	Pig	
Water	Yes	Yes	Yes	Required for biochemical reactions and whole-body homeostasis; high requirement in all animals
Energy	Yes	Yes	Yes	Required for survival, growth, development, and health; obesity occurs with excess deposition of white fats
Amino acids (AAs)[a]	Yes	Yes	Yes	Building blocks of proteins and precursors of bioactive molecules; Dietary requirement: Pigs < Dogs < Cats
Fatty acids (FAs)[b]	Yes	Yes	Yes	Fatty acids serve as major metabolic fuels and other functions
Starch[c]	No	No	No	Major source of glucose in domestic dogs and pigs after weaning but not cats; excess causes gut dysfunction in cats
Glycogen[c]	No	No	No	Present in meat; a source of glucose in dogs and cats but insignificant in domestic pigs
Dietary fiber	No	No	No	Not required as an essential nutrient; plays important roles in intestinal health
Glucose[d]	No	No	No	Exclusive metabolic fuel for brain and red blood cells; high rate of synthesis in cats
Sweet substances[e]	No	No	No	Dogs and pigs select for, but cats avoid, sweet substances
Arginine[f]	Yes	Yes	Yes	Essential for the urea cycle to detoxify ammonia; Pigs < Dogs < Cats (dietary requirement)
Taurine[g]	Yes	Yes	No	A deficiency of taurine causes DCM and retinal damage; Dogs < Cats (dietary requirement)
Felinine	No	No	No	Synthesized from cysteine; used for communication
Phe plus Tyr	Yes	Yes	Yes	Required for eumelanin synthesis for black hair color and skin health
α-Linolenic acid	Yes	Yes	Yes	Precursor of EPA and DHA; nutritionally essential for all animals
Linoleic acid	Yes	Yes	Yes	Precursor of arachidonic acid; nutritionally essential for all animals
Arachidonic acid[h]	No	Yes	No	Essential for the synthesis of 2-PGs and 4-LKs; nutritionally essential for cats
EPA[i]	No	Yes	No	Essential for the synthesis of 3-PGs and 5-LKs; nutritionally essential for cats

(continued)

Table 4.2 (continued)

Nutrients and energy	Dietary requirement			Comments on metabolism, function, and dietary requirements
	Dog	Cat	Pig	
DHA[i]	No	Yes	No	Essential for the growth and development of brain; nutritionally essential for cats
Vitamin A[j]	Yes	Yes	Yes	Essential for the integrity and function of retinal cells, reproduction, and health
Vitamin D_3	Yes	Yes	Yes	Effectively used in dogs and cats; essential for the intestinal absorption of Ca and P
Vitamin D_2	No	No	No	Dogs, but not cats, can utilize vitamin D_2 from sun-dried plants
Vitamin C	No	No	No	Essential for many enzyme-catalyzed reactions and the integrity of connective tissue
Niacin	Yes	Yes	Yes	

AAs, amino acids, AD, age-dependent (high at birth; inadequate at 7- to 21-days of age, as well as during pregnancy and lactation); DCM, dilated cardiomyopathy; DHA, docosahexaenoic acid; EPA, eicosapentaenoic acid; 4-LKs and 5-LKs, series 4 and 5 leukotrienes, respectively; 2- and 3-PGs, series 2 and 3 prostaglandins, respectively.

[a] Cats have greater endogenous losses of nitrogen and greater requirements for dietary protein than dogs. Cats select high-protein diets over low-protein diets, whereas dogs select medium-level protein diets over high-protein diets

[b] Fatty acids are esterified with glycerol to form fat. Dogs select for high-fat diets over low-fat diets; whereas cats prefer animal (e.g., beef) fats to plant fats. In both dogs and cats, chronic consumption of high-fat diets increases risk for the development of overweight and obesity (particularly when the animals lack sufficient exercise) and must be avoided

[c] Adequate intake of starch and glycogen by dogs may reduce their use of amino acids for gluconeogenesis. Cats have a low ability to digest dietary starch and glycogen. Excess intake of starch and glycogen may be detrimental for the digestive tract and may increase risk for metabolic disorders in cats. The saliva of dogs and cats has no or little α-amylase activity. Pancreatic α-amylase activity is high in the small intestine of dogs but is limited in cats

[d] Cats have a greater ability to synthesize glucose from amino acids and other glucogenic substrates than dogs. The liver of cats lacks glucokinase

[e] The tongue of cats lacks receptors for sweet substances

[f] Dogs can synthesize arginine to maintain nitrogen balance but the endogenously synthesized arginine is inadequate to meet functional needs such as ammonia detoxification via the urea cycle and possibly nitric-oxide synthesis. Cats have a very low rate of arginine synthesis via the intestinal-renal axis. Both dogs and cats require dietary arginine for health and survival

[g] Most breeds of dogs synthesize sufficient taurine but some dogs of certain breeds do not. Healthy pigs have no requirement for dietary taurine. Animal-sourced food contain sufficient taurine, but this amino acid is absent from plant-sourced foods

[h] Arachidonic acid is synthesized from linoleic acid via Δ^6 desaturase in dogs, but the activity of this enzyme is very low in cats. Dogs have no requirement for dietary arachidonic acid if linoleic acid intake is sufficient

[i] Eicosapentaenoic acid (EPA) and docosahexaenoic acid are synthesized from α-linolenic acid via Δ^6 desaturase in dogs but the activity of this enzyme is very low in cats. Dogs have no requirement for dietary eicosapentaenoic acid and docosahexaenoic acid if α-linolenic acid intake is sufficient

[j] Dogs can synthesize vitamin A from β-carotene in the enterocytes of their small intestine via β-carotene 15,15-dioxygenase, but this enzyme is absent from cats. Diets for dogs can use β-carotene to replace vitamin A, and this vitamin must be included in diets for cats

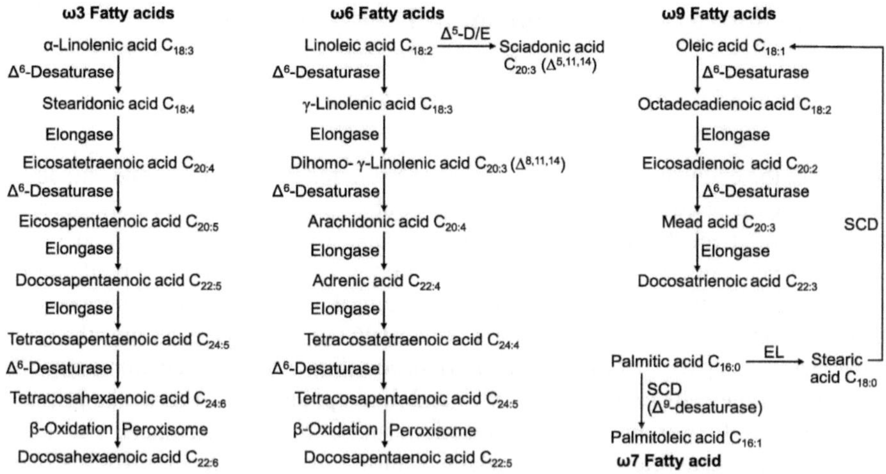

Fig. 4.1 Elongation of the unsaturated fatty acid chain beyond C_{18}, as well as the formation of monosaturated fatty acids in the smooth endoplasmic reticulum. This metabolic pathway also requires the addition of acetyl-CoA as malonyl-CoA to the hydrocarbon chain, as well as NADPH. The shortening of a polyunsaturated fatty acid chain from C_{24} to C_{22} through β-oxidation in the peroxisome is also shown. All desaturases are localized in the smooth endoplasmic reticulum. Δ^5-D/E = Δ^5-desaturase and elongase; EL = elongase; SCD = stearoyl-CoA desaturase (Δ^9 desaturase)

4.3.3 *Lipid Metabolism in Cats*

In contrast to dogs, cats have a very limited Δ^6 desaturase activity (Pawlosky et al. 1994; Rivers et al. 1975). Thus, when cats are fed a diet rich in linoleic acid, there is a modest increase in: (1) tissue concentrations of *eicosadienoic acid* ($C_{20:2}$, ω6; $\Delta^{11,14}$) due to chain elongation, and (2) the formation of sciadonic acid ($C_{20:3}$, ω6; $\Delta^{5,11,14}$) due to the action of Δ^5 desaturase activity (Sinclair et al. 1981). These metabolic pathways are shown in Fig. 4.1. Sciadonic acid is a biomarker for limited Δ^6 desaturase activity but active Δ^5 desaturase in animal tissues (Bauer 1997). Accordingly, cats generate only a limited amount of arachidonic acid from linoleic acid (Sinclair et al. 1981), but dogs and omnivores (e.g., humans and pigs) readily convert linoleic acid to arachidonic acid as noted previously. Cats have a greater requirement for dietary arachidonic acid than dogs. In addition, a low activity of Δ^6 desaturase in cats limits the conversion of α-linolenic acid into EPA and DHA. Thus, diets for cats must contain EPA and DHA. Daily supplementation of DHA and EPA is recommended for pregnant and lactating cats for the proper brain and nervous system development of their offspring (Vuorinen et al. 2020).

4.4 Carbohydrate Metabolism in Dogs and Cats

4.4.1 Overview of Carbohydrate Metabolism

Glucose is the exclusive energy source for the brain, red blood cells, kidney medulla, and retinal cells of dogs and cats in fed and post-absorptive states (Wu 2018). Because the brain is relatively large in newborn dogs and cats (accounting for 2.74% and 3.61% of BW, respectively; Latimer 1967), a large amount of glucose is used by the brain of puppies and kittens. For comparison, the brain of adult dogs and cats represents 0.85% and 0.80% of BW, respectively (Latimer 1967), which is greater than the brain of cattle and sheep (0.2% of BW, Wu 2018). In addition, glucose is a major metabolic fuel for immune cells (e.g., lymphocytes and macrophages), as well as the primary source of NADPH for anti-oxidative reactions and nitric-oxide synthesis. As noted previously, dogs, unlike cats, have adapted during evolution to omnivorous diets that contain both animal- and plant-sourced foods (Baker and Czarnecki-Maulden 1991). Gluconeogenesis (formation of glucose from non-glucose substrates such as AAs, glycerol, lactate, and odd-carbon number fatty acids) plays an important role in maintaining glucose homeostasis and carbon balances in all mammals, including both dogs and cats (Wu 2018). There are no established requirements of dogs and cats for dietary carbohydrates.

4.4.2 Carbohydrate Metabolism in Dogs

When diets do not provide sufficient starch, glycogen, or glucose, dogs must synthesize glucose from glucogenic AAs, lactate, and glycerol in their liver and kidneys (Belo et al. 1976). In the canine liver and kidneys, phosphoenolpyruvate carboxykinase (PEPCK) is localized to both the cytosol and mitochondria (e.g., equal distribution in the liver but 50% and 65%, respectively, in the kidney under fed conditions) (Croniger et al. 2002; Wolf and Mehlman 1972). Such an intracellular compartmentation of PEPCK allows for the use of all glucogenic substrates (including glucogenic AAs) for glucose synthesis (Feng et al. 1996; Wolf and Mehlman 1972). Note that if PEPCK is exclusively localized to the mitochondria of the liver and kidneys, there is no formation of glucose from AAs due to the absence of NADH provision, as reported for the liver of fed or fasted chickens (Watford 1985). Interestingly, the activities of mitochondrial PEPCK (converting oxaloacetate into phosphoenolpyruvate) and pyruvate carboxylase (converting pyruvate into oxaloacetate) in the liver and kidneys of dogs are enhanced by the feeding of a carbohydrate-free diet (Feng et al. 1996). Intakes of dietary carbohydrates (including starch and fiber) affect postprandial concentrations of glucose and insulin in the plasma of dogs (Carciofi et al. 2008).

In dogs, water-soluble fibers (e.g., plant pectin, fructans, gums, psyllium, and β-glucan) and water-insoluble fibers (e.g., plant cellulose and hemicellulose) play an important role in their intestinal health (Nogueira et al. 2019). Soluble fibers in oats, peas, beans, apples, citrus fruits, carrots, barley, and psyllium form a viscous solution upon contact with water and are highly fermentable in the large intestine (NRC 2006; Silvio et al. 2000). Insoluble fibers in plant cell walls, wheat bran, vegetables, and whole grains retain water, do not form a viscous solution, and are much less fermentable than soluble fibers in the gut. Insoluble fibers stimulate intestinal peristalsis, increase fecal mass, and decrease the transit time of food through the GIT (Schneeman 1994). Some fermentable fibers (e.g., fructooligosaccharides) are beneficial prebiotics for the intestinal health of dogs (Montserrat-Malagarriga et al. 2024; Nogueira et al. 2019).

4.4.3 Carbohydrate Metabolism in Cats

As noted previously, cats consume meat (containing low glycogen content, < 5% on a DM basis) and no starch. Cats, just like dogs, have little or no salivary α-amylase activity (McGeachin and Akin 1979). In addition, pancreatic α-amylase activity is low in cats as compared with dogs. For example, pancreatic α-amylase activity in adult cats is only 2.3% and 2% of that in adult dogs and pigs, respectively (Kienzle 1993a), and the abundance of glucose transporters in the small intestine of cats is much lesser than that in dogs (Batchelor et al. 2011). Therefore, cats have a much lower ability to digest and use dietary starch in the small intestine than dogs, and their consumption of carbohydrate-rich diets can result in intestinal inflammation (Morris et al. 1977; Verbrugghe and Hesta 2017).

Because of a limited intake of glycogen, starch, and sugars in their natural foods, cats primarily depend on constant gluconeogenesis (mainly from AAs) to provide glucose (Legrand-Defretin 1994). This metabolic pathway is particularly significant for newborn cats with relatively a large brain (accounting for 3.62% of BW; Latimer 1938), where a large amount of glucose is used via glycolysis and the Krebs cycle. For comparison, the brain of adult cats represents 0.80% of BW (Latimer 1967), which is similar to that for adult dogs. Unlike dogs, glucokinase activity is absent from the feline liver (Washizu et al. 1999) just like the ruminant liver (Wu 2018). By contrast, the activities of other glycolytic enzymes (e.g., hexokinase I, phosphofructokinase, and pyruvate kinase) in the feline liver are substantially greater than those in the canine liver (Washizu et al. 1999). Furthermore, the activities of rate-controlling enzymes for gluconeogenesis (e.g., pyruvate carboxylase, fructose-1,6-bisphosphatase, and glucose-6-phosphatase) in the feline liver are also much greater than those in the canine liver (Washizu et al. 1999). This is consistent with a greater rate of whole-body glucose production in cats than in dogs (Legrand-Defretin 1994). Interestingly, hepatic PEPCK activity is not altered in cats that are fasted after consuming high-protein diets (Kettelhut et al. 1980) or after the protein content of their diet is increased from 17.5% (low) to 70% (high) (Rogers et al. 1977). These results support the view that cats express high basal activities of hepatic gluconeogenic enzymes to ensure the rapid conversion of excess dietary AAs into glucose.

In cats, gluconeogenesis plays an essential role in the provision of glucose to their brain, red blood cells, and immunocytes and therefore their survival.

Just like dogs, carbohydrates (including water-soluble and insoluble fibers) are also extensively fermented by microbes in the large intestine of cats (Brosey et al. 2000; Legrand-Defretin 1994; Morris et al. 1977). For example, in adult cats, the prececal apparent digestibility of starch (e.g., cooked corn starch, 72%; raw corn starch, 46%; raw potato starch, 0%) is lower than its total apparent (fecal) digestibility (cooked corn starch was digested to nearly 100%, raw corn starch to 78%, raw potato starch to 36%), but the pattern of apparent digestibility of raw starch is similar to that for the cooked starch (Kienzle 1993b). This may help to explain the observation that the concentration of glucose in adult cats fed a 31.7%-starch diet was continuously elevated from 4.6 mM at 3 h after the meal, compared with the pre-meal baseline (4.8 mM) to 6.2 mM at 19 h after feeding, whereas the concentration of glucose in adult cats fed a 23%-starch diet did not differ from the pre-meal level or during the period of 3 and 19 h after the meal (Hewson-Hughes et al. 2011). Due to its microbial fermentation, starch led to an acidification of the large bowel chyme and feces, and the effect of dietary undigested starch can be alleviated by oral antibiotics in cats (Kienzle 1993b). This may explain, in part, why cats should not be fed dog food and require carefully formulated diets. Commercial diets for cats generally contain a small amount of water-soluble and insoluble fibers (a total of 3–5% fibers, DM basis) to maintain intestinal health (Zoran 2023).

4.5 Protein and AA Metabolism in Dogs and Cats

4.5.1 Overview of Protein and AA Metabolism

Both dogs and cats have a high ability to digest dietary protein in the gastrointestinal tract and degrade dietary AAs in a tissue-specific manner (Galim et al. 1980; Beliveau and Freedland RA 1982; Hammer et al. 1996; Hu et al. 2021; Li and Wu 2023a; Wu 1998). These mammals, just like all other animals, cannot form de novo the carbon skeletons of Cys, His, Ile, Leu, Lys, Met, Phe, Thr, Trp, Tyr, and Val, but can convert Phe into Tyr, Met into Cys, and Trp to serotonin (Wu 2022). Some metabolites of tyrosine (melanins) are crucial for the color of hair and skin health (Anderson et al. 2002; Biourge and Sergheraert 2002; Yu et al. 2001), whereas serotonin regulates animal behavior (Landsberg et al. 2017) and intestinal function (Chiocchetti et al. 2021). In addition, these two animal species, young or adult, are not capable of synthesizing sufficient arginine from glutamine/glutamate and proline (Burns et al. 1981; Czarnecki and Baker 1984; Weber et al. 1977; Yu et al. 1996). This is likely due to a deficiency of pyrroline-5-carboxylate synthase, proline oxidase, carbamoylphosphate synthetase I, and/or N-acetylglutamate synthetase as well as ornithine aminotransferase in enterocytes of the canine and feline small intestines (Dillion and Wu 2021; Morris 2002; Rogers and Phang 1985; and Wu and Morris 1998). In addition, most breeds of dogs,with possible exceptions of a few, can synthesize sufficient taurine when fed methionine- and cysteine-adequate diets (Jacobsen and Smith 1968). Interestingly, golden retrievers, American Cocker Spaniels, and a small population (1.3–2.5%) of the Newfoundlands may be genetically predisposed to taurine deficiency possibly due to gene mutations (Backus et al. 2003, 2006; Kaplan et al. 2018; Kittleson et al. 1997). By contrast, all breeds of cats have a very limited ability to convert cysteine into taurine due to low activities of cysteine dioxygenase and cysteinesulphinic acid decarboxylase in the liver (Knopf et al. 1978). Thus, Arg, His, Ile, Leu, Lys, Met, Phe, Thr, Trp, and Val must be included in canine and feline diets; taurine must be provided to all cats and some breeds of dogs. As reported for pigs, rats, fish and crustaceans (He and Wu 2022; He et al. 2023; Hou et al. 2015, 2016; Hou and Wu 2015; Li et al. 2020, 2021b,c, 2023; Wu 2009; Wu and Li 2022), dogs and cats may have dietary requirements for AAs (e.g., glutamate, glutamine, glycine, and proline) that are synthesized de novo and also use dietary AAs as major metabolic fuels. Their quantitative amounts have recently been suggested for both animal species during the growth period and in adulthood (Li and Wu 2023a). Growing evidence shows that AAs are signaling molecules to activate the mechanistic target of rapamycin pathway for initiating protein synthesis in animal cells (Rezaei and Wu 2022), as well as regulating energy sensing and metabolism (Jobgen et al. 2022a, b, 2023). Both dogs and cats have dietary requirements for AAs but not protein or nitrogen. In the following sections, we focus on arginine and cysteine in canine and feline nutrition, as well as AA imbalances and antagonisms, metabolic adaptation to AA intakes, and AA requirements.

4.5.2 Arginine Metabolism in Dogs and Cats

In most mammals, including dogs and cats, the small intestine synthesizes de novo and releases citrulline, which is taken up by extraintestinal tissues (primarily the kidneys and, to a lesser extent, all other tissues) for arginine synthesis (Levillain et al. 1996; Wu and Morris 1998). However, endogenous synthesis of arginine is limited in both animal species, as noted previously. All mammalian tissues contain arginase for the hydrolysis of arginine into urea and ornithine, with ornithine being either converted into proline, glutamate, polyamines, agmatine, and NO or oxidized to CO_2 in a cell-specific manner (Wu et al. 2016b). Thus, the dietary intake, endogenous synthesis, and catabolism of arginine regulate its homeostasis in the body.

Consistent with the lower rate of arginine synthesis in cats than in dogs, the milk of cats contains more arginine than the milk of dogs (Davis et al. 1994a, b) to support feline survival, growth, and development. In addition, the content of arginine in the milk of dogs is greater than that in the milk of herbivores (e.g., cows) and omnivores (e.g., humans and pigs) to compensate for a lower rate of arginine synthesis in the dogs (Heinze et al. 2014). Current commercial vegan foods may not provide sufficient arginine for cats (Zafalon et al. 2020b).

A deficiency of arginine results in decreased food intake, impaired ureagenesis, hyperammonemia, abnormal hemodynamics, hypertension, severe emesis, frothing at the mouth, muscle tremors, and cataract in dogs (Burns et al. 1981; Czarnecki and Baker 1984; Ranz et al. 2002). Cats have a lower rate of endogenous arginine synthesis, ingest more AAs, and are more sensitive to a deficiency of dietary arginine than dogs. Thus, a lack of dietary arginine rapidly causes hyperammonemia (occurring within 1–3 h after feeding), vomiting, neurological signs, severe emesis, ataxia, tetanic spasms, and death in cats (Morris and Rogers 1978). Dietary supplementation with arginine, citrulline, or ornithine to dogs and cats can prevent hypoargininemia-induced hyperammonemia and mortality. Dietary or intravenous administration of arginine or citrulline, but not ornithine, can restore growth in young cats fed arginine-free or deficient diets (Czarnecki and Baker 1984). This is due to the complex compartmentalization of ornithine metabolism in enterocytes, as extracellular ornithine is preferentially channeled into the formation of proline but not citrulline in these cells (Wu and Morris 1998). Greater protein requirements by cats than dogs may be explained, in part, by a lower ability of cats to synthesize arginine than dogs and a need for the greater provision of arginine from feline than canine diets (Li and Wu 2023a).

4.5.3 Cysteine Metabolism in Dogs and Cats

All cats and some breeds of dogs have a limited ability to synthesize taurine from cysteine (a metabolite of methionine; Fig. 4.2), as noted previously. This metabolic pathway occurs in the liver of dogs and cats, with dietary cysteine being capable to replace up to 50% dietary methionine (Che et al. 2021; Oberbauer and Larsen 2021). In all animals, the synthesis of cysteine requires methionine and serine, as the sulfur atom of cysteine is derived from methionine, whereas serine provides both the amino group and the whole carbon skeleton of cysteine (Wu 2022). Of particular note, taurine is abundant in animals-sourced foodstuffs but is absent from plants (Hou et al. 2017, 2019; Li et al. 2021a). In dogs and cats, the primary bile acids are conjugated exclusively and almost exclusively with taurine, respectively, to form bile salts, which are required for the digestion and absorption of dietary lipids (including long-chain fatty acids and lipid-soluble vitamins). In healthy dogs, the primary bile acids are cholic acid, deoxycholic acid, and chenodeoxycholic acid; their tauro-conjugated forms account for approximately 73%, 20%, and 6%, respectively, of the total bile acid pool in the gallbladder (Washizu et al. 1994); and there are no glycine-conjugated bile acids in the liver or bile (Wildgrube et al. 1986). In healthy cats, the major primary bile acids are cholic acid and deoxycholic acid, and their distribution percentages (%) are 98.4% taurine-conjugated form (79.1% cholic acid and 19.3% deoxycholic acid), 1.1% glycine-conjugated cholic acid, and 0.5%of unconjugated cholic acid (Rabin et al. 1976).

Because the content of methionine and cysteine in plant-sourced ingredients is generally lower than that in animal-sourced ingredients, an inadequate intake of methionine and cysteine in plant-based foods can result in a deficiency of taurine in both dogs and cats (Cavanaugh et al. 2021; Knopf et al. 1978; McCauley et al. 2020). A deficiency of taurine in these companion animals can result in dilated cardiomyopathy that is characterized by thin heart muscle and enlarged chambers (Cavanaugh et al. 2021; McCauley et al. 2020; Morris 2002). Additional symptoms in cats include heart failure, central retinal degeneration, blindness, deafness, poor reproduction, and impaired immune responses (Morris 2002; Pion et al. 1992). All foods for cats and some breeds of dogs must provide sufficient taurine to maintain the integrity and health of their tissues. Morris et al. (1990) reported that the content of 0.12% taurine (on a DM basis) in commercial expanded (dry) cat foods was sufficient for cats, but the content of taurine in canned diets should be increased to 0.20–0.25% (DM basis) because heating during the canning process forms products that promotes the loss of taurine via the enterohepatic circulation.

Another metabolite of cysteine via cooperation of the liver and kidneys in both dogs and cats is isovalthine, the formation of which also requires leucine (Wu 2022). Interestingly, male cats produce more felinine than females (Hendriks et al. 2004). In cats but not dogs, cysteine is also used to generate felinine and isobuteine from isopentenyl pyrophosphate (formed from acetyl-CoA) and valine, respectively, via cooperation of the liver and kidneys (Che et al. 2021; Hendriks et al. 2008; Rutherfurd-Markwick et al. 2005). Cysteine participates in all these reactions in the form of glutathione. These synthetic pathways may contribute to greater requirements of cats for cysteine, as compared with dogs (Li and Wu 2023a). Physiological functions of felinine, isovalthine, and isobuteine remain unknown but may serve as pheromones in cats for the purpose of territorial marking and intra-species communications (e.g., chemical signals to attract females) (Miyazaki et al. 2008).

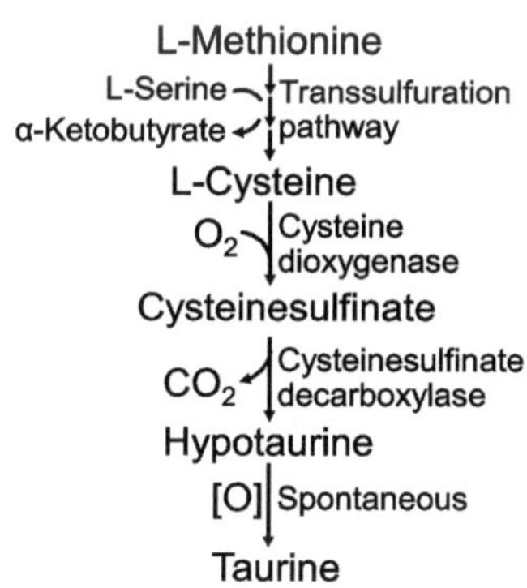

Fig. 4.2 Synthesis of taurine from L-cysteine in the liver of dogs and cats. In the presence of serine, L-methionine is degraded on the liver via the transsulfuration pathway to generate cysteine and α-ketobutyrate. Subsequently, L-cysteine is oxidized to hypotaurine via reactions catalyzed by cysteine dioxygenase and cysteinesulfinate decarboxylase. Hypotaurine is spontaneously oxidized to taurine. Most breeds of dogs can synthesize sufficient taurine when fed L-methionine- or L-cysteine-adequate diets. However, this synthetic pathway is limited in cats due to low activities of cysteine dioxygenase and cysteinesulfinate decarboxylase

4.5.4 AA Imbalances and Antagonisms, as Well as Metabolic Adaptation to AA Intakes

Compared with non-collagen animal proteins, plant proteins (particularly those in cereals) generally contain low amounts of nearly all AAs, particularly the AAs (e.g., lysine, tryptophan, threonine, methionine, and cysteine) that are not synthesized de novo and the AAs (glycine and proline + hydroxyproline) that are most abundant in the body (Li et al. 2021d). Dietary imbalances of AAs (defined as improper ratios of AAs in diets) can occur in dogs and cats fed plant-based diets containing a small amount of or no animal-sourced ingredients (Li and Wu 2023a). Antagonisms among AAs with similar chemical structures (e.g., BCAAs) or net electric charges (e.g., arginine and lysine) that share the same transmembrane transporters can occur in response to their improper ratios in diets or tissue (Wu 2022). Effects of dietary AA imbalances or antagonisms in dogs are more similar to those in rats than in cats based on alterations in food intake, plasma metabolic patterns, and hepatic enzyme activities (Morris 2002). Animal-sourced foods are abundant providers of all proteinogenic AAs as well as taurine and 4-hydroxyproline and play an important role in balancing AAs in diets for dogs and cats (Che et al. 2021; Li and Wu 2018; 2020; Li et al. 2021a; Wu et al. 2016a).

Dogs and cats have a high ability to use and oxidize dietary AAs, with the rate of AA catabolism in cats being greater than that in dogs under both fed and post-absorptive conditions (Morris 2002; Russell et al. 2002; Wester et al. 2015). For example, dogs and cats can tolerate at least 32% and 60% dietary protein, respectively (DeNapoli et al. 2000; Dodman et al. 1996; Green et al. 2008; Laflamme and Hannah 2013). These animals, just like most of other mammals (e.g., pigs and rats) can adapt to: (1) low-protein diets by increasing food intake and reducing whole-body AA catabolism, and (2) high-protein diets by reducing food intake and up-regulating the expression of AA-catabolic enzymes (Harper et al. 1984; Rogers et al. 1998). However, dogs are not as efficient as rats in metabolic adaptation to low- or high-protein intakes. In support of this view, studies with adult dogs have shown that the rate of whole-body protein degradation is not affected by low dietary protein intake and that the rate of whole-body leucine oxidation is not influenced by high dietary protein intake (e.g., increasing from 32 to 148 g CP/Mcal ME) (Humbert et al. 2002). In contrast, cats rapidly respond to protein-free diets by decreasing the rate of whole-body AA oxidation (Hendricks et al. 1996) and to elevated protein intake by increasing the rate of whole-body AA oxidation (Hendricks et al. 1997; Wester et al. 2015).

4.5.5 AA Requirements of Dogs and Cats

Compared with dogs, cats have greater a loss of endogenous nitrogen (urinary plus fecal nitrogen) when fed nitrogen-free diets (Hendriks et al. 1996 and 2002; Kendall et al. 1982). The whole-body degradation of protein provides AAs (e.g., arginine) for metabolic utilization. Accordingly, cats have greater requirements for dietary AAs than dogs (Mansilla et al. 2018; 2020a,b; NRC 2006). Quantitative requirements of dogs and cats for dietary AAs may be estimated from experiments involving nitrogen balance, the factorial analysis of use of AAs for growth and product (e.g., milk and hair) formation, and the oxidation of direct and indicator AAs (Lambie et al. 2024; Singh et al. 2024; Wu 2022). Interestingly, differences in dietary requirements for EAAs between these two animal species do not appear to be substantial (Rogers and Morris 1979). It is possible that cats degrade NEAAs at greater rates in a tissue-specific manner, as compared with dogs, with some of NEAAs being used for hepatic and renal gluconeogenesis. This is physiologically and nutritionally important because feline diets contain only a small amount of glycogen or starch (Washizu et al. 1999).

Skeletal muscle is the largest tissue in healthy animals. Like other mammals (e.g., humans), elderly dogs and cats experience sarcopenia (aging-related progressive losses of skeletal muscle mass and strength) because the rate of protein synthesis is lower than the rate of proteolysis (Kealy 1999; Laflamme 2008a,b). Sarcopenia results in not only physical weakness and disability but also tremendous increases in risks for falls, fractures, morbidity, and mortality, as well as health care costs. Exercise and AAs (e.g., arginine, branched-chain AAs, glutamine, glycine, and tryptophan) are co-stimuli to activate the mechanistic target of rapamycin cell signaling pathway and protein synthesis in skeletal muscle, thereby mitigating the loss of muscle protein with aging in humans (Glynn et al 2010; Holowaty et al. 2023; Mitchell et al. 2015). This is true for dogs and cats (Laflamme and Hannah 2013; Laflamme and Danièlle 2014; Williams et al. 2001), although the underlying signal transduction mechanisms have not yet been elucidated. There are suggestions that dietary protein requirements be increased by approximately 50% and 2.5-fold in older dogs and cats, respectively, compared with young adults (Laflamme 2008b; Laflamme and Hannah 2013; Wannemacher and McCoy 1966). This is equivalent to 2.55 and 5 g protein/kg BW/day for healthy adult dogs and cats, respectively (Churchill and Eirmann 2021). Considering long-term kidney health, we (Li and Wu 2023a) recommended adequate intakes of high-quality protein (i.e., 32% and 40% animal protein in diets for aging dogs and cats, respectively; DM basis) to alleviate sarcopenia. This goal can be achieved through the inclusion of pet-food grade animal-sourced foodstuffs in canine and feline diets (Li et al. 2021a, d; Li and Wu 2022, 2023b).

4.6 Vitamin Metabolism in Dogs and Cats

4.6.1 Vitamin Metabolism in Dogs

Like most animals, dogs can convert: (1) dietary β-carotene into vitamin A (all-*trans*-retinol) in enterocytes of the small intestine (Schweigert 1998), (2) tryptophan into niacin in the liver (Krehl 1981), and (3) glucose into vitamin C in the liver (McDowell 1989). Thus, β-carotene and tryptophan can replace some vitamin A and niacin in canine diets, respectively, and dogs do not have a dietary requirement for vitamin C. However, in contrast to other mammals (e.g., sheep, cattle, horses, pigs, rats, and humans), dogs have a limited ability to synthesize vitamin D_3 (cholecalciferol) from 7-dehydrocholesterol in the skin exposed to sunlight and, therefore, depend on the dietary intake of vitamin D_3 for bone health (How et al. 1994; Morris 1999). This is because the skin of dogs has a low concentration of 7-dehydrocholesterol and a high activity of 7-dehydrocholesterol Δ^7 reductase to rapidly convert 7-dehydrocholesterol into cholesterol (Zafalon et al. 2020a). For example, concentrations of 7-dehydrocholesterol in the skin of dogs are only about 10% of those in the skin of rats (How et al. 1994). Cholecalciferol is generally added to the commercial diets for dogs, although dogs can effectively use the plant-sourced vitamin D_2 (ergocalciferol). Synthetic vitamin K_3 (menadione sodium bisulfite; water-soluble) is used in the commercial diets of dogs. Dogs do not synthesize other vitamins required for cell metabolism (NRC 2006). Lipid- and water-soluble vitamins are excreted from the body mainly via bile (feces) and urine, respectively (Wu 2018).

4.6.2 Vitamin Metabolism in Cats

Vitamin B_6, Vitamin B_{12}, folate, and choline, along with methionine, serine, and histidine, play an important role in one-carbon metabolism in cats, as in other animals (Verbrugghe and Bakovic 2013). Like dogs, cats synthesize vitamin C and do not require a dietary source of this vitamin under normal feeding and environmental conditions (McDowell 1989), and also have a very limited ability to synthesize vitamin D_3 from 7-dehydrocholesterol in response to sunlight (How et al. 1994). This is because of a very low concentration of 7-dehydrocholesterol and a high activity of 7-dehydrocholesterol Δ^7 reductase to rapidly convert 7-dehydrocholesterol into cholesterol (How et al. 1994; Morris 1999), as reported for dogs. Cholecalciferol is generally added to the commercial diets for cats. Unlike dogs, cats cannot convert dietary β-carotene into vitamin A due to the absence of β-carotene 15,15-dioxygenase (Gershoff et al. 1957), although β-carotene is absorbed by the enterocytes of the small intestine (Schweigert et al. 2002). In addition, cats cannot use dietary vitamin D_2 (the vitamin D in sun-dried plants). Furthermore, cats cannot synthesize niacin from tryptophan due to (a) a high activity of picolinate carboxylase to divert 2-amino-3-carboxymuconate semialdehyde into α-ketoadipate, and (b) a low activity of quinolinate phosphoribosyltransferase for converting 2-amino-3-carboxymuconate semialdehyde into nicotinate mononucleotide (Badawy 2017). Therefore, feline diets must provide vitamin A, vitamin D_3, and niacin. Other vitamins are not synthesized by cats and are required in their diets for cell metabolism (NRC 2006).

4.6.3 Anti-vitamin Factors in Foods

Some foods contain anti-vitamin factors. For example, thiaminase is present in the viscera of some freshwater fish [e.g., common bream (*Abramis brama*), central stoneroller (*Campostoma anomalum*), goldfish (*Carassius auratus*), and common carp (*Cyprinus carpio*)], some marine fish [e.g., broad-striped anchovy (*Anchoa hepsetus*), Atlantic herring (*Clupea harrengus*), Atlantic cod (*Gadus morhua*), and Yellowfin tuna (*Neothunnus macropterus*)], and some bacteria (Greig and Gnaedinger 1971). This enzyme is heat-labile and is inactivated by cooking. Among the 32 species of freshwater fish tested, 18 of them were found to contain thiaminase. Among the 61 marine fish tested, 32 were found to contain thiaminase (Lichtenberger 2021). In other words, 56% of the freshwater fish examined contained thiaminase as compared with 51% of the marine fish species. Interestingly, there is little or a small difference in thiaminase activity between freshwater and marine fish. Furthermore, raw egg white contains avidin (a glycoprotein) that binds biotin very tightly. Fortunately, avidin is heat-labile and can be inactivated by cooking. Caution should be taken when feeding dogs and cats with raw fish or raw eggs.

4.7 Mineral Metabolism in Dogs and Cats

4.7.1 Mineral Metabolism in Dogs

In dogs, nutritionally essential macrominerals are Ca, P, Na, Cl, K, and Mg, whereas nutritionally essential microminerals are Fe, Zn, Cu, Mn, I, and Se (NRC 2006). Possibly essential minerals for these animals are Mo, boron, and Cr. Methionine and cysteine provide the sulfur atom required for the structures of protein and other molecules in dogs. Dietary inorganic sulfur is not required by these animals. Minerals are essential for nutrient transport, bone growth and development, the activities of proteins (including enzymes), and acid–base balance (Wu 2018). However, excessive Ca intake (e.g., > 0.31% of dietary DM) results in skeletal abnormalities in growing dogs (Hazewinkel et al. 1991; Hedhammar et al. 1974). Likewise, excessive Ca and P intake (0.31% Ca and 0.28% P of dietary DM) during early maturation (3 to 17 weeks of age) in dogs alters Ca and P balances by as early as 9 weeks of age, but dietary normalization (0.1% Ca and 0.08% P of dietary DM) during 17 to 27 weeks of age can effectively mitigate any long-term adverse effects on Ca and P balance (Schoenmakers et al. 1999).

Dogs ingesting 4 g NaCl/kg BW/day showed no sign of appreciable salt retention for 6 days if they have free access to drinking water (Ladd and Raisz 1949). However, seizures, hypernatremia [serum sodium concentration (211 mM; reference range, 145–158 mM) and serum chloride concentration (180 mM; reference range, 105–122 mM), and death occur after consuming an excessive amount of sodium from a salt–flour mixture (finely ground salt; Khanna et al. 1997). Dogs consuming excessive salts should have free access to sufficient drinking water.

4.7.2 Mineral Metabolism in Cats

Cats have the same requirements for minerals as dogs but can tolerate a higher intake of dietary salt than dogs (NRC 2006). Cats are commonly beset with the problem of urolithiasis, namely the formation of sediment (consisting of one or more poorly soluble urine crystalloids) within the urinary tract (Houston et al. 2003). Magnesium ammonium phosphate (also known as struvite) and calcium oxalate (CaOx) are often present in the majority of such deposits in alkaline urine (Robertson et al. 2002). In North America, struvite and CaOx stones are the most and second most common mineral types found in the feline uroliths. Thus, caution must be exercised to avoid an excess of dietary magnesium and calcium. In practice, sufficient water intake can reduce risk for urolithiasis in cats. In addition, a high dietary intake of phosphorus and a low ratio of calcium-to-phosphorus in diets increases risks for kidney damage in cats (Summers et al. 2020). Furthermore, the content of minerals in the diet can affect its palatability for cats and must be monitored carefully. Of note, there is evidence that current commercial vegan foods for dogs and cats may not provide: (a) adequate calcium, potassium, and sodium for dogs; or (b) sufficient potassium and proper Ca/P ratio for cats (Zafalon et al. 2020b). This nutritional problem can be prevented by the inclusion of animal-sourced foodstuffs, which contain relatively a large amount of minerals (Li and Wu 2022, 2023b).

4.8 Water Metabolism in Dogs and Cats

4.8.1 Overview of Water Metabolism

Water is the quantitatively most important and most abundant nutrient for all animals, and its loss occurs via evaporation, respiration, urine, and feces (Wu 2018). Drinking and dietary (moist food) water are by far the major sources of water in animals, as the production of metabolic water from the oxidation of nutrients is limited in most mammals including dogs and cats (e.g., 1.09, 0.60, and 0.41 g water for the oxidation of 1 g fat, glucose, and protein, respectively (Wu 2018). Thirst results from both intracellular and extracellular dehydration (e.g., a loss of 0.5–1% of body water), as well as an increase in plasma osmolality, leading to the stimulation of water consumption (NRC 2006). It usually takes only days for animals (including dogs and cats) to be dehydrated particularly in an environment with elevated temperatures, but weeks or longer before clinical signs of a deficiency of a non-water nutrient are evident. In both dogs and cats, clinical signs of water deficiency include the loss of elasticity of the skin, reduced saliva secretion, reduced appetite, reduced blood volume, impaired blood circulation, the formation of bladder and kidney stones, and the dysfunction of tissues. These abnormalities appear when water stores in dogs are decreased by 5%, and death may occur when the body loses about 15% water (Kirk and Bistner 1981).

4.8.2 Water Metabolism in Dogs

Thirsty dogs can drink sufficient water within minutes to restore the losses of water, whereas starved dogs reduce water consumption by 67–80%. By contrast, increasing protein or salt intake increases water consumption in dogs. Voluntary water consumption by dogs is affected by the moisture of ingested foods. The ratio of total water intake to DM intake is 2.33 ml water for 1 g DM food. At 22–25 °C, a dog needs 50–60 ml water/kg BW/day (Schaer 1989). Water intake is increased when dogs exercise or are exposed to heat stress. Exposure to cold does not appear to affect water consumption.

4.8.3 Water Metabolism in Cats

Like dogs, cats require water as the most abundant nutrient and can maintain body water balance when fed meat containing ~ 70% water. At 22–25 °C, a cat needs 45–55 ml water/kg BW/day (NRC 2006). Cats have a lower physiological thirst drive than dogs and concentrate their urine (through the reabsorption of water in the lumen of kidney tubules into the blood) to a greater degree than dogs, which helps to conserve water (Anderson 1982; Chew 1965). In addition, the dense hair of cats can minimize the evaporative loss of water through the skin. Thus, cats may have a lower requirement for exogenous water than dogs under similar feeding conditions and can tolerate mild dehydration better than dogs (NRC 2006).

A loss of 20% water from cats can be fatal (Anderson 1982). Compared with continuous feeding, periodic feeding reduces the daily consumption of food and water by cats (Finco et al. 1986). In both dogs and cats, drinking too much water may result in adverse effects. As important as it is to avoid dehydration, water intoxication in dogs due to their excessive water intake may occur during swimming, diving, or water-retrieving (Flaim 2019). A cat drinking an excessive amount of water (e.g., > 100 ml/kg BW/day) may appear thirsty, but may actually have potential health problems (e.g., diabetes mellitus, chronic kidney disease, and hyperthyroidism; Sparkes et al. 2015).

4.9 Feeding Behavior and Management of Dogs and Cats

4.9.1 General Considerations

The evolution of domestic dogs and cats has resulted in the idiosyncrasies of their digestive tract, metabolism, and nutrient requirements. Both dogs and cats can develop learned taste aversions (Mugford 1977), avoid arginine-deficient diets (Morris 2002), prefer flavor components in meat and peptides (e.g., protein hydrolysates) and free AAs (e.g., alanine, histidine, leucine, lysine, and proline, as well as monosodium glutamate plus 5'-nucleotides (Hargrove et al. 1994; Kumazawa and Kurihara 1990a,b). In both animal species, age and the percentage of lean or fat body mass can influence food intake (Hall et al. 2018a, b). Their appetite is regulated by signals produced in the hypothalamus and peripheral organs (e.g., the digestive tract and white adipose tissue), as well as by the concentrations of glucose, fatty acids, amino acids, and their metabolites in the blood (Wu 2018). In addition, overweight or obesity occurs in dogs and cats if they are allowed to eat as much as they want at all times during the day without adequate exercise (Zoran 2010). Furthermore, spaying (the removal of ovaries from females) and neutering (the removal of testes from males) are performed as a management procedure before reaching sexual maturity (e.g., at or before 5 and 6 months of age in cats and dogs, respectively) with adequate nutritional support to prevent any unwanted behaviors (Hart et al. 2020; Vendramini et al. 2020). In both dogs and cats, care must be taken to prevent gastrointestinal parasites, as well as infectious diseases [e.g., canine and feline parvovirus (highly contagious viral diseases of dogs and cats, respectively) that cause acute gastrointestinal illness] (Laflamme and Hannah 2010). Of particular note, cow or goat milk is not a suitable replacement milk for both puppies and kittens, because the ruminant milk contains much less protein, arginine, taurine, methionine, fat, ME, calcium, phosphorus, sodium, copper, iron, and oleic acid than dog and cat milk (Debraekeleer et al. 2010; Gross et al. 2010), as shown in Table 4.3. Orphaned puppies and kittens can be fed commercial milk replacers especially produced for them. However, despite the above shared similarities, dogs and cats differentially select diets containing different compositions of macronutrient when given choices to eat foods with similar palatability (Hall et al. 2018a). These peculiar aspects of canine and feline nutrition are highlighted in the following paragraphs. Such knowledge will aid in our understanding of the different feeding behaviors of dogs and cats. This, in turn, can help to develop optimal management methods to raise, nurture, and care for them.

Table 4.3 Concentrations of nutrients in the milks of cats, dogs, cows, goats, and pigs[a]

Nutrient	Cat	Dog	Cow	Goat	Pig[n]
Water (g/100 g milk)	79	77.3	87.7	87.0	79.9
Dry matter (g/100 g milk)	21	22.7	12.3	13.0	20.1
Crude protein (g/100 g milk)	7.5	7.5	3.3	3.6	4.8
Arginine (mg/100 g milk)	347	420	119	119	186
Taurine (mg/100 g milk)	27	33[b]	0.13	9.8[d]	19.0
Methionine (mg/100 g milk)	188	194[c]	82	80	91.8
Crude fat (g/100 g milk)	8.5	9.5	3.6	4.1	8.0
Palmitate (C16:0; g/100 g milk)	1.3[e]	2.5[e]	0.94[e]	0.91[g]	2.0
Oleic acid (C18:1; ω9; Δ^9; g/100 g milk)	2.3[e]	3.8[e]	1.0[e]	0.98[g]	2.2
α-Linolenic acid (C18:3; ω3; $\Delta^{9,12,15}$; g/100 g milk)	0.16[e]	0.03[f]	0.05[h]	0.04[g]	0.09
Linoleic acid (C18:2; ω6; $\Delta^{9,12}$; g/100 g milk)	1.3	1.1	0.10[h]	0.11[g]	1.1
Arachidonic acid (C20:4; ω6; $\Delta^{5,8,11,14}$; g/100 g milk)	0.11	0.04[f]	0.09	0.17	0.04
Lactose (g/100 g milk)	4.0	3.3	4.7	4.0	5.2
Calcium (mg/100 g milk)	180	240	119	133	200
Phosphorus (mg/100 g milk)	162	180	93	111	142
Potassium (mg/100 g milk)	103	120	150	204	89
Sodium (mg/100 g milk)	90	80	53[i]	56[k]	42
Magnesium (mg/100 g milk)	9	11	14	14	11
Zinc (mg/100 g milk)	0.66[e]	0.95	0.44[i]	0.15[m]	0.68
Copper (mg/100 g milk)	0.11	0.33	0.05[j]	0.07[m]	0.09
Iron (mg/100 g milk)	0.35	0.70	0.05	0.05	0.20
Vitamin A (IU/100 g milk)	651	867[l]	156[k]	207[k]	160
Metabolizable energy (kcal/100 g milk)	121	146	64	69	110
Metabolizable energy (kJ/100 g milk)	506	610	268	288	462

[a] Adapted from Gross et al. (2010) and Debraekeleer et al. (2010) for the milk of cats, dogs, cows, and goats, unless indicated otherwise
[b] Rassin et al. (1978)
[c] Heinze et al. (2014)
[d] Prosser (2021)
[e] Dobenecker et al. (1998)
[f] Heinze et al. (2014)
[g] Kompan and Komprej (2012)
[h] Wu (2018)
[i] Rodrıguez et al. (2001)
[j] Supplee and Bellis (1922)
[k] Bruhn (2017)
[l] Anderson et al. (1940)
[m] Sawaya et al. (1984)
[n] Reza et al. (2016) for the content of protein, fat, and lactose; Wu (2022) for the content of amino acids; calcium; and Hurley (2015) for the content of fatty acids and minerals
"–", data are not available

4.9.2 Weaning of Dogs and Cats

Weaning refers to the full transition of the puppy's diet from its mother's liquid milk to the solid growth diet. This is a significant and stressful period for young dogs. In dogs, weaning can begin at 3–4 weeks of age (Case et al. 2011). A small amount of canned pet food (usually consisting of ingredients of animal origin, such as meat, poultry, fish, and animal by-products) mixed with water (1:1, g/g) can be introduced to puppies 1 week before weaning so that they can adapt more easily to solid foods. Care must be taken to protect the intestinal microflora and prevent intestinal disorders in post-weaning dogs, such as gastritis, abnormal stool, gut atrophy, diarrhea, enteritis, leaky gut syndrome, inflammatory bowel disease, and vomiting (Case et al. 2011). The weanling diet must contain sufficient proteinogenic AAs (particularly arginine, glutamate, glutamine, glycine, methionine, proline, and tryptophan) and taurine (Wu 2018).

Weaning of cats usually begins when kittens are ~ 4 weeks of age by separating them from their mother for a few hours at a time each day and is completed at ~ 8 weeks of age (Case et al. 2011). Moistened kitten food and clean water bowls can be provided to kittens before the nursing phase ends, so as to stimulate the development of their digestive tract and facilitate their transition from the mother's liquid milk to solid weaning diets. As for dogs, weaning allows kittens to gradually progress from the sole maternal care to social independence. This is a critical period in neonatal life when environmental stresses should be minimized or prevented. During weaning, cats must be provided with diets containing high-quality protein and taurine to ensure optimal nutrition for the small intestine and the whole body.

4.9.3 Food Selection of Dogs and Cats

Dogs are similar to other carnivores in the GIT anatomy but to omnivores in the metabolism of most nutrients. This characteristic is useful to guide canine feeding. Dogs have evolved in their ability to digest a relatively large amount of cooked starch (Axelsson et al. 2013; Félix et al. 2012) and select a diet lower in protein (30% of ME from protein) than a high-protein (52% of ME from protein) diet of wild wolves (Buff et al. 2014). Thus, dogs have more latitude in the selection of food ingredients and more flexible adaptability to both animal- and plant-sourced diets (NRC 2006). In addition, dogs have receptors for sweet substances and, therefore, have preference for 10% sucrose solution, compared to water. When offered complete and balanced diets with varying levels of protein, fat, and carbohydrate, dogs will choose a high-fat (63% of ME from fat) diet over a low-fat (15% of ME from fat) diet (Bradshaw 2006; Buff et al. 2014; Hewson-Hughes et al., 2016). Regarding dietary protein choices, dogs are more like rats than cats. When offered various levels of protein in diets, dogs will select the diets containing 25–30% of energy as protein (Romos and Ferguson

1983; Tôrres et al. 2003). Dogs can eat powdered, dry, semi-moist, or canned diets. Food should be stored in under air-tight, dry, and cool conditions.

Cats are picky eaters (Pekel et al. 2020). Preferring meat with palatability factors (e.g., fats, AAs, peptides, and nucleotides) as food, cats eat small prey, such as rats, mice, birds, lizards, and insects. Odor, taste, texture, particle size, and temperature of diets affect the food preference of cats. Cats may avoid a particular diet because of the textural difference but not the difference in its nutrient composition. Clearly, cats prefer (1) moist foods to dry foods, (2) warm foods to cold or hot foods, (3) high-protein (e.g., 50% protein on a DM basis) foods to low-protein (e.g., 20% protein on a DM basis) or protein-free foods, (4) animal fat to plant fat, and (5) diets containing 25% total lipids to diets containing 10% or 50% total lipids (Cook et al. 1985, 1996; Eyre et al. 2022; Morris 2002; Rogers et al. 2004). Thus, cats would select beef tallow over butter and chicken fats, but have similar preference for beef tallow, lard, and partially hydrogenated vegetable oil (Kane et al. 1981). A diet consisting of high fat content and 5% hydrogenated beef tallow is particularly palatable to cats (Rogers et al. 2004). By contrast, medium-chain fatty acids, present in either the free form (e.g., 0.1% caprylic acid) or TAG (e.g., 5% medium-chain TAG), and 25% hydrogenated coconut oil in diets can reduce food intake by cats (MacDonald et al. 1985). Of particular note, cats cannot taste sweetness as indicated previously, because of the lack of sweet taste receptors due to the deletion of the *Tas1r* gene and, unlike dogs, do not select sweet substances such as sucrose (Li et al. 2005, 2006). For this reason, cats show no preference for sugar-rich foods such as fruits and juice.

When cats are offered complete and balanced diets with varying levels of protein, fat, and carbohydrate (without controlling palatability), cats will choose high-protein diets (Bradshaw 2006; Hewson-Hughes et al. 2016). Thus, high-starch diets over a prolonged period of time could be a major contributing to poor feline health and must be avoided in feeding both young and adult cats particularly those with overweight or diabetes (Verbrugghe and Hesta 2017). Likewise, domestic cats select a macronutrient profile (52% and 48% of ME from protein and fat, respectively) similar to the diet (52%, 46%, and 2% of ME from protein, fat, and glycogen, respectively) of wild cats. With regards to AAs, cats have preference for proline, cysteine, ornithine, lysine, histidine, alanine, and leucine (Bradshaw et al. 1996), but against arginine, isoleucine, phenylalanine, and tryptophan (Oliveira et al. 2016; Zaghini and Biagi 2005). This is consistent with the notion that cats are more sensitive to bitter taste than dogs. Exposure of kittens to numerous different flavors and textures early in life can result in wider preference for such foods later in life (Kuo 1967). However, adult cats can learn to select new and better flavors in addition to the ones early in life (Mugford 1977).

Healthy cats do not select sodium even if they are deficient in sodium, but acidotic cats will select diets containing excess sodium (Yu et al. 1997). Furthermore, cats can develop a learned taste aversion to diets that causes metabolic acidosis (Cook et al. 1996). Healthy cats are normally neophilic, namely selecting a new food (or flavor) if offered a variety of foods (or flavors) throughout their lives (Beauchamp et al. 1977). Furthermore, cats do not consume a powdered diet but will ingest the same diet in a pelleted or gel form. If a diet is too dry or powdery, cats will eat very little even when no alternative food is offered (NRC 2006).

4.9.4 Meal Frequency of Dogs and Cats

Dogs eat small and large prey, as well as plant-sourced foods. These animals can consume a large meal within 10 min that is sufficient to meet daily nutrient requirements, but they usually eat 4 to 8 meals/day (depending on breed; Bradshaw 2006). Growing puppies should have free access to food or be fed 2–3 times daily. Young and adult dogs generally ingest food and drink water during the light period and may eat at night if they are hungry or dehydrated, but some breeds of dogs also eat and drink during the dark period (Mugford and Thorne 1980). On average, dogs drink more water than cats per kg BW per day. Although adult dogs may do well on a regimen of eating a meal per day with possible health benefits (Bray et al. 2021), these animals should be offered foods at least twice daily to ensure adequate nutrition and welfare (Brooks et al. 2014). Multiple meals per day may help dogs to alleviate boredom, while reducing the risk for the problem of gastric dilatation-volvulus in susceptible breeds of dogs (Bataller 1995). If overweight or obese, they should be fed either less or lower energy-dense diets. Caloric restriction may reduce oxidative stress and delay aging in dogs (Kealy et al 2002; Lawler et al. 2008). Of particular note, intermittent fasting on a ketogenic diet containing medium-chain triglycerides may confer both metabolic and immunological effects in healthy dogs (Leung et al. 2020). Dogs with special health issues or dietary needs may require individualized feeding programs.

In contrast to dogs, cats do not have clear-cut circadian rhythms (Hawking et al. 1971; Randall et al. 1985). Some cats are nocturnal (e.g., resting during the daytime and going out or walking at night), and their existing rhythms can be altered by minor disturbances such as human noises (Macdonald and Apps 1978). Cats generally eat frequent, small meals and drink water throughout the day and night and can voluntarily eat 12–20 meals/day, but fresh food should be offered daily (Kane et al. 1981). After they adapt to a new diet for 1 week, the size and number of meals are not affected by the type of diet. When cats eat a big portion of food at one time, they may immediately vomit it due to gastric irritation. Therefore, appropriate and frequent meals are critical for feline health. Pet owners can decide meal size and frequency based on breed, age, health condition, ambient environment, and practicality for their dogs and cats rather than simply food availability.

Camara et al. (2020) recently reported that compared with the feeding of the same canned food four times/day, adult cats fed only once daily had greater plasma concentrations of glucagon-like protein-1, gastric inhibitory protein, and some AAs within the initial postprandial hours (as expected because of more nutrient intakes at a single time), but a lower value of respiratory quotient (RQ) in the fasting state. These authors interpreted the data as that a reduced RQ value may result from a greater rate of whole-body fatty acid oxidation and that feeding once daily may be a beneficial feeding management strategy for indoor cats to promote satiation and lean body mass. However, a reduced RQ value in the fasting state may also result from a greater rate of the conversion of AAs (released from the breakdown of muscle protein; possibly an adverse effect) into glucose, because this process has an RQ value of 0.333 (Wu 2018). In the work (Study 1) of Camara et al. (2020), neither body weight nor feed (energy) intake differed between the two groups of cats. Clearly, more studies are warranted to assess whether feeding once daily may have health benefits for cats.

4.10 Conclusion and Perspectives

Both dogs and cats naturally prefer and effectively digest meat, but can also digest appropriate amounts of properly (e.g., suitable temperature and period) processed and formulated plant-sourced ingredients. These two animal species differ markedly in some aspects of nutrition and metabolism: including (1) the ability to digest the quantities (i.e., small vs large) of dietary cooked starch in the small intestine; (2) thermoneutral zones and rates of basal energy metabolism; (3) qualitative (i.e., presence or absence) and quantitative (i.e., amounts) requirements for many dietary nutrients, particularly protein, certain AAs (arginine, taurine, methionine, and cysteine, as well as NEAAs), ω3 and ω6 PUFAs, and vitamins (e.g., vitamin A and niacin); (4) sensitivity to dietary AA imbalances and antagonisms; (5) felinine synthesis from cysteine; (6) feeding behavior and meal frequencies; (7) food preference and refusal; (8) the selection for foods containing various amounts of protein and fats; (9) the form of diets; and (10) the occurrence of gastrointestinal and metabolic disorders (Table 4.1). Recommended minimum requirements and allowances of dietary nutrients other than AAs for growing and adult dogs and cats are summarized in Table 4.4. Recommended values for their dietary AA requirements have been recently summarized (Li and Wu 2023a, b). Unlike cats, dogs have adapted by expressing a high activity of pancreatic α-amylase activity during evolution to omnivorous diets containing taurine-rich meat and plant ingredients with relatively a large amount of starch. Thus, a high intake of starch (e.g., 40% on a DM basis) can result in metabolic disorders in cats but not in dogs, whereas dietary fiber is beneficial for the intestinal health of both cats and dogs. Given the increasing incidences of obesity or overweight in both dogs and cats, dietary supplementation with arginine may be a promising solution to prevent metabolic syndromes in these animals, as reported for rats (Jobgen et al. 2009, 2022a, b, 2023). In practice, cats should not be fed canine foods. Because the composition of the milk of both dogs and cats differ from that of farm mammals, the young pets should not be fed replacer diets formulated based on goat or cow milk. For both dogs and cats, there may be breed differences in dietary requirements for nutrients and thermoneutral zones even among different sizes of the same breed. The fundamental knowledge of the idiosyncrasies of metabolism in dogs and cats is essential for guiding their feeding and care, as well as food manufacturing.

Of particular note, current commercial vegan petfoods may be nutritionally inadequate for dogs (low content of calcium, potassium, sodium, and methionine) and cats (low content of protein, arginine, taurine, and potassium, as well as an improper Ca/P ratio). Animal-sourced foods contain nutritionally significant amounts of AAs, lipids, and minerals and therefore play an important role in balancing AAs in diets for dogs and cats play an important role in optimizing the nutrition and health of these companion animals.

Table 4.4 Recommended allowances of dietary fatty acids, minerals, and vitamins for dogs, cats, and pigs

Nutrient	Dogs (Amt/kg DM)		Cats (Amt/kg DM)		Pigs (Amt/kg DM)	
	Growing[a]	Adult	Growing	Adult	6-kg BW	110-kg BW
	NRC (2006)		NRC (2006)		(NRC 2012)[b]	
Metabolizable energy (kcal)	~4000	~4000	~4000	~4000	3778	3667
Crude protein (g)	225	100	225	200	252	116
Total fat (g)	85	55	90	90	–	–
Fatty acid (g)						
Linoleic acid	13	11	5.5	5.5	1.1	1.1
α-Linolenic acid	0.8	0.44	0.2	ND	–	–
Arachidonic acid	0.3	–	0.2	0.06	–	–
EPA + DHA	0.5	0.44	0.1	0.1	–	–
Minerals						
Calcium (g)	12	4.0	8.0	2.9	9.4	5.1
Phosphorus (g)	10	3.0	7.2	2.6	7.8	4.8
Potassium	4.4	4.0	4.0	5.2	3.3	1.9
Sodium (g)	2.2	0.80	1.4	0.68	4.4	1.1
Chloride (g)	2.9	1.2	0.90	0.96	5.6	0.89
Magnesium (g)	0.40	0.60	0.40	0.40	0.44	0.44
Iron (mg)	88	30	80	80	111	44
Copper (mg)	11	6.0	8.4	5.0	6.7	3.3
Zinc (mg)	100	60	75	74	111	56
Manganese (mg)	5.6	4.8	4.8	4.8	4.4	2.2
Iodine (mg)	0.88	0.88	1.8	1.4	0.16	0.16
Selenium (mg)	0.35	0.35	0.30	0.30	0.33	0.17

(continued)

(continued)

Nutrient	Dogs (Amt/kg DM)		Cats (Amt/kg DM)		Pigs (Amt/kg DM)	
	Growing[a]	Adult	Growing	Adult	6-kg BW	110-kg BW
	NRC (2006)		NRC (2006)		(NRC 2012)[b]	
Vitamins and choline						
All-*trans*-retinol (vitamin A; μg)	1515	1515	1000	1000	733	433
Cholecalciferol (vitamin D_3; μg)	13.8	13.8	5.6	7.0	6.1	4.2
α-Tocopherol (vitamin E; mg)	30	30	38	38	12	8.2
Menadione (vitamin K_3; mg)	1.64	1.63	1.0	1.0	0.56	0.56
Thiamin (mg)	1.38	2.25	5.5	5.6	1.7	1.1
Riboflavin (mg)	5.25	5.25	4.0	4.0	4.4	2.2
Pyridoxine (mg)	1.5	1.5	2.5	2.5	7.8	1.1
Niacin (mg)	17	17	40	40	33	33
Pantothenic acid (mg)	15	15	5.7	5.75	13	7.8
Cobalamin (vitamin B_{12}; μg)	35	35	22.5	22.5	22	5.6
Folic acid (mg)	0.27	0.27	0.75	0.75	0.33	0.33
Biotin (μg)	ND	ND	75	75	89	56
Choline (g)	1.7	1.7	2.55	2.55	0.67	0.33

[a] The recommended allowance of dietary crude protein for $\geq$ 14-week-old growing dogs is 175 g/kg DM

[b] Different breeds of swine have different body weights at young and adult ages. For the pig offspring (Yorkshire × Landrace sows and Duroc × Hampshire boars) with a normal birth weight, 6- and 110-kg body weights correspond to approximately 3 weeks and 6 months of age. At about 18 months of age (adult), the lean-tissue weight curve of swine is flattened (Reiland 1978). The dietary requirement of adult swine for crude protein is approximately 100 g/kg dry matter

Amt = amount; BW = body weight; DHA = docosahexaenoic acid; DM = dry matter of diet; EAAs = traditionally classified nutritionally essential amino acids; EPA = eicosapentaenoic acid; ND = not determined; NEAAs = traditionally classified nutritionally nonessential amino acids; "–" = data are not available

Acknowledgements We thank Professor Nancy Ing (DVM and Ph.D., Department of Animal Science at Texas A&M University) for critically reviewing and providing helpful comments on this manuscript. This work was supported by Texas A&M AgriLife Research (H-8200).

Conflict of Interest The authors declare no conflict of interest.

References

Anderson HD, Johnson BC, Arnold A (1940) The composition of dog's milk. Am J Physiol 129:631–634

Anderson RS (1982) Water balance in the dog and cat. J Small Anim Pract 23:588–598

Anderson PJ, Rogers QR, Morris JG (2002) Cats require more dietary phenylalanine or tyrosine for melanin deposition in hair than for maximal growth. J Nutr 132:2037–2042

Axelsson E, Ratnakumar A, Arendt M-L et al (2013) The genomic signature of dog domestication reveals adaptation to a starch-rich diet. Nature 495:360–364

Backus RC, Cohen G, Pion PD, Good KL, Rogers QR, Fascetti AJ (2003) Taurine deficiency in Newfoundlands fed commercially available complete and balanced diets. J Am Vet Med Assoc 223:1130–1136

Backus RC, Ko KS, Fascetti AJ, Kittleson MD, MacDonald KA, Maggs DJ, Berg JR, Rogers QR (2006) Low plasma taurine concentration in Newfoundland dogs is associated with low plasma methionine and cyst(e)ine concentrations and low taurine synthesis. J Nutr 136:2525–2533

Badawy AA (2017) Kynurenine pathway of tryptophan metabolism: Regulatory and functional aspects. Int J Tryptophan Res 10:1178646917691938

Baker DH, Czarnecki-Maulden GL (1991) Comparative nutrition of cats and dogs. Annu Rev Nutr 11:239–263

Bartges JM, Cook M (1999) Influence of feeding conjugated linoleic acid on body composition in healthy adult cats. In: Proceedings 17th ACVIM. pp 729

Bataller N (1995) Risk factors and the debate of diet in canine gastric dilatation-volvulus. Vet Clin Nutr 2:87

Batchelor DJ, Al-Rammahi M, Moran AW, Brand JG, Li X, Haskins M, German AJ, Shirazi-Beechey SP (2011) Sodium/glucose cotransporter-1, sweet receptor, and disaccharidase expression in the intestine of the domestic dog and cat: two species of different dietary habit. Am J Physiol 300:R67-75

Bauer JE (1997) Fatty acid metabolism in domestic cats (Felis catus) and cheetahs (Acinonyx jubatas). Proc Nutr Soc 56:1013–1024

Bauer JE (2007) Responses of dogs to dietary omega-3 fatty acids. J Am Vet Med Assoc 231:1657–1661

Beauchamp GK, Maller O, Rogers JG Jr (1977) Flavor preferences in cats *(Felis catus* and *Panthera sp*). J Comp Physiol Psychol 91:1118–1127

Beliveau GP, Freedland RA (1982) Metabolism of serine, glycine and threonine in isolated cat hepatocytes Felis domestica. Comp Biochem Physiol B 71:13–18

Belo PS, Romsos DR, Leveille GA (1976) Influence of diet on glucose tolerance, on the rate of glucose utilization and on gluconeogenic enzyme activities in the dog. J Nutr 106:1465–1474

Bermingham EN, Thomas DG, Morris PJ, Hawthorne AJ (2010) Energy requirements of adult cats. Br J Nutr 103:1083–1093

Besch EL, Woods JE (1977) Heat production biorhythms of laboratory animals. Lab Anim Sci 27:54–59

Biourge V, Sergheraert R (2002) Hair pigmentation can be affected by diet in dogs. Proc Comp Nutr Soc 4:103–104

Bordoni A, Biagi PL, Turchetto E, Hrelia S (1988) Aging influence on delta-6-desaturase activity and fatty acid composition of rat liver microsomes. Biochem Int 17:1001–1009

Bradshaw JWS (2006) The evolutionary basis for the feeding behavior of domestic dogs (*Canis familiaris*) and cats (*Felis catus*). J Nutr 136:1927S-1931S

Bradshaw JWS, Goodwin D, Legrand-Defretin V, Nott HMR (1996) Food selection by the domestic cat, an obligate carnivore. Comp Biochem Physiol A 114:205–209

Bray EE, Zheng Z, Tolbert MK, McCoy BM, Dog Aging Project Consortium, Kaeberlein M, Kerr KF (2021) Once-daily feeding is associated with better health in companion dogs: results from the Dog Aging Project. bioRxiv preprint. https://doi.org/10.1101/2021.11.08.467616

Brody S, Procter RC, Ural S Ashworth US (1934) Growth and development with special reference to domestic animals. XXXIV, Basal metabolism, endogenous nitrogen, creatinine and neutral sulphur excretions as functions of body weight. Agric Exp Stn Res Bull No 220. University of Missouri, Columbia

Brooks D, Churchill J, Fein K et al (2014) AAHA weight management guidelines for dogs and cats. J Am Anim Hospital Assoc 50:1–11

Brosey BP, Hill RC, Scott KC (2000) Gastrointestinal volatile fatty acid concentrations and pH in cats. Am J Vet Res 61:359–361

Bruhn JC (2017) Dairy goat milk composition. https://drinc.ucdavis.edu/goat-dairy-foods/dairy-goat-milk-composition. Accessed 5 Oct 2021

Bryan DL, Hart P, Forsyth K, Gibson R (2001) Incorporation of alpha-linolenic acid and linoleic acid into human respiratory epithelial cell lines. Lipids 36:713–717

Buff PR, Carter RA, Bauer JE, Kersey JH (2014) Natural pet food: a review of natural diets and their impact on canine and feline physiology. J Anim Sci 92:3781–3791

Burns RA, Milner JA, Corbin JE (1981) Arginine: An indispensable amino acid for mature dogs. J Nutr 111:1020–1024

Camara A, Verbrugghe A, Cargo-Froom C, Hogan K, DeVries TJ, Sanchez A, Robinson LE, Shoveller AK (2020) The daytime feeding frequency affects appetite-regulating hormones, amino acids, physical activity, and respiratory quotient, but not energy expenditure, in adult cats fed regimens for 21 days. PLoS ONE 15:e0238522

Carciofi AC, Takakura FS, de-Oliveira LD, Teshima E, Jeremias JT, Brunetto MA, Prada F (2008) Effects of six carbohydrate sources on dog diet digestibility and post-prandial glucose and insulin response. J Anim Physiol Anim Nutr 92:326–336

Case LP, Daristotle L, Hayek MG, Raasch MF (2011) Canine and feline nutrition, 3rd edn. Mosby, Maryland Heights, Missouri

Cavanaugh SM, Cavanaugh RP, Gilbert GE, Leavitt EL, Ketzis JK, Vieira AB (2021) Shortterm amino acid, clinicopathologic, and echocardiographic findings in healthy dogs fed a commercial plant-based diet. PLoS ONE 16:e0258044

Che DS, Nyingwa PS, Ralinala KM, Maswanganye GMT, Wu G (2021) Amino acids in the nutrition, metabolism, and health of domestic cats. Adv Exp Med Biol 1285:217–231

Chew RM (1965) Water metabolism of mammals. In: Mayer WV, van Gelder RG (eds) Physiological mammalogy: mammalian reactions to stressful environments. Academic Press, New York, pp 43–178

Chiocchetti R, Galiazzo G, Giancola F, Tagliavia C, Bernardini C, Forni M, Pietra M (2021) Localization of the serotonin transporter in the dog intestine and comparison to the rat and human intestines. Front Vet Sci 8:802479

Churchill JA, Eirmann L (2021) Senior pet nutrition and management. Vet Clin North Am Sm Anim Pract 51:635–651

Clark MH, Ferguson DC, Bunick D, Hoenig M (2013) Molecular and histological evidence of brown adipose tissue in adult cats. Vet J 195:66–72

Cook NE, Kane E, Rogers QR, Morris JG (1985) Self selection of dietary casein and soy-protein by the cat. Physiol Behav 34:583–594

Cook NE, Rogers QR, Morris JG (1996) Acid-base balance affects dietary choice in cats. Appetite 26:175–192

Croniger CM, Olswang Y, Reshef L, Kalhan SC, Tilghman SM, Hanson RW (2002) Phosphoenolpyruvate carboxykinase revisited: insight into its metabolic role. Biochem Mol Bil Educ 30:14–20

Czarnecki GL, Baker DH (1984) Urea cycle function in the dog with emphasis on the role of arginine. J Nutr 114:581–590

Davis TA, Nguyen HV, Garcia-Bravo R, Fiorotto ML, Jackson EM, Lewis DS, Lee DR, Reeds PJ (1994a) Amino acid composition of human milk is not unique. J Nutr 124:1126–1132

Davis TA, Nguyen HV, Garcia-Bravo R, Fiorotto ML, Jackson EM, Reeds PJ (1994b) Amino acid composition of the milk of some mammalian species changes with stage of lactation. Br J Nutr 72:845–853

De Bruijne JJ, van den Brom WE (1986) The effect of long-term fasting on ketone body metabolism in the dog. Comp Biochem Physiol B 83:391–395

Debraekeleer J, Gross KL, Zicker SC (2010) Feeding nursing and orphaned puppies from birth to weaning. In: Thatcher CD, Remillard RL, Roudebush P (eds) Small Animal clinical nutrition Hand MS. Mark Morris Institute, Topeka, KS, pp 295–309

DeNapoli JS, Dodman NH, Shuster L, Rand WM, Gross KL (2000) Effect of dietary protein content and tryptophan supplementation on dominance aggression, territorial aggression, and hyperactivity in dogs. J Am Vet Med Assoc 217:504–508

Dillon EL, Wu G (2021) Cortisol enhances ctrulline synthesis from proline in enterocytes of suckling piglets. Amino Acids 53:1957–1966

Dobenecker B, Zottmann B, Kienzle E, Zentek J (1998) Investigations on milk composition and milk yield in queens. J Nutr 128:2618S-2619S

Dodman NH, Reisner I, Shuster L, Rand W, Luescher UA, Robinson I, Houpt KA (1996) Effect of dietary protein content on behavior in dogs. J Am Vet Med Assoc 208:376–379

Dunbar BL, Bauer JE (2002) Conversion of essential fatty acids by delta-6 desaturase in dog liver microsomes. J Nutr 132:1701S-1703S

Eyjolfson V, Spriet LL, Dyck DJ (2004) Conjugated linoleic acid improves insulin sensitivity in young, sedentary humans. Med Sci Sports Exerc 36:814–820

Eyre R, Trehiou M, Marshall E, Carvell-Miller L, Goyon A, McGrane S (2022) Aging cats prefer warm food. J Vet Behav 47:86–92

Félix AP, Gabeloni LR, Brito CBM, Oliveira SG, Silva AVF, Maiorka A (2012) Effect of β-mannanase on the digestibility of diets with different protein sources in dogs determined by different methodologies. J Anim Sci 90:3060–3067

Feng BC, Li J, Kliegman RM (1996) Transcription of hepatic cytosolic phosphoenolpyruvate carboxykinase gene in newborn dogs. Biochem Mol Med 59:13–19

Finco DR, Adams DD, Crowelle WA, Stattelman AJ, Brown SA, Barsanti JA (1986) Food and water intake and urine composition in cats: influence of continuous versus periodic feeding. Am J Vet Res 47:1638–1642

Flaim D (2019) Can dogs drink too much water? The dangers of water intoxication. American Kennel Club. https://www.akc.org/expert-advice/health/can-dogs-drink-much-water-dangers-water-intoxication. Accessed 12 Jan 2022

Galim EB, Hruska K, Bier DM, Matthews DE, Haymond MW (1980) Branched-chain amino acid nitrogen transfer to alanine in vivo in dogs. J Clin Invest 66:1295–1304

Gershoff SN, Andrus SB, Hegsted DM, Lentini EA (1957) Vitamin A deficiency in cats. Lab Invest 6:227–240

Gil F, Arencibia A, García V, Ramírez G, Vázquez JM (2018) Anatomic and magnetic resonance imaging features of the salivary glands in the dog. Anat Histol Embryol 47:551–559

Glynn EL, Fry CS, Drummond MJ, Timmerman KL, Dhanani S, Volpi E, Rasmussen BB (2010) Excess leucine intake enhances muscle anabolic signaling but not net protein anabolism in young men and women. J Nutr 140:1970–1976

Green AS, Ramsey JJ, Villaverde C, Asami DK, Wei A, Fascetti AJ (2008) Cats are able to adapt protein oxidation to protein intake provided their requirement for dietary protein is met. J Nutr 138:1053–1060

Greig RA, Gnaedinger RH (1971) Occurrence of thiaminase in some common aquatic animals of the United States and Canada. Special Scientific Report—Fish. US Dept Commer Natl Mar Fish Serv 631:1–7

Gross KL, Becvarova I, Debraekeleer J (2010) Feeding nursing and orphaned kittens from birth to weaning. In: Thatcher CD, Remillard RL, Roudebush P (eds) Small Animal clinical nutrition Hand MS. Mark Morris Institute, Topeka, KS, pp 415–427

Hall JA, Vondran JC, Vanchina MA, Jewell DE (2018a) When fed foods with similar palatability, healthy adult dogs and cats choose different macronutrient compositions. J Exp Biol 221:jeb173450

Hall JA, Jackson MI, Vondran JC, Vanchina MA, Jewell DE (2018b) Comparison of circulating metabolite concentrations in dogs and cats when allowed to freely choose macronutrient intake. Biol Open 7:bio036228

Hammer VA, Rogers QR, Freedland RA (1996) Threonine is catabolized by L-threonine 3-dehydrogenase and threonine dehydratase in hepatocytes from domestic cats (*Felis domestica*). J Nutr 126:2218–2226

Hanson RW, Garber AJ (1972) Phosphoenolpyruvate carboxykinase. I. Its role in gluconeogenesis. Am J Clin Nutr 25:1010–1021

Hargrove DM, Morris JG, Rogers QR (1994) Kittens choose a high leucine diet even when isoleucine and valine are the limiting amino acids. J Nutr 124:689–693

Harper AE, Miller RM, Block KP (1984) Branched-chain amino acid metabolism. Annu Rev Nutr 4:409–454

Hart BL, Hart LA, Thigpen AP, Willits NH (2020) Assisting decision-making on age of neutering for 35 breeds of dogs: associated joint disorders, cancers, and urinary incontinence. Front Vet Sci 7:388

Hawking F, Lobban MC, Gammage K, Worms MJ (1971) Circadian rhythms (activity, temperature, urine and microfilariae) in dog, cat, hen, duck, thamnomys and gerbillus. Int J Cycle Res 2:455–473

Hazewinkel HAW, van den Brom WE, van't Klooster A, Voorhout G, van Wees A (1991) Calcium metabolism in Great Dane dogs fed diets with various calcium and phosphorus contents. J Nutr 121:S99–S106

He W, Connolly ED, Wu G (2024) Characteristics of the digestive tract of dogs and cats. Adv Exp Med Biol 1446:15–38. https://doi.org/10.1007/978-3-031-54192-6_2

He WL, Wu G (2022) Oxidation of amino acids, glucose, and fatty acids as metabolic fuels in enterocytes of developing pigs. Amino Acids 54:1025–1039

He WL, Posey EA, Steele CC, Savell JW, Bazer FW, Wu G (2023) Dietary glycine supplementation enhances postweaning growth and meat quality of pigs with intrauterine growth restriction. J Anim Sci 101:skadskad354

Hedhammar A, Wu FM, Krook L, Schryver HF, de LaHunta A, Wahlen JP, Kallfelz FA, Nunez EA, Hintz HF, Sheffy BE, Ryan GD (1974) Overnutrition and skeletal disease. An experimental study in growing Great Dane dogs. Cornell Vet 64 (suppl 5):1–160

Heinze CR, Larsen JA, Kass PH, Fascetti AJ (2009) Plasma amino acid and whole blood taurine concentrations in cats eating commercially prepared diets. Am J Vet Res 70:1374–1382

Heinze CR, Freeman LM, Martin CR, Power ML, Fascetti AJ (2014) Comparison of the nutrient composition of commercial dog milk replacers with that of dog milk. J Am Vet Med Assoc 244:1413–1422

Hendriks WH, Moughan PJ, Tarttelin MF (1996) Gut endogenous nitrogen and amino acid excretions in adult domestic cats fed a protein-free or an enzymatically hydrolyzed casein-based diet. J Nutr 126:955–962

Hendricks WH, Moughan PJ, Tarttelin MF (1997) Urinary excretion of endogenous nitrogen metabolites in adult domestic cats using a protein-free diet and the regression technique. J Nutr 127:623–629

Hendriks WH, Sritharan K, Hodgkinson SM (2002) Comparison of the endogenous ileal and faecal amino acid excretion in the dog (*Canis familiaris*) and the rat (*Rattus rattus*) determined under protein-free feeding and peptide alimentation. J Anim Physiol Anim Nutr 86:333–341

Hendriks WH, Vather R, Rutherfurd SM, Weidgraaf K, Rutherfurd-Markwick KJ (2004) Urinary isovalthine excretion in adult cats is not gender dependent or increased by oral leucine supplementation. J Nutr 134:2114S-2116S

Hendriks WH, Rutherfurd-Markwick KJ, Weidgraaf K, Morton RH, Rogers QR (2008) Urinary felinine excretion in intact male cats is increased by dietary cystine. Br J Nutr 100:801–809

Hewson-Hughes AK, Gilham MS, Upton S, Colyer A, Butterwick R, Miller AT (2011) The effect of dietary starch level on postprandial glucose and insulin concentrations in cats and dogs. Br J Nutr 106:S105–S109

Hewson-Hughes AK, Colyer A, Simpson SJ, Raubenheimer D (2016) Balancing macronutrient intake in a mammalian carnivore: disentangling the influences of flavour and nutrition. R Soc Open Sci 3:160081

Hill RC (1998) The nutritional requirements of exercising dogs. J Nutr 128:2686S-2690S

Hill RC, Scott KC (2004) Energy requirements and body surface area of cats and dogs. JAVMA 225:689–694

Holloway B, Stribling D, Freeman S, Jamieson L (1985) The thermogenic role of adipose tissue in the dog. Int J Obesity 9:423–432

Holowaty MNH, Lees MJ, Abou Sawan S, Paulussen KJM, Jäger R, Purpura M, Paluska SA, Burd NA, Hodson N, Moore DR (2023) Leucine ingestion promotes mTOR translocation to the periphery and enhances total and peripheral RPS6 phosphorylation in human skeletal muscle. Amino Acids 55:253–261

Hou YQ, Wu G (2017) Nutritionally nonessential amino acids: a misnomer in nutritional sciences. Adv Nutr 8:137–139

Hou YQ, Yin YL, Wu G (2015) Dietary essentiality of "nutritionally nonessential amino acids" for animals and humans. Exp Biol Med 240:997–1007

Hou YQ, Yao K, Yin YL, Wu G (2016) Endogenous synthesis of amino acids limits growth, lactation and reproduction of animals. Adv Nutr 7:331–342

Hou YQ, Wu ZL, Dai ZL, Wang GH, Wu G (2017) Protein hydrolysates in animal nutrition: Industrial production, bioactive peptides, and functional significance. J Anim Sci Biotechnol 8:24

Hou YQ, He WL, Hu SD, Wu G (2019) Composition of polyamines and amino acids in plant-source foods for human consumption. Amino Acids 51:1153–1165

Houston DM, Moore AEP, Favrin MG, Brent Hoff B (2003) Feline urethral plugs and bladder uroliths: a review of 5484 submissions 1998–2003. Can Vet J 44:974–977

How KL, Hazewinkel HAW, Mol JA (1994) Dietary vitamin D dependence of cat and dog due to inadequate cutaneous synthesis of vitamin D. Gen Comp Endocrinol 96:12–18

Hu SD, He WL, Wu G (2021) Hydroxyproline in animal metabolism, nutrition, and cell signaling. Amino Acids 54:513–528

Humbert B, Martin L, Dumon H, Darmaun D, Nguyen P (2002) Dietary protein level affects protein metabolism during the postabsorptive state in dogs. J Nutr 132:1676S – 1678

Hurley WL (2015) Composition of sow colostrum and milk. In: Farmer C (ed) The gestating and lactating sow. Wageningen Academic Publishers, Wageningen, The, Netherlands, pp 193–229

Iyer MS, Paszkiewicz RL, Bergman RN, Richey JM, Woolcott OO, Asare-Bediako I, Wu Q, Kim SP, Stefanovski D, Kolka CM, Clegg DJ, Kabir M (2019) Activation of NPRs and UCP1-independent pathway following CB1R antagonist treatment is associated with adipose tissue beiging in fat-fed male dogs. Am J Physiol 317:E535-547

Jacobsen JG, Smith LH (1968) Biochemistry and physiology of taurine and taurine derivatives. Physiol Rev 48:424–511

Jobgen WJ, Meininger CJ, Jobgen SC, Li P, Lee M-J, Smith SB, Spencer TE, Fried SK, Wu G (2009) Dietary L-arginine supplementation reduces white-fat gain and enhances skeletal muscle and brown fat masses in diet-induced obese rats. J Nutr 139:230–237

Jobgen WS, Wu G (2022a) L-Arginine increases AMPK phosphorylation and the oxidation of energy substrates in hepatocytes, skeletal muscle cells, and adipocytes. Amino Acids 54:1553–1568

Jobgen WS, Wu G (2022b) Dietary L-arginine supplementation increases the hepatic expression of AMP-activated protein kinase in rats. Amino Acids 54:1569–1584

Jobgen WS, Lee M-J, Fried SK, Wu G (2023) L-Arginine supplementation regulates energy-substrate metabolism in skeletal muscle and adipose tissue of diet-induced obese rats. Exp Biol Med 248:209–216

Kane E, Rogers QR, Morris JG, Leung PMB (1981) Feeding behavior of the cat fed laboratory and commercial diets. Nutr Res 1:499–507

Kaplan JL, Stern JA, Fascetti AJ, Larsen JA, Skolnik H, Peddle GD, Kienle RD, Waxman A, Cocchiaro M, Gunther-Harrington CT, Klose T, LaFauci K, Lefbom B, Lamy MM, Malakoff R, Nishimura S, Oldach M, Rosenthal S, Stauthammer C, O'Sullivan L, Visser LC, Williams R, Ontiveros E (2018) Taurine deficiency and dilated cardiomyopathy in golden retrievers fed commercial diets. PLoS ONE 13:e0209112

Kealy RD (1999) Factors influencing lean body mass in aging dogs. Comp Cont Edu Small Anim Pract 2:34–37

Kealy RD, Lawler DF, Ballam JM et al (2002) Effects of diet restriction on life span and 358 age-related changes in dogs. J Am Vet Med Assoc 220:1315–1320

Kendall PT, Blaze SE, Holme DW (1982) Assessment of endogenous nitrogen output in adult dogs of contrasting size using a protein-free diet. J Nutr 112:1281–1286

Kettelhut IC, Foss MC, Migliorini RH (1980) Glucose homeostasis in a carnivorous animal (cat) and in rats fed a high-protein diet. Am J Physiol 239:R437-444

Khanna C, Boermans HJ, Wilcock B (1997) Fatal hypernatremia in a dog from salt ingestion. J Am Anim Hosp Assoc 33:113–116

Kienzle E (1993a) Carbohydrate metabolism of the cat. 1. Activity of amylase in the gastrointestinal tract of the cat. J Anim Physiol Anim Nutr 69:92–101

Kienzle E (1993b) Carbohydrate metabolism of the cat. 2. Digestion of starch. J Anim Physiol Anim Nutr 69:102–114

Kirk RW, Bistner SI (1981) Handbook of veterinary procedures and emergency treatment. W.B. Saunders, Philadelphia, p 582

Kittleson MD, Keene B, Pion PD, Loyer CG (1997) Results of the multicenter spaniel trial (MUST): taurine- and carnitine-responsive dilated cardiomyopathy in American cocker spaniels with decreased plasma taurine concentration. J Vet Intern Med 11:204–211

Kleiber M (1961) The fire of life. Wiley & Sons, New York

Knopf K, Sturman JA, Armstrong M, Hayes KC (1978) Taurine: an essential nutrient for the cat. J Nutr 108:773–778

Kompan D, Komprej A (2012) The effect of fatty acids in goat milk on health. In: Chaiyabutr N (ed) Milk production, IntechOpen, London, UK

Krehl WA (1981) Discovery of the effect of *tryptophan* on niacin deficiency. Fed Proc 40:1527–1530

Kumazawa T, Kurihara K (1990a) Large enhancement of canine taste responses to sugars by salts. J Gen Physiol 95:1007–1018

Kumazawa T, Kurihara K (1990b) Large synergism between monosodium glutamate and 59–nucleotides in canine taste nerve responses. Am J Physiol 259:R420–R426

Kuo ZY (1967) The dynamics of behaviour development: an epigenetic view. Random House, New York

Ladd M, Raisz LG (1949) Response of the normal dog to dietary sodium chloride. Am J Physiol 159:149–152

Laflamme D (2008a) Effect of diet on loss and preservation of lean body mass in aging dogs and cats. Companion Animal Nutrition Summit, May 3–5, 2008, Charleston, South Carolina. pp 41–46

Laflamme DP (2008b) Pet food safety: dietary protein. Top Companion Anim Med 23:154–157

Laflamme DP, Hannah SS (2010) Pet dogs and vats. In: Pond KR (ed) Introduction to animal science (pond WG. John Wiley & Sons, New York, pp 526–552

Laflamme DP, Hannah SS (2013) Discrepancy between use of lean body mass or nitrogen balance to determine protein requirements for adult cats. J Feline Med Surg 15:691–697

Laflamme D, Danièlle G-M (2014) Nutrition of aging cats. Vet Clin North Am Sm Anim Pract 44:761–774

Lambie JG, Pezzali JG, Richards TL, Ellis JL, Verbrugghe A, Shoveller AK (2024) Phenylalanine requirements using the direct amino acid oxidation technique, and the effects of dietary phenylalanine on food intake, gastric emptying, and macronutrient metabolism in adult cats. J Anim Sci 102:skae009

Landsberg G, Milgram B, Mougeot I, Kelly S, de Rivera C (2017) Therapeutic effects of an alpha-casozepine and L-tryptophan supplemented diet on fear and anxiety in the cat. J Feline Med Surg 19:594–602

Larsen TM, Toubro A, Astrup A (2003) Efficacy and safety of dietary supplements containing CLA for the treatment of obesity: evidence from animal and human studies. J Lipid Res 44:2234–2241

Larsen JA, Oberbauer AM (2023) Dog nutrition. In: Phillips CJC (ed) Encyclopedia of Animal Nutrition, 2nd edition. CABI, Wallingford, pp 176–179

Latimer HB (1938) The weights of the brain and of its parts, of the spinal cord and of the eyeballs in the adult cat J Comp Neurol 68:395–404

Latimer HB (1967) Variability in body and organ weights in the newborn dog and cat compared with that in the adult. Anat Red 157:449–456

Lawler DF, Larson BT, Ballam JM, Smith GK, Biery DN, Evans RH, Greeley EH, Segre M, Stowe HD, Kealy RD (2008) Diet restriction and ageing in the dog: major observations over two decades. Br J Nutr 99:793–805

Legrand-Defretin V (1994) Differences between cats and dogs: a nutritional view. Proc Nutr Soc 53:15–24

Lemaitre RN, King IB, Sotoodehnia N, Rea TD, Raghunathan TE, Rice KM, Lumley TS, Knopp RH, Cobb LA, Copass MK, Siscovick DS (2009) Red blood cell membrane alpha-linolenic acid and the risk of sudden cardiac arrest. Metabolism 58:534–540

Leung YB, Cave NJ, Heiser A, Edwards PJB, Godfrey AJR, Wester T (2020) Metabolic and immunological effects of intermittent fasting on a ketogenic diet containing medium-chain triglycerides in healthy dogs. Front Vet Sci 6:480

Levillain O, Parvy P, Hus-Citharel A (1996) Arginine metabolism in cat kidney. J Physiol 491:471–477

Li P, Wu G (2018) Roles of dietary glycine, proline and hydroxyproline in collagen synthesis and animal growth. Amino Acids 50:29–38

Li P, Wu G (2020) Composition of amino acids and related nitrogenous nutrients in feedstuffs for animal diets. Amino Acids 52:523–542

Li P, Wu G (2022) Functional molecules of intestinal mucosal products in animal nutrition and health. Adv Exp Med Biol 1354:263–277

Li P, Wu G (2023a) Amino acid nutrition and metabolism in domestic cats and dogs. J Anim Sci Biotechnol 14:19

Li P, Wu G (2023b) Composition of nutrients in rendered animal-sourced feedstuffs. In: Phillips C (ed) Encyclopedia of animal nutrition, 2nd edn. CABI, Wallingford, Oxon, UK, pp 576–579

Li X, Li W, Wang H, Cao J, Maehashi K, Huang L, Bachmanov AA, Reed DR, Legrand-Defretin V, Beauchamp GK (2005) Pseudogenization of a sweet-receptor gene accounts for cats' indifference toward sugar. PLoS Genet 1:e27-35

Li X, Li W, Wang H, Bayley DL, Cao J, Reed DR, Bachmanov AA, Huang L, Legrand-Defretin V, Beauchamp GK, Brand JG (2006) Cats lack a sweet taste receptor. J Nutr 136(Suppl 7):1932S-1934S

Li XL, Zheng SX, Jia SC, Song F, Zhou CP, Wu G (2020) Oxidation of energy substrates in tissues of largemouth bass (*Micropterus salmoides*). Amino Acids 52:1017–1032

Li P, He WL, Wu G (2021a) Composition of amino acids in foodstuffs for humans and animals. Adv Exp Med Biol 1332:189–210

Li XY, Zheng SX, Wu G (2021b) Nutrition and functions of amino acids in fish. Adv Exp Med Biol 1285:133–168

Li XY, Han T, Zheng SX, Wu G (2021c) Nutrition and functions of amino acids in aquatic crustaceans. Adv Exp Med Biol 1285:169–197

Li P, He WL, Wu G (2021d) Composition of amino acids in foodstuffs for humans and animals. Adv Exp Med Biol 1332:189–209

Li XY, He WL, Wu G (2023) Dietary glycine supplementation enhances the growth performance of hybrid striped bass (*Morone saxatilis* ♀× *Morone chrysops* ♂) fed soybean meal-based diets. J Anim Sci 101:skad345

Lichtenberger M (2021) Thiaminase and its role in predatory pet fish (and other piscivores) nutrition. http://www.wetwebmedia.com/ca/volume_6/volume_6_1/thiaminase.htm: Accessed 19 Sept 2021

Loftus JP, Yazwinski M, Milizio JG, Wakshlag JJ (2014) Energy requirements for racing endurance sled dogs. J Nutr Sci 3:e34

Loncar D, Afzelius BA (1989) Ontogenetical changes in adipose tissue of the cat: convertible adipose tissue. J Ultrastruct Mol Struct Res 102:9–23

Macdonald DW, Apps PJ (1978) The social behaviour of a group of semi-dependent farm cats, *Felis catus*: a progress report. Carnivore Genet Newslett 3:256–268

MacDonald ML, Rogers QR, Morris JG (1985) Aversion of the cat to dietary medium-chain triglycerides and caprylic acid. Physiol Behav 35:371–375

Mansilla WD, Gorman A, Fortener L, Shoveller AK (2018) Dietary phenylalanine requirements are similar in small, medium, and large breed adult dogs using the direct amino acid oxidation technique. J Anim Sci 96:3112–3120

Mansilla WD, Templeman JR, Fortener L, Shoveller AK (2020a) Minimum dietary methionine requirements in Miniature Dachshund, Beagle, and Labrador Retriever adult dogs using the indicator amino acid oxidation technique. J Anim Sci 98:skaa324

Mansilla WD, Fortener L, Templeman JR, Shoveller AK (2020b) Adult dogs of different breed sizes have similar threonine requirements as determined by the indicator amino acid oxidation technique. J Anim Sci 98:skaa066

McCauley SR, Clark SD, Quest BW, Streeter RM, Oxford EM (2020) Review of canine dilated cardiomyopathy in the wake of diet-associated concerns. J Anim Sci 98:skaa155

McCue MD (2010) Starvation physiology: reviewing the different strategies animals use to survive a common challenge. Comp Biochem Physiol A 156:1–18

McDowell LR (1989) Vitamin C: vitamins in animal nutrition. San Diego, Academic Press, pp 365–387

McGeachin RL, Akin JR (1979) Amylase levels in the tissues and body fluids of the domestic cat (*Felis catus*). Comp Biochem Physiol B 63:437–439

Middleton RP, Lacroix S, Scott-Boyer MP, Dordevic N, Kennedy AD, Slusky AR, Carayol J, Petzinger-Germain C, Beloshapka A, Kaput J (2017) Metabolic differences between dogs of different body sizes. J Nutr Metab 2017:4535710

Mitchell WK, Phillips BE, Williams JP, Rankin D, Lund JN, Wilkinson DJ, Smith K, Atherton PJ (2015) The impact of delivery profile of essential amino acids upon skeletal muscle protein synthesis in older men. Clinical efficacy of pulse versus bolus supply. Am J Physiol 309:E450-457

Miyazaki M, Yamashita T, Taira H, Suzuki A (2008) The biological function of cauxin, a major urinary protein of the domestic cat. In: Hurst JL, Beynon RJ, Roberts SC, Wyatt TD (eds) Chemical signals in vertebrates, vol 11. Springer, New York, pp 51–60

Montserrat-Malagarriga M, Castillejos L, Salas-Mani A, Torre C, Martín-Orúe SM (2024) The impact of fiber source on digestive function, fecal microbiota, and immune response in adult dogs. Animals 14:196

Morris JG (1999) Ineffective vitamin D synthesis in cats is reversed by an inhibitor of 7-dehydrocholestrol-Δ^7-reductase. J Nutr 129:903–908

Morris JG, Rogers QR, Pacioretty LM (1990) Taurine: an essential dietary nutrient for cats. J Small Anim Pract 31:502–509

Morris JG (2002) Idiosyncratic nutrient requirements of cats appear to be diet-induced evolutionary adaptations. Nutr Res Rev 15:153–168

Morris JG, Rogers QR (1978) Arginine: an essential amino acid for the cat. J Nutr 108:1944–1953

Morris JG, Trudell J, Pencovic T (1977) Carbohydrate digestion by the domestic *cat* (*Felis catus*). Br J Nutr 37:365–373

Mugford RA (1977) External influences on the feeding of carnivores. In: Maller O (ed) The chemical senses and nutrition (Kare MR. Academic Press, New York, pp 25–50

Mugford RA, Thorne C (1980) Comparative studies of meal patterns in pet and laboratory housed dogs and cats. In: Anderson RS (ed) Nutrition of the dog and cat. Pergamon Press, Oxford, UK, pp 3–14

National Research Council (NRC (2006) Nutrient requirements of dogs and cats. National Academies Press, Washington DC

National Research Council (NRC (2012) Nutrient requirements of swine. National Academies Press, Washington DC

Nogueira JPDS, He F, Mangian HF, Oba PM, de Godoy MRC (2019) Dietary supplementation of a fiber-prebiotic and saccharin-eugenol blend in extruded diets fed to dogs. J Anim Sci 97:4519–4531

Oberbauer AM, Larsen JA (2021) Amino acids in dog nutrition and health. Adv Exp Med Biol 1285:199–216

Oliveira R, Haesel D, Kill JL, Lima A, Malini PV, Thompson GR (2016) Palatability of cat food with sodium pyrophosphate and yeast extract. Ciência Rural 46:2202–2205

Pawlosky R, Barnes A, Salem N Jr (1994) Essential fatty acid metabolism in the feline: relationship between liver and brain production of long-chain polyunsaturated fatty acids. J Lipid Res 35:2032–2040

Pekel AY, Mülazımoğlu SB, Acar N (2020) Taste preferences and diet palatability in cats. J Appl Anim Res 48:281–292

Pion PD, Kittleson MD, Rogers QR, Morris JG (1987) Myocardial failure in cats associated with low plasma taurine: a reversible cardiomyopathy. Science 237:764–768

Pion PD, Kittleson MD, Thomas WP, Skiles ML, Rogers QR (1992) Clinical findings in cats with dilated cardiomyopathy and relationship of findings to taurine deficiency. J Am Vet Med Assoc 201:267–274

Prosser CG (2021) Compositional and functional characteristics of goat milk and relevance as a base for infant formula. J Food Sci 86:257–265

Rabin B, Nicolosi RJ, Hayes KC (1976) Dietary influence on bile acid conjugation in the cat. J Nutr 106:1241–1246

Randall W, Johnson RF, Randall S, Cunningham JT (1985) *Circadian rhythms* in food intake and activity in domestic cats. Behav Neurosci 99:1162–1175

Ranz D, Gutbrod F, Corinna Eule C, Kienzle E (2002) Nutritional lens opacities in two litters of newfoundland dogs. J Nutr 132:1688S-1689S

Rassin DK, Sturman JA, Gaull GE (1978) Taurine and other free amino acids in milk of man and other mammals. Early Hum Dev 2:1–13

Reiland S (1978) Growth and skeletal development of the pig. Acta Radiol 358(Suppl):15–22

Rezaei R, Wu G (2022) Branched-chain amino acids regulate intracellular protein turnover in porcine mammary epithelial cells. Amino Acids 54:1491–1504

Rezaei R, Wu ZL, Hou YQ, Bazer FW, Wu G (2016) Amino acids and mammary gland development: nutritional implications for neonatal growth. J Anim Sci Biotechnol 7:20

Riond JL, Stiefel M, Wenk C, Wanner M (2003) Nutrition studies on protein and energy in domestic cats. J Anim Physiol Anim Nutr (berl) 87:221–228

Rivera NLM, Félix AP, Ferreira FM, da Silva AVF, Maiorka A (2011) Body measurements and serum lipid profile of overweight adult dogs fed diet with containing conjugated linoleic acid. Ciência Rural 41:2020–2025

Rivers JPW, Sinclair AJ, Crawford MA (1975) Inability of the cat to desaturate essential fatty acids. Nature 258:171–173

Robertson WG, Jones JS, Heaton MA, Stevenson AE, Markwell PJ (2002) Predicting the crystallization potential of urine from cats and dogs with respect to calcium oxalate and magnesium ammonium phosphate (struvite). J Nutr 132:1637S-1641S

Rodrıguez EMR, Alaejos MS, Romero CD (2001) Mineral concentrations in cow's milk from the canary Island. J Food Compost Anal 14:419–430
Rogers QR, Morris JG (1979) Essentiality of amino acids for the growing kitten. J Nutr 109:718–723
Rogers QR, Phang JM (1985) Deficiency of pyrroline-5-carboxylate synthase in the intestinal mucosa of the cat. J Nutr 115:146–150
Rogers QR, Taylor TP, Morris JG (1998) Optimizing dietary amino acid patterns at various levels of crude protein for cats. J Nutr 128:2577S-2580S
Rogers QR, Morris JG, Freedland RA (1977) Lack of hepatic enzymatic adaptation to low and high levels of dietary protein in the adult cat. Enzyme 22:348–356
Rogers QR, Wigle AR, Laufer A, Castellanos VH, Morris JG (2004) Cats select for adequate methionine but not threonine. J Nutr 134:2046S-2049S
Romsos DR, Ferguson D (1983) Regulation of protein intake in adult dogs. J Am Vet Med Assoc 182:41–43
Romsos DR, Palmer HJ, Muiruri KL, Bennink MR (1981) Influence of a low carbohydrate diet on performance of pregnant and lactating dogs. J Nutr 111:678–689
Russell K, Murgatroyd PR, Batt RM (2002) Net protein oxidation is adapted to dietary protein intake in domestic cats (*Felis silvestris catus*). J Nutr 132:456–460
Rutherfurd-Markwick KJ, Rogers QR, Hendriks WH (2005) Mammalian isovalthine metabolism. J Anim Physiol Anim Nutr 89:1–10
Sawaya WN, Khalil JK, A F Al-Shalhat AF (1984) Mineral and vitamin content of goat's milk. J Am Diet Assoc 84:433–435
Schaer M (1989) General principles of fluid therapy in small animal medicine. Vet Clin N Am Sm Anim Pract 19:203–213
Schoenherr W, Jewell J (1999) Effect of conjugated linoleic acid on body composition of mature obese Beagles. FASEB J 13:A262
Schneeman BO (1994) Carbohydrates: significance for energy balance and gastrointestinal function. J Nutr 124(Suppl 9):1747S-1753S
Schoenmakers I, Hazewinkel HAW, van den Brom WE (1999) Excessive Ca and P intake during early maturation in dogs alters Ca and P balance without long-term effects after dietary normalization. J Nutr 129:1068–1074
Schweigert FJ (1998) Metabolism of carotenoids in mammals. In: Liaaen-Jensen S, Pfander H (eds) Carotenoids (Britton G. Birkhäuser Verlag, Boston, pp 249–284
Schweigert FJ, Raila J, Wichert B, Kienzle E (2002) Cats absorb β-carotene, but it is not converted to vitamin A. J Nutr 132:1610S-1612S
Silvio J, Harmon DL, Gross KL, McLeod KR (2000) Influence of fiber fermentability on nutrient digestion in the dog. Nutrition 16:289–295
Sinclair AJ, Slattery W, McLean JG, Monger EA (1981) Essential fatty acid deficiency and evidence for arachidonate synthesis in the cat. Br J Nutr 46:93–96
Singh P, Banton S, Bosch G, Hendriks WH, Shoveller AK (2024) Beyond the bowl: understanding amino acid requirements and digestibility to improve protein quality metrics for dog and cat foods. Adv Exp Med Biol 1446:99–134. https://doi.org/10.1007/978-3-031-54192-6_5
Sparkes AH, Cannon M, Church D, Fleeman L, Harvey A, Hoenig M, Peterson ME, Reusch CE, Taylor S, Rosenberg D, ISFM (2015) ISFM consensus guidelines on the practical management of diabetes mellitus in cats. J Feline Med Surg 17:235–250
Summers SC, Stockman J, Larsen JA, Zhang L, Rodriguez AS (2020) Evaluation of phosphorus, calcium, and magnesium content in commercially available foods formulated for healthy cats. J Vet Intern Med 34:266–273
Supplee GC, Bellis B (1922) The coper content of cow's milk. J Dairy Sci 5:455–467
Tôrres CL, Hickenbottom SJ, Quinton R. Rogers QR (2003) Palatability affects the percentage of metabolizable energy as protein selected by adult beagles. J Nutr 133:3516–3522
Trevizan L, de Mello KA, Bigley KE, Anderson WH, Waldron MK, Bauer JE (2010) Effects of dietary medium-chain triglycerides on plasma lipids and lipoprotein distribution and food aversion in cats. Am J Vet Res 71:435–440

Ugawa T, Kurihara K (1993) Large enhancement of canine taste responses to amino acids by salts. Am J Physiol 264:R1071-1076

Van Kruiningen HJ, Wojan LD, Stake PE, Lord PF (1987) The influence of diet and feeding frequency on gastric function in the dog. J Am Anim Hosp Assoc 23:146–153

Vendramini THA, Amaral AR, Pedrinelli V, Zafalon RVA, Rodrigues RBA, Brunetto MA (2020) Neutering in dogs and cats: current scientific evidence and importance of adequate nutritional management. Nutr Res Rev 33:134–144

Verbrugghe A, Bakovic M (2013) Peculiarities of one-carbon metabolism in the strict carnivorous cat and the role in feline hepatic lipidosis. Nutrients 5:2811–2835

Verbrugghe A, Hesta M (2017) Cats and carbohydrates: the carnivore fantasy? Vet Sci 4:55

Vhile SG, Skrede A, Ahlstrøm Ø, Hove K (2005) Comparative apparent total tract digestibility of major nutrients and amino acids in dogs (Canis familiaris), blue foxes (*Alopex lagopus*) and mink (*Mustela vison*). Anim Sci 81:141–148

Vinay P, Lemieux G, Gougoux A, Halperin M (1986) Regulation of glutamine metabolism in dog kidney in vivo. Kidney Int 29:68–79

Vuorinen A, Bailey-Hall E, Karagiannis A, Yu S, Roos F, Sylvester E, Wilson J, Irina Dahms I (2020) Safety of algal oil containing EPA and DHA in cats during gestation, lactation and growth. J Anim Physiol Anim Nutr (berl). 104:1509–1523

Wannemacher RW, McCoy JR (1966) Determination of optimal dietary protein requirements of young and old dogs. J Nutr 88:66–74

Washizu, T, Ishida T, Washizu M, Tomoda I, Kaneko JJ (1994) Changes in bile acid composition of serum and gallbladder bile in bile duct ligated dogs. J Vet Med Sci 56:299–303

Washizu T, Tanaka A, Sako T, Washizu M, Arai T (1999) Comparison of the activities of enzymes related to glycolysis and gluconeogenesis in the liver of dogs and cats. Res Vet Sci 67:205–206

Watford M (1985) Gluconeogenesis in the chicken: regulation of phosphoenolpyruvate carboxykinase gene expression. Fed Proc 44:2469–2074

Weber FL Jr, Maddrey WC, Walser M (1977) Amino acid metabolism of dog jejunum before and during absorption of keto analogues. Am J Physiol 232:E263-269

Weber FL Jr, Veach GL (1979) The importance of the small intestine in gut ammonium production in the fasting dog. Gastroenterology 77:235–240

Weber FL Jr, Friedman DW, Fresard KM (1988) Ammonia production from intraluminal amino acids in canine jejunum. Am J Physiol 254:G264-268

Weber MP, Stambouli F, Martin LJ, Dumon HJ, Biourge VC, Nguyen PG (2002) Influence of age and body size on gastrointestinal transit time of radioopaque markers in healthy dogs. Am J Vet Res 63:677–682

Weber MP, Martin LJ, Biourge VC, Nguyen PG, Dumon HJ (2003) Influence of age and body size on orocecal transit time as assessed by use of the sulfasalzine method in healthy dogs. Am J Vet Res 64:1105–1109

Wester TJ, Weidgraaf K, Hekman M, Ugarte CE, Forsyth SF, Tavendale MH (2015) Amino acid oxidation increases with dietary protein content in adult neutered male cats as measured using [1-^{13}C]leucine and [$^{15}N_2$]urea. J Nutr 145:2471–2478

White TD, Boudreau JC (1975) Taste preferences of the cat for neurophysiologically active compounds. Physiol Psychol 3:405–410

Widdowson EM (1985) Development of the digestive system: comparative animal studies. Am J Clin Nutr 41(Suppl):384–390

Wildgrube HJ, Stockhausen H, Petri J, Füssel U, Lauer H (1986) Naturally occurring conjugated bile acids, measured by highperformance liquid chromatography, in human, dog, and rabbit bile. J Chromatogr 353:207–213

Williams CC, Cummins KA, Hayek MG, Davenport GM (2001) Effects of dietary protein on whole-body protein turnover and endocrine function in young-adult and aging dogs. J Anim Sci 79:3128–3136

Wolf LGG, Mehlman MA (1972) Subcellular distribution of pyruvate carboxylase and phosphoenolpyruvate carboxykinase in dog liver and kidney. Proc Soc Exp Biol Med 141:532–535

Woods JE, Besch EL (1971) Dissipation of body heat in dogs in a controlled environment. Physiologist 14:252
Wu G (1998) Amino acid metabolism in the small intestine. Trends Comp Biochem Physiol 4:39–74
Wu G (2009) Amino acids: metabolism, functions, and nutrition. Amino Acids 37:1–17
Wu G (2016) Dietary protein intake and human health. Food Funct 7:1251–1265
Wu G (2018) Principles of animal nutrition. CRC Press, Boca Raton, Florida
Wu G (2020a) Metabolism and functions of amino acids in sense organs. Adv Exp Med Biol 1265:201–217
Wu G (2020b) Important roles of dietary taurine, creatine, carnosine, anserine and 4-hydroxyproline in human nutrition and health. Amino Acids 52:329–360
Wu G (2022) Amino acids: biochemistry and nutrition, 2nd edn. CRC Press, Boca Raton, Florida
Wu G, Morris SM Jr (1998) Arginine metabolism: nitric oxide and beyond. Biochem J 336:1–17
Wu G, Li P (2022) The "ideal protein" concept is not ideal in animal nutrition. Exp Biol Med 247:1191–1201
Wu G, Borbolla AG, Knabe DA (1994) The uptake of glutamine and release of arginine, citrulline and proline by the small intestine of developing pigs. J Nutr 124:2437–2444
Wu G, Bazer FW, Satterfield MC, Li XL, Wang XQ, Johnson GA, Burghardt RC, Dai ZL, Wang JJ, Wu ZL (2013) Impacts of arginine nutrition on embryonic and fetal development in mammals. Amino Acids 45:241–256
Wu G, Cross HR, Gehring KB, Savell JW, Arnold AN, McNeill SH (2016a) Composition of free and peptide-bound amino acids in beef chuck, loin, and round cuts. J Anim Sci 94:2603–2613
Wu ZL, Hou YQ, Hu SD, Bazer FW, Meininger CJ, McNeal CJ, Wu G (2016b) Catabolism and safety of supplemental L-arginine in animals. Amino Acids 48:1541–1552
Xenoulis PG, Palculict B, Allenspach K, Steiner JM, Van House AM, Suchodolski JS (2008) Molecular-phylogenetic characterization of microbial communities imbalances in the small intestine of dogs with inflammatory bowel disease. FEMS Microbiol Ecol 66:579–589
Yu YM, Wagner DA, Tredget EE, Walaszewski JA, Burke JF, Young VR (1990) Quantitative role of splanchnic region in leucine metabolism: L-[1-^{13}C,^{15}N]leucine and substrate balance studies. Am J Physiol 259:E36-51
Yu YM, Burke JF, Tompkins RG, Martin R, Young VR (1996) Quantitative aspects of interorgan relationships among arginine and citrulline metabolism. Am J Physiol 271:E1098-1109
Yu S, Rogers QR, Morris JG (1997) Absence of a salt (NaCl) preference or appetite in sodium-replete or depleted kittens. Appetite 29:1–10
Yu S, Rogers QR, Morris JG (2001) Effect of low levels of dietary tyrosine on the hair colour of cats. J Small Anim Pract 42:176–180
Zafalon RVA, Risolia LW, Pedrinelli V, Vendramini THA, Rodrigues RBA, Amaral AR, Kogika MM, Brunetto MA (2020a) Vitamin D metabolism in dogs and cats and its relation to diseases not associated with bone metabolism. J Anim Physiol Anim Nutr 104:322–342
Zafalon RVA, Risolia LW, Vendramini THA, Ayres Rodrigues RB, Pedrinelli V, Teixeira FA et al (2020b) Nutritional inadequacies in commercial vegan foods for dogs and cats. PLoS ONE 15:e0227046
Zaghini G, Biagi G (2005) Nutritional peculiarities and diet palatability in the cat. Vet Res Commun 29(Suppl 2):39–44
Zhai H, Adeola O (2011) Apparent and standardized ileal digestibilities of amino acids for pigs fed corn-and soybean meal-based diets at varying crude protein levels. J Anim Sci 89:3626–3633
Zhang Q, Hou YQ, Bazer FW, He WL, Posey EA, Wu G (2021) Amino acids in swine nutrition and production. Adv Exp Med Biol 1285:81–107
Zoran DL (2010) Obesity in dogs and cats: a metabolic and endocrine disorder. Vet Clin North Am Small Anim Pract 40:221–239
Zoran DL (2023) Cat (Felis catus), essential nutrients. In: Phillips CJC (ed) Encyclopedia of Animal Nutrition, 2nd edition. CABI, Wallingford, pp 100–102

Chapter 5
Beyond the Bowl: Understanding Amino Acid Requirements and Digestibility to Improve Protein Quality Metrics for Dog and Cat Foods

Pawanpreet Singh, Sydney Banton, Guido Bosch, Wouter H. Hendriks, and Anna K. Shoveller

Abstract The determination of amino acid (AA) requirements for mammals has traditionally been done through nitrogen (N) balance studies, but this technique underestimates AA requirements in adult animals. There has been a shift toward researchers using the indicator amino acid oxidation (IAAO) technique for the determination of AA requirements in humans, and recently in dogs. However, the determination of AA requirements specific to adult dogs and cats at maintenance is lacking and the current requirements outlined by the National Research Council are based on a dearth of data and are likely underreporting the requirements of indispensable AA (IAA) for the population. To ensure the physiological requirements of our cats and dogs are met, we need methods to accurately and precisely measure digestibility. In vivo methods, such as ileal cannulation, are most commonly used, however, due to ethical considerations, we are moving away from animal models and toward in vitro methods. Harmonized static digestion models have the potential to replace in vivo methods but work needs to be done to have these methods more accurately represent the gastrointestinal tract (GIT) of cats and dogs. The Digestible IAA Score (DIAAS) is one metric that can help define protein quality for individual ingredients or mixed diets that uses AA SID estimates and ideally those can be replaced with in vitro AA digestibility estimates. Finally, we need accurate and reliable laboratory AA analyses to measure the AA present in complete diets, especially those used to quantify methionine (Met) and cysteine (Cys), both often limiting AAs in cat and dog diets. Together, this will guide accurate feed formulation for our companion animals to satisfy requirements while avoiding over-supplying protein, which inevitably contributes to excess N excretion, affecting both the environment and feed sustainability.

Pawanpreet Singh and Sydney Banton are the co-first authors

P. Singh · S. Banton · A. K. Shoveller (✉)
Department of Animal Biosciences, University of Guelph, Guelph, ON, Canada
e-mail: ashovell@uoguelph.ca

G. Bosch · W. H. Hendriks
Animal Nutrition Group, Wageningen University, Wageningen, The Netherlands

G. Wu (ed.), *Nutrition and Metabolism of Dogs and Cats*, Advances in Experimental Medicine and Biology 1446, https://doi.org/10.1007/978-3-031-54192-6_5

Keywords Requirements · Amino acids · Indicator amino acid oxidation · Digestibility

Abbreviations

AA	Amino acid
AAFCO	Association of American feed control officials
AI	Adequate intake
ATD	Apparent ileal digestibility
ATTD	Apparent total tract digestibility
BW	Body weight
CI	Confidence interval
CP	Crude protein
DAA	Dispensable amino acid
DIAAS	Digestible indispensable amino acid score
FEDIAF	European pet food industry federation
GIT	Gastrointestinal tract
GSH	Glutathione
IAA	Indispensable amino acid
IAAO	Indicator amino acid oxidation
MR	Minimal requirement
RA	Recommended allowance
SAA	Sulfur amino acid
SID	Standardized ileal digestibility
Tau	Taurine
TID	True ileal digestibility
Trp	Tryptophan
Tyr	Tyrosine
Val	Valine

5.1 Introduction

In 2018, approximately 471 million dogs and 373 million cats were kept as pets globally (Euromonitor 2019). Assuming that most of those pets are being fed a commercial diet, feeding this many animals creates a major demand for protein. Given the danger of not complying with regulations and the lack of sufficient data on amino acid (AA) requirements, it is likely that most of the pet food industry over-formulates protein. For example, the average crude protein (CP) content of adult canine maintenance dry foods available from two major retailers in Canada is 28% (range from 20–42%), 10% greater than the recommended minimum suggested by the

Association of American Feed Control Officials (AAFCO 2016) and European Pet Food Industry Federation (FEDIAF 2021). Additionally, consumers rely on protein claims when buying pet food. In a survey spanning Europe, Canada, and the United States, 24% of respondents stated they seek "protein #1 ingredient" on the label (Banton et al. 2021). This elucidates the rising demand for higher protein pet diets, with pet food companies vying for quality ingredients in competition with other nutrition sectors, including human. Traditionally, pet food relied on animal-based protein ingredients, but the industry is now utilizing new sources like plant, microbial, and synthetic proteins to meet the growing demand. For environmental and economic reasons, it is becoming increasingly important to be more precise with protein sources for pet food, which necessitates characterization of protein and AA composition, digestibility of AAs, and eventually comparison to AA requirements. The pet food industry continues to introduce new ingredients, but accurate data on the digestibility of the "classic" ingredients remains limited.

This chapter aims to discuss the current understanding of assessing AA requirements in adult pet dogs and cats, as well as the methodology to assess digestibility of AAs in ingredients used in pet foods. It will also explore the associated limitations and the necessary advancements in these methodologies to more accurately measure requirements, digestibility, protein quality, and dietary AA concentrations in pet foods. By comprehensively considering both the animal and its diet, we can better comprehend the quality of protein we provide to our pets and potentially improve the effective use of the limited amount of ingredients and lower the environmental impact of companion animal ownership.

5.2 Current Understanding

There are 22 AAs used for body protein synthesis (Wu 2009; Lopez and Mohiuddin 2023), that can be categorized into three groups; indispensable, conditionally indispensable, and dispensable (Reeds 2000). Protein requirements for mammals at maintenance can be defined by the necessary amount of indispensable AAs (IAAs) in combination with dispensable AAs (DAAs) (NRC 1989) to support whole-body protein turnover and nitrogen (N) balance in the animal. Indispensable AAs are characterized as those that the body cannot synthesize in sufficient quantities to meet the physiological needs of the individual (Borman et al. 1946). There are 10 IAAs for dogs and cats: arginine (Arg), histidine (His), isoleucine (Ile), leucine (Leu), lysine (Lys), methionine (Met), phenylalanine (Phe), threonine (Thr), tryptophan (Trp), and valine (Val) (NRC 2006). Additionally, cats have a dietary requirement for taurine, which although not incorporated into protein, is a metabolic product of Cys oxidation and required for a variety of functions such as osmoregulation, bile acid conjugation, retinal development, and cardiac function (Morris et al. 1990; Novotny et al. 1991). Conditionally, IAAs are synthesized by the body at efficient rates but may become essential to receive from the diet under circumstances when there are insufficient IAA precursors to meet the animal's metabolic demands (Reeds 2000). In such situations, it becomes necessary to increase the dietary intake of these AAs to support metabolic requirements. Some conditionally IAAs have sparing effects on IAAs, such that Cys can spare the Met requirement by up to 40–50% by weight (Finkelstein and Mudd 1967; DiBuono et al. 2001; Shoveller et al. 2003) and Tyr can spare up to 50% of the Phe requirement by weight (Sasse and Baker 1972; Kim 1993). Consequently, they are collectively referred to as total sulfur and total aromatic AAs, respectively. Dispensable AAs are categorized as AAs that are adequately synthesized within the body to support optimal health. However, there is limited data supporting that DAAs are sufficiently synthesized de novo in monogastric mammals (Wu 2014; Ennis and Elango 2021). As such, each DAA is considered unnecessary in the diet unless their precursors are restricted or there is an insufficient supply of dietary N.

Amino acids are utilized for a plethora of functions within the body beyond protein synthesis such as cell signaling, gene expression, synthesis of hormones, gluconeogenesis, and provision of structural components for essential compounds, including other DAAs (Wu 2009). Amino acids also play a crucial role in immune function. Cysteine, for example, serves as a direct precursor to glutathione (GSH), an antioxidant that protects cells from reactive oxygen species. For instance, supplementation of Cys, through N-acetylcysteine, in HIV patients, promoted lymphocyte proliferation and prolonged survival (Breitkreutz et al. 2000). As Met can influence Cys status, both would have implications for GSH synthesis (Wu et al. 2004). Clinical observations have indicated lower erythrocyte GSH concentrations in ill compared to healthy cats and dogs (Viviano et al. 2009), suggesting an increased requirement for GSH and coincidingly sulfur AAs (SAAs) during states of compromised health. Since AAs cannot be stored in the body, a consistent dietary intake is required to replace the direct losses through urine, feces, sweat, and other bodily secretions, as well as indirect losses occurring through catabolism or synthesis into other vital N-containing compounds (NRC 1989). A substantial portion, 30–50% of dietary IAAs are also catabolized in the small intestine through first-pass metabolism (Stoll et al. 1998). In-depth reviews by Wu (2009) and Li and Wu (2023) further summarize the metabolic and functional roles of AAs and emphasize that for these reasons, the requirements for AAs are not set but variable based on the fluctuating physiological demands of the individual.

5.3 Methods to Determine AA Requirements

Traditional methods for determining protein and AA requirements have been focused on measures such as N balance, growth, and plasma or serum AA and urea concentrations where animals were adapted to diets over weeks/months or in the case of growth, longer. Requirements can be assessed through factorial or empirical methods. The factorial method defines the protein requirements as the sum needed to replace obligatory losses (maintenance) and protein accretion of tissue (growth, lactation, and gestation; Fuller and Chamberlain 1985). Meanwhile, the empirical method often involves graded levels of a specific AA, above and below the estimated requirement, with a control group containing all nutrients above this requirement. This method is preferred when establishing the optimal intake levels of specific AAs to meet physiological needs. This is primarily done through the construction of dose–response curves depicting the impact of varying levels of AA intake on specific biological or physiological outcomes, such as protein synthesis, N balance or muscle growth (Pencharz and Ball 2003). To obtain accurate results, all other aspects of the experimental diet need to be held constant (Baker 1986; Pencharz and Ball 2003).

Historically, N balance studies have been used to determine AA requirements in mammals (FAO 2013). Nitrogen balance represents the difference between N intake, from protein or non-protein N sources, and N excretion (fecal, urinary, and integument). When a stepwise increase in the limited/test AA is provided, there will be a progressive increase in N balance until the point where the N balance equates to zero, which is considered the requirement (Waterlow 1986). There are a few considerations to be mindful of when using N balance studies because the body metabolically adapts to modulate AA metabolism to its maximum capacity to maintain N balance. As such, it is possible that AA requirements can be incorrectly reported using this method. As an example, one consideration is making sure experimental diets meet the energy requirements of the animal, because when diets fail to do so, it may result in the oxidation of AAs for energy, therefore elevating the dietary protein requirement (Campbell et al. 1985; Moughan 2003). Overestimation of AA requirements using N balance is unlikely, instead this method is more likely to underestimate AA requirements by up to 40% (Zello et al. 1995; Elango et al. 2008). This is because N balance can be maintained even with inadequate amounts of dietary protein due to the adaptation of mammals lowering protein turnover to maintain N homeostasis (Young and Marchini 1990). In adult cats, N balance was maintained with the provision of lower dietary protein (1.5 g protein/kg), but to maintain lean body mass as observed through dual energy x-ray absorptiometry, more than triple that amount (5.2 g protein/kg) was required (Laflamme and Hannah 2013). A reduction in lean body mass can be indicative of inadequate dietary protein (Huh and Son 2022), thus N balance may not be sensitive enough to determine physiological AA needs for optimal health (Baker 1986; Young and Marchini 1990; Pencharz and Ball 2003).

In contrast, the indicator AA oxidation (IAAO) method for determining AA requirements is more accurate, less invasive, requires fewer animals, shorter adaptation period, and each animal can be in their own control (Pencharz and Ball 2003). The accuracy of N balance studies is adversely affected by the relatively long (minimum 5 days and typically a period of 7–10 days; Motil et al. 1994) adaptation period for dietary protein, in comparison to 2 days needed for the IAAO method regardless of the animal's age (Moehn et al. 2004; Elango et al. 2009). Long adaptation periods down-regulate N metabolism (Millward 2012). In contrast, because tRNA is charged shortly after consumption of a meal, IAAO only requires a single diet with an adaptation of total protein intake before that.

The IAAO method is based on the concept that when one IAA is lacking for protein synthesis, all other AAs, including the indicator AA (typically L-[1-^{13}C]Phe), will be oxidized (Pencharz and Ball 2003). As the intake of the limiting AA increases, the oxidation of the indicator AA decreases, representing more of it is being used for protein synthesis. Once the requirement for the test AA is met, the oxidation of the indicator AA remains stable. The point at which the indicator AA oxidation levels off is called the breakpoint, determined using bi-phase linear regression analysis, and represents the mean or estimated average requirement of the test AA (Pencharz and Ball 2003). The concept of the most limiting AA governing protein metabolism underlies the IAAO method. This method can be used to determine the bioavailability (defined below) of AAs in ingredients as well, however, it is the preferred method of choice for determining requirements over bioavailability.

While IAAO methods are more precise than N balance, there are some limitations. For example, IAAO experimental diets must be carefully formulated to evaluate requirements for one AA at a time, necessitating the use of synthetic AAs/protein sources, which, when combined with the isotope and equipment necessary, can be expensive in comparison to N balance. Furthermore, the use of synthetic AAs in IAAO test diets may underestimate requirements because synthetic AAs are absorbed more readily than intact protein sources typically found in diets, resulting in a lower required concentration to reach the breakpoint (Wolfe et al. 2022). As such, determining a definitive breakpoint can be difficult. Additionally, due to the small concentrations of CO_2 produced from a single IAA, it can be difficult to accurately determine the isotope-enriched CO_2 exhaled. Thus, the exhaled CO_2 is multiplied by the total volume of CO_2 to calculate the $^{13}CO_2$ excretion, measured by indirect calorimetry or isotope dilution techniques. However, this has an error margin of $\pm$ 5% or higher which can exceed the $^{13}CO_2$ enrichment variation observed at different rates of AA oxidation, making it difficult to define the breakpoint in AA oxidation relative to intake (Wolfe et al. 2016). Variations in breakpoint definitions and forms of AAs used in experimental diets would thus result in discrepancies in the determination of an AA requirement.

To determine AA requirements accurately, careful attention is required to experimental design, statistical analysis, and interpretation of results. Requirements for AA should be determined by ideally providing six or more graded levels of test AA (Baker 1986) across a range of intakes around the predicted level of requirement for a specific response variable (Pencharz and Ball 2003; Wolfe et al. 2016). When comparing concentrations, lack of statistical difference between two test levels of consecutive AA provision is not relevant (Baker 1986). As such, utilizing the broken line model or quadratic regression is more appropriate to determine AA requirements in these studies (Baker 1986; Pencharz and Ball 2003). Furthermore, Latin square designs can be used to try to get inter-individual responses to graded AAs, which are a stronger predictor of the 95% confidence interval of the target population. The inclusion of random variables in the model to account for other sources of error is also possible when using a Latin square design. Lastly, body weight (BW) should always be considered and to avoid perceiving that heavier animals require greater total amounts of nutrients; requirements should be presented on a g per kg BW per day basis [(g AA/kg BW)/d] and similar metabolizable energy content (Baker 1986).

5.4 Factors Affecting AA Requirements

Amino acid requirements for dogs and cats vary based on inter-individual nuances. The size and BW of various dog species can differ vastly, and the energy and protein requirements are also disparate among these phenotypes (Burger and Johnson 1991). In combination with BW, the distinction among breeds due to body composition, type of coat and shape of dog, all play a role in AA requirements (Burger and Johnson 1991). Moreover, with evident phenotypic differences, there are implications of genotype between different sizes of dogs. For example, larger dogs have a lower rate of taurine biosynthesis and greater loss through bile acid secretion than smaller dogs regardless of consuming similar amounts of SAAs (Ko et al. 2007; Tôrres et al. 2022). While even less is known about feline idiosyncrasies, phenotypic differences are known to also impact AA demands such as darker coat colors increasing Phe and Tyr requirements (Anderson et al. 2002).

Protein requirements are also affected by dietary factors such as fiber content, anti-nutritional factors (ANFs), energy density, and protein structure and processing. Pulse ingredients, for example, are a popular and widely used protein source in human, agricultural, and companion animal feeds. Pulses are rich sources of fiber, which is known to reduce lipid absorption (Schwesinger et al. 1999; Hoenig et al. 2001). This is because bile acids are synthesized from cholesterol, and fiber can bind with bile acids to increase their excretion. As dogs and cats preferentially conjugate bile acids with taurine (Rabin et al. 1976; Falany et al. 1994), this can also augment the demand for bile acids and subsequently taurine, and in the case of dogs, its precursor SAAs (Che et al. 2021; Oberbauer and Larsen 2021). Furthermore, pulses contain ANFs, like trypsin inhibitors, which can reduce protein digestion. Nevertheless, the elimination of specific ANFs can be achieved through the application of heat and pressure during the cooking process of pet foods (Rehman and Shah 2005; Cargo-Froom et al. 2020), while others are reduced through careful ingredient selection. Heat treatment can, however, have adverse effects on AA bioavailability (Hendriks et al. 1999; Bhat et al. 2021), thereby influencing the quantity of food necessary to fulfill AA requirements.

Amino acids (dietary and endogenous) are utilized more efficiently for protein synthesis when consumption is reduced, thereby increasing the overall efficiency of AA retention (Harper 1965; Young and Marchini 1990). This concept of efficient conservation of AAs has been depicted through lower oxidation of AAs when dietary intake is marginal or restricted (Young and Marchini 1990). Young adult men who were provided with deficient, marginal or excess levels of dietary protein in reference to maintenance requirements were observed to have greater levels of L-[1-^{13}C]Leu and L-[alpha^{15}N]Lys oxidation when protein exceeded requirements compared to those consuming a protein-restricted diet (Motil et al. 1981). Similarly, other studies with adult humans, in work mostly mediated by Dr. Vernon Young, demonstrate that limited intakes of a certain IAAs or protein result in an improved AA efficiency compared to when dietary protein is provided in excess (i.e., Meguid et al. 1986a,b).

Additionally, there is evidence suggesting that AA intake surpassing levels necessary for achieving protein retention can enhance several physiological functions, including protein turnover (Young and Marchini 1990) and transmethylation reactions (Robinson et al. 2016). This is an important consideration when looking forward to ascertaining AA requirements of adult animals at different physiological stages such as compromised health or increased physical activity, for example, as greater dietary protein in human athletes can improve exercise performance and protein metabolism (Williamson et al. 2019).

5.5 Nutritional Guidelines for Formulating Pet Foods

5.5.1 The National Research Council (2006)

The Nutrient Requirements of Dogs and Cats, published by the National Research Council (NRC 2006), presents a comprehensive compilation of data available at the time, outlining the recommended nutrient requirements of growing, pregnant, lactating and adult dogs and cats. The requirement estimates were primarily derived from studies conducted on growing animals, with additional consideration given to the bioavailability of nutrients. Bioavailability is defined as the proportion of those ingested AAs that are in a chemically available form for metabolism and protein synthesis (digestible, absorbable, and chemically available; Batterham 1992). The NRC released specifications for cats and dogs in 2006 with distinction to the specific physiological stages on a dry matter, g/kg metabolic BW (g/kg $BW^{0.75}$ for dogs, g/kg $BW^{0.67}$ for cats), and metabolizable energy basis (g/Mcal). The NRC specifies a minimal requirement (MR) for each stage, defined as the "minimal concentration or amount of a bioavailable nutrient that will support a defined physiological state". The recommended allowance (RA) is based on MR with consideration to a bioavailability factor, which in the case of AAs and CP is 80%. Hence, the RA for AAs and CP is calculated as MR × 1.25 (i.e., 1/0.8). The bioavailability of AAs in food is affected by composition such as fiber, ANF, etc., the physical characteristics of the ingredients (e.g., particle size), processes applied (e.g., heat, moisture), and the physiological state of the animal (e.g., life stage, activity). Analyses of data in the literature on ileal AA digestion from varying food sources (experimental and commercial diets) in adult dogs indicate that the bioavailability factor of 80% used by the NRC for most AAs and CP is an overestimate (Hendriks et al. 2015).

The AA requirements for adult dogs at maintenance outlined in the NRC (2006) are predicted from studies done in immature dogs, but based largely on two observational studies that elucidated the lowest amount of AAs to avoid deficiency (Ward 1976; Sanderson et al. 2001; Table 5.1). In terms of adult feline requirements, only Lys (Burger and Smith 1987) and SAA (Burger and Smith 1984) requirements were determined using dose–response studies aimed to maintain N balance. The requirements for all other AAs are based on commercial diets observed to avoid a deficiency.

In 1994, Morris and Rogers stressed the lack of available data in regard to requirements of adult animals at maintenance and since then, there has remained a dearth of dose–response studies providing MR data on the AA requirements in adult dogs and cats. Tables 5.1 and 5.2 summarize all data used to establish the requirements for puppies/kittens and adult dogs/cats MR at maintenance.

5.5.2 AAFCO (2016) and FEDIAF (2021)

In North America and Europe, there are regulatory bodies that present guidelines for canine and feline nutrient requirements based on, but disparate from recommendations outlined by the NRC. The North American (AAFCO 2016) and European (FEDIAF 2021) regulatory bodies set minimum dietary recommendations for protein and IAAs in which the digestibility, bioavailability, and processing of pet foods on dietary protein are accounted for. The recommendations for IAAs are based on RA estimations from the NRC, with specific safety margins to allow for the adequacy of meeting standards (2006). Moreover, in the case of FEDIAF (2021) recommendations for adult dogs, an additional increase of 20% is applied to account for the lower estimated energy requirements of household dogs compared to the average energy intake assumed by the NRC (2006). This is a critical distinction between AAFCO and FEDIAF because it acknowledges a heightened risk for low-activity or overweight dogs to have an AA deficiency when commercial diets are restricted by owners for the purpose of weight loss (Gaylord et al. 2018; Kaplan et al. 2018; Grant et al. 2020), which can be a concern as nearly 50% of the pet population are overweight (German 2016). Diets are formulated with nutrient levels relative to the energy density of the overall food in mind; consequently, a dietary restriction in commercial diets that are not designed for weight loss may risk a deficiency in nutrients. As an example, retrospective data with adult dogs elucidate that a restriction in diet due to veterinary recommendations to mitigate weight loss resulted in IAAs being provided below the requirement for Met and Cys (Gaylord et al. 2018; Kaplan et al. 2018), Trp (Gaylord et al. 2018), and Arg in cats (Grant et al. 2020). Due to limited data on adult dogs and cats' maintenance requirements and the bioavailability of AAs in ingredients and complete foods, well-designed experiments are needed to determine AA requirements and ingredient bioavailability. This will ensure nutrient adequacy, preventing both deficiencies and excess AAs.

5.5.3 Limitations

There is a lack of dose–response data to accurately determine the AA requirements of adult cats and dogs. The NRC (2006) has outlined safety margins, suggesting that the requirements outlined will meet the requirements of 95% of the canine and feline population needs. However, the requirements have not been revised, unlike for

Table 5.1 Daily protein and amino acid requirements of > 14 week puppies and adult dogs at maintenance as presented by NRC, AAFCO, FEDIAF, and IAAO (g/100 g DM, 4.0 kcal ME/g) with dose–response data used to establish them

Nutrient	Animals	Diet approach	Outcome	NRC MR puppies > 14 wks	NRC MR adults	NRC RA adults	AAFCO MR	FEDIAF MR (110 kcal/kg^0.75)	Upper 95% CL determined by IAAO
Crude protein (g/kg)					8	10	18	18	–
Arginine	English pointer puppies[1]	Simple diet, Arg titration	Max weight gain, 4 g/kg & minimize orotic acid, 5.3 g/kg	0.53					
	Mature female pointers[2]	Simple titration with purified L-AA diets	Avoid signs of deficiency; emesis, increased plasma ammonia and orotic acid		0.28	0.35	0.51	0.52	–
Histidine	Weanling beagles[3]	Simple titration, His titration	Max N retention at 2.1 g/kg	0.20					
	–	–	–		0.15	0.19	0.19	0.23	–
Isoleucine	Weanling beagles[4]	Simple diet, Ile titration	Max N retention at 4 g/kg	0.40					
	–	–	–		0.30	0.38	0.38	0.46	–
Methionine	Immature beagles[5] Immature beagles[6] English pointer puppies[7]	Simple diet[5] Simple diet, Met titration[6] Simple diet, SAA titration[7]	Weight gain, 2.1 g/kg Met & 3.5 g/kg Cys[5] Weight gain, 2 g/kg Met, 1.9 g/kg Cys[6] Weight gain, 2.3 g/kg Met, 2.2 g/kg Cys[7]	0.21					

(continued)

Table 5.1 (continued)

Nutrient	Animals	Diet approach	Outcome	NRC MR puppies > 14 wks	NRC MR adults	NRC RA adults	AAFCO MR	FEDIAF MR (110 kcal/kg^0.75)	Upper 95% CL determined by IAAO
	Adult beagles[8] Adult dachshunds, Beagles, labradors[9] Adult labrador retrievers[10]	Intact protein, protein restricted[8] Simple titration, crystalline AA, IAAO[9] Simple titration using semi–synthetic diet, IAAO[10]	Avoid taurine deficiency and maintain healthy cardiac function (echocardiography)[8] Reach steady state to determine breakpoint[9,10]		0.26	0.33	0.33	0.40	0.48
Methionine & Cysteine	Immature beagles[5] Immature beagles[6] English pointer puppies[7]	Simple diet[5] Simple diet, Met titration[6] Simple diet, SAA titration[7]	Weight gain, 2.1 g/kg Met & 3.5 g/kg Cys[5] Weight gain, 2 g/kg Met, 1.9 g/kg Cys[6] Weight gain, 2.3 g/kg Met, 2.2 g/kg Cys[7]	0.42					
	–	–	–		0.52	0.65	0.65	0.76	–
Leucine	Weanling beagles[4] 8–14 wk beagles[11]	Simple diet, Leu titration[4] 140 or 280 g/kg CP, titrated Leu[11]	Max N retention at 6 g/kg[4] Max weight gain at 7 g/kg[11]	0.65					
	–	–	–		0.54	0.68	0.68	0.82	–

(continued)

Table 5.1 (continued)

Nutrient	Animals	Diet approach	Outcome	NRC MR puppies > 14 wks	NRC MR adults	NRC RA adults	AAFCO MR	FEDIAF MR (110 kcal/ kg^0.75)	Upper 95% CL determined by IAAO
Lysine	Weanling pointer[12]	Intact protein diet, titrated Lys[12]	Weight gain, simple AA = 7 g/kg Intact protein = 8 g/kg[12]	0.56					
	Dachshunds, Beagles, labradors[13]	Simple titration using crystalline AA, IAAO[13]	Reach steady state to determine breakpoint[13]		0.28	0.35	0.63	0.42	0.53
Phenylalanine	Dachshunds, Beagles, labradors[14]	Simple titration using crystalline AA, IAAO[14]	Reach steady state to determine breakpoint[14]	0.40	0.36	0.45	0.45	0.54	0.46
Phenylalanine & Tyrosine	10-wk weanling beagles[15]	Simple diet, Phe titration[15]	Max N retention at 8 g/kg[15]	0.80					
	–	–	–		0.59	0.74	0.74	0.89	–
Threonine	15-wk weanling beagles[3]	Simple diet, Thr titration	Max N retention, 5 g/kg	0.50					
	Dachshunds, Beagles, labradors[16]	Simple titration using crystalline AA, IAAO[16]	Reach steady state to determine breakpoint[16]	–	0.34	0.43	0.48	0.52	0.56

(continued)

Table 5.1 (continued)

Nutrient	Animals	Diet approach	Outcome	NRC MR puppies > 14 wks	NRC MR adults	NRC RA adults	AAFCO MR	FEDIAF MR (110 kcal/kg^0.75)	Upper 95% CL determined by IAAO
Tryptophan	7-wk beagles[5] 6–14 wk English pointer puppies[17]	Simple diet, Trp titration[5] Simple diet[17]	Max N retention, 1.5 g/kg[5] Max weight gain, 1.2 g/kg[17]	0.14	–	–	–	–	–
	Dachshunds, Beagles, labradors[18]	Simple titration using crystalline AA, IAAO[18]	Reach steady state to determine breakpoint[18]	–	0.11	0.14	0.16	0.17	0.20
Valine	12-wk weanling beagles[4]	Simple diet, Val titration	Max weight gain and N retention, 4.5 g/kg	0.45					
	–	–	–		0.39	0.49	0.49	0.59	–

[1](Czarnecki and Baker 1984). [2](Burns et al. 1981). [3](Burns and Milner 1982). [4](Burns et al. 1984). [5](Milner, 1979b). [5](Burns and Milner 1981). [7](Hirakawa and Baker 1985). [8](Sanderson et al. 2001). [9](Mansilla et al. 2020a). [10](Harrison et al. 2020). [11](Delaney et al. 2001). [12](Hirakawa and Baker 1986). [13](Sutherland et al. 2020). [14](Mansilla et al. 2018). [15](Milner et al. 1984). [16](Mansilla et al. 2020b). [17](Czarnecki and Baker 1982). [18](Templeman et al. 2019)

Table 5.2 Daily crude protein and amino acid requirements of kittens and adult cats at maintenance as presented by NRC, AAFCO, and FEDIAF, (g/100 g DM, 4.0 kcal ME/g) with dose–response data used to establish them

Nutrient	Animals	Diet approach	Outcome	NRC MR Kittens	NRC MR adult	NRC RA adult	AAFCO MR	FEDIAF MR (110 kcal/ kg^0.75)
Crude protein (g)	–	–	–	180	16	20	26	25
Arginine	Weanling kittens[1] Weanling kittens[2] Weanling kitten[3]	Simple diet[1] Simple diet, Arg titration[2] Simple diet[3]	Max growth, 8–8.3 g/kg (4.7 kcal ME/g)[1] Min urinary orotic aciduria, 10.5 g/kg[2] Min orotic aciduria, 7.5 g/kg diet of 155 g CP/diet (4.0 kcal ME/g)[3]	0.77				
	–	–	–		–	0.77	1.04	1.00
Histidine	Young kittens[4] Young kittens[5]	Simple diet[4] Simple diet[5]	Max weight gain, 3 g/kg[4] Max weight gain and N balance, 2 g/kg but to avoid cataract development 3 g/kg (4.7 kcal ME/g)[5]	0.26				
	–	–	–		–	0.26	0.31	0.26
Isoleucine	Weanling kittens[6] Kitten[7]	Simple diet[6] Simple diet[7]	Max weight gain, 3 g/kg (4.7 kcal ME/g)[6] Max weight gain and N balance, 5 g/kg (4.7 kcal ME/g)[7]	0.43				
	–	–	–		–	0.43	0.52	0.43
Methionine	Weanling kittens[8] Kittens[9]	Purified AA diets[8] Purified AA diets, titrated Met[9]	Max weight gain, 4.5 g g/kg Met, 6 g/kg Cys[8] Max growth, 4.5 g/kg Met, 4.5 g/kg Cys[8] Max weight gain, N retention, 3.9 g/kg Met, 6 g/kg Cys[9]	0.35				
	–	–	–		0.135	0.17	0.20	0.17

(continued)

Table 5.2 (continued)

Nutrient	Animals	Diet approach	Outcome	NRC MR Kittens	NRC MR adult	NRC RA adult	AAFCO MR	FEDIAF MR (110 kcal/kg^0.75)
Methionine & Cysteine	**same as Met*	**same as Met*	**same as Met*	0.70				
	Adult cats[10]	–	N balance, 3 g/kg (4.5 kcal ME/diet)[10]		0.27	0.34	0.40	0.34
Leucine	Weanling kittens[6] Kitten[11]	Simple diet[6] Simple diet, Leu titration[11]	Max weight gain, 12 g/kg (4.7 kcal ME/g)[6] Max weight gain, 9 g/kg, and Max N retention, 12 g/kg (4.7 kcal ME/g)[11]	1.02				
	–	–	–		–	1.02	1.24	1.02
Lysine	Kittens[5.1] Kittens[12]	Simple diet, Lys titration[5.1] Simple diet[12]	Max weight gain, 8 g/kg[5.1] Max weight gain, N retention, 8 g/kg (4.7 kcal ME/g)[12]	0.68				
	Adult cats[10]	Simple diet, Lys titration[10]	N balance, 3 g/kg (4.5 kcal ME/diet)[10]	–	–	0.34	0.83	0.34
Phenylalanine	Kittens[13]	Purified L-AA diet, Phe titration[13]	Max weight gain, N retention, 7.5 g/kg (No Tyr), 3.5 g/kg (with 10 g/kg Tyr)[13]	0.40				
	–	–	–	–	–	0.40	0.42	0.40
Phenylalanine & Tyrosine	Young kittens[4] Kitten^s13 Kitten[14]	Simple diet[4] Purified L-AA diet, Phe titration[13] Purified AA diets, Phe + Tyr titration[14]	Max weight gain, 10 g total (5 g/kg Phe, 5 g/kg Tyr)[4] Max weight gain, N retention, 3.5 g/kg Phe, 10 g/kg Tyr[13] Max black color in coat, 15.3 g/kg[14]	1.53				
	–	–	–		–	1.53	1.53	1.53

(continued)

Table 5.2 (continued)

Nutrient	Animals	Diet approach	Outcome	NRC MR Kittens	NRC MR adult	NRC RA adult	AAFCO MR	FEDIAF MR (110 kcal/ kg^0.75)
Threonine	Kittens[15] Kittens[16]	Purified L-AA diets, Thr titration[15] Simple diet, Thr titration[16]	Max growth, 7 g/kg[15] Max growth, N retention, 6 g/kg (4.6 kcal ME/diet)[16]	0.52				
	–	–	–		–	0.52	0.73	0.52
Tryptophan	Young kittens[4] Kittens[16]	Simple diet[4] Simple diet, Trp titration[16]	Max weight gain, 1.5 g/kg[4] Weight gain, N balance, 1.1 g/kg[16]	0.13				
	–	–	–		–	0.13	0.16	0.13
Valine	Weanling kittens[17] Weanling kittens[6]	Semi-purified diet, Val titration[17] Simple diet[6]	Weight gain, 6 or 18 g/kg[17] Weight gain, 6 g/kg[6]	0.51				
	–	–	–		–	0.51	0.62	0.51
Taurine				0.032	0.032	0.04	0.10	0.10

[1](Anderson et al. 1979). [2](Costello et al. 1980). [3](Taylor et al. 1997). [4](Anderson et al. 1980a). [5](Quam et al. 1987). [6](Anderson et al. 1980b). [7](Hargrove et al. 1994). [8](Teeter et al. 1978). [9](Smalley et al. 1983). [10](Burger and Smith 1987). [11](Hargrove et al. 1984). [12](Morris et al. 2004). [13](Williams et al. 1987). [14](Anderson et al. 2002). [15](Titchenal et al. 1980). [16](Hargrove et al. 1983). [17](Hardy et al. 1977)

example, human requirements which have steadily increased upon reassessment from using N balance studies to isotopic methodology (WHO/FAO/UNO 2007; Elango et al. 2009; Zello et al. 1993). A series of empirical studies using the IAAO technique to determine the AA requirements of small, medium, and large breed adult dogs fed for weight maintenance indicate that the NRC 95% confidence interval (CI) does not account for diversity in AA requirements in this population. As an example, medium and large breed dogs have a seemingly greater requirement for Met (Mansilla et al. 2020a) and Lys (Sutherland et al. 2020), whereas medium breed dogs may have greater Trp requirements than small and large breed dogs (Templeman et al. 2019). However, in all cases, when the data for all breed sizes were pooled, the estimate of the 95% CI was greater than NRC, AAFCO, and FEDIAF (Table 5.1). These IAAO studies elucidate that the requirements outlined in the NRC, based on N balance studies of the beagle or English pointers, likely underestimate the requirements of medium and large breed dogs. Other safety margins are yet to be supported with data encompassing the bioavailability of ingredients to provide adequate nutrients and shelf-life stability implications on requirements. Hendriks (2003) reviewed and quantified a number of physiological characteristics that impact AA demands of cats and dogs such as gut endogenous losses, losses in hair and urine, and AAs needed to synthesize N compounds in the body. Unique to both dogs and cats compared to other mammals with less hair/fur on skin, such as humans or swine, is the varying AA requirements during annual periods of main hair growth (e.g., seasonal periods of shedding coat). The most abundant AA in hair is Cys (Hendriks et al. 1998), as such the determination of SAAs during the main hair growth period in the year would likely depict a higher SAA requirement compared to the low hair growth period. There is a dearth of scientific literature in this regard for cats and dogs and utilization of the IAAO method may be beneficial to understand these fluctuations in requirements. Hendriks (2003) proposes a factorial model to assess the requirements of adult cats and dogs as a tool that considers various physiological processes for a more detailed understanding of the animal's physiological requirements from which requirements can be derived pertaining to varying conditions for different breeds. First, we must better understand physiological AA requirements at maintenance, and then define how the factors described above, underlying inter-individual nuances, can impact AA requirements to guide optimal formulation and feeding practices of companion animals in specific physiological states. The development of additive models will aid in the more accurate establishment of requirements for specific groups or individuals.

5.6 Formulation in Practice

Ideally, nutrient requirements of the animal are accurately quantified within their physiological state and then nutrients are appropriately balanced in the diet through using digestibility and bioavailability data of ingredients with consideration to processing factors contributing to the final product. In pet food, there is a growing trend to increase absolute CP in foods. However, the focus should be mitigated toward specific AAs in the diet to improve protein utilization efficiency and reduce waste. In this sense, the concept of the response criterion is important such that the desired response will influence the amount of AAs required. For example, if maximal immune function is the response criterion, greater concentrations of Cys may be needed to maintain antioxidant activity (Li et al. 2007). Adequate AAs are required to maintain health (Reeds 2000; Milner 1979a) and there are extensive reviews about the effects of balanced ratios between IAA:DAA on N balance and on meeting protein requirements (Harper et al. 1970; Heger 2003). The IAA:DAA ratio may affect longevity and could be useful to assess dietary AA requirements to prevent deficiency (Heger 2003). The ideal ratio of AAs with differing total N concentrations has been investigated in other mammalian species, largely originating from rats (Womack and Rose 1947) but explored in humans (Cooper et al. 2020) and swine (Heger et al. 1998; Lenis et al. 1999). However, the optimal ratio of IAA to total N for cats and dogs at maintenance has not been defined and is yet to be investigated considering factors such as age, physiological state, health status, dietary factors such as IAA:N ratio, and type of diet such as intact/hydrolyzed protein vs. crystalline AAs.

Any IAA or sum of DAAs that are insufficiently provided in the diet (limiting) will result in all other absorbed AAs exceeding requirements to be oxidized rather than utilized for protein synthesis. (Humbert et al. 2001; Moehn et al. 2005). The practice of over-formulation to ensure nutrient requirements are met can lead to waste of valuable resources. This concept has been shown in pigs where it was observed that the efficiency with which absorbed AAs are utilized for body protein accretion decreases as AA intake levels increase (Gatel et al. 1992; Heger et al. 1998). Similarly in pets, increasing dietary protein can further exacerbate the effects of limiting AAs and increase N waste (Harper 1965). Milner (1981) demonstrated that doubling dietary CP in canine diets also increased requirements for Lys. If there is an AA imbalance in the diet, then increasing overall CP actually increases the requirement for the limiting AAs (Harper et al. 1970). Furthermore, consideration should be given to the adaptation period on high protein diets as an increase in dietary CP also increases AA requirements in part due to the upregulation of degrading enzymes in the liver (Keene and Austic 2001). This warrants consideration for an optimal dietary AA ratio and possibly complementary ingredients because if a protein source is limiting in one AA, adding more of that protein to increase dietary CP will further contribute to the oxidation of all other AAs in excess.

A balanced ratio of IAA:DAA improves net protein deposition by reducing endogenous protein breakdown (Fuller et al. 1987). There is a plethora of data in other species elucidating how imbalances in AAs can lower the quality of protein being provided. For example, Romano et al. (2019) fed growing mice different ratios of IAA:DAA and observed that even a mildly imbalanced diet, about < 20% poorer in IAA compared to typical protein-rich diets like casein, could result in rapid loss of muscle mass and premature death in mice. Similarly, maximal growth was observed in pigs when fed increased ratios of AAs, specifically Lys proportionally to CP, in contrast to supplementing excess AAs without consideration to the overall IAA:N ratios of the diet (Gatel and Fekete 1989). Extreme ratios of IAAs to CP have been shown to have adverse effects on various physiological variables (e.g., growth), possibly due to the limited availability of DAA in the diet (Heger 2003). A recent review by Li and Wu (2023) delves into the physiological and dietary factors that affect AA requirements such as these ratios and specifically the need for DAA (as mentioned earlier) when there is a limitation on endogenous production. For adults at maintenance, less is understood about these ratios and how they mitigate losses. Thus, there is an urgent need for research to better understand protein and AA metabolism and how to optimize dietary IAA:N ratios to support the health of dogs and cats with consideration to their impact on the global ingredient supply and the environment.

5.7 Digestion and Absorption of Protein in Monogastric Species

In order for a food to be able to satisfy the IAA and total AA requirements of animals, the protein first needs to be digested into smaller components to be absorbed (Fig. 5.1). Digestion is the process of enzymatic hydrolysis of proteins, after which the AAs and di and tripeptides can be absorbed from the lumen into circulation via enterocytes (Stein et al. 2007a). The greatest source of variation in bioavailability is digestibility, thus we must first understand digestion and how to measure it (FAO 2013). Briefly, protein digestion begins in the stomach, where proteolysis is initiated by gastric pepsins resulting in large polypeptides, smaller oligopeptides, and free AAs (Erickson and Kim 1990). Next, the pancreas releases several proteolytic enzymes that are activated in the duodenum resulting in proteins, polypeptides, and oligopeptides being cleaved into smaller oligopeptides and a smaller proportion as free AAs (Erickson and Kim 1990). The final stage of digestion occurs via peptidases in the luminal fluid and at the brush border membrane of the small intestine resulting in di and tripeptides and free AAs that are available for transport via specialized carriers (Erickson and Kim 1990). Undigested and unabsorbed food proteins and peptides flow into the large intestine together with endogenous proteins, such as sloughed cell proteins, mucoproteins, pancreatic secretions, and bile acids containing greater amounts of taurine, proline (Pro), glycine (Gly), serine (Ser), and Thr (Yen 2001). Little to no AA absorption occurs in the large intestine of monogastrics, but non-protein N is readily absorbed (Fuller and Reeds 1998; Hendriks et al. 2012). Large intestinal microbiota can degrade most endogenous and undigested food proteins and AAs, changing the AA profile of the large intestinal digesta. Threonine and Trp are often degraded while Met and Lys may be synthesized by bacteria (Moughan et al. 2014). When digestibility is measured at the end of the large intestine (named "apparent total tract digestibility" (ATTD)), the microbial contributions can lead to considerable overestimation of digestibility values. For example, the mean difference in apparent ileal and fecal N digestibility across 141 diets in dogs was 9.4% and varied from –4.1 to 31.3% (Hendriks et al. 2012). For evaluation of AA bioavailability in ingredients or foods, it is therefore incorrect to use the ATTD approach.

Therefore, measures that estimate AA digestibility at the end of the small intestine ("ileal nutrient digestibility") should be applied.

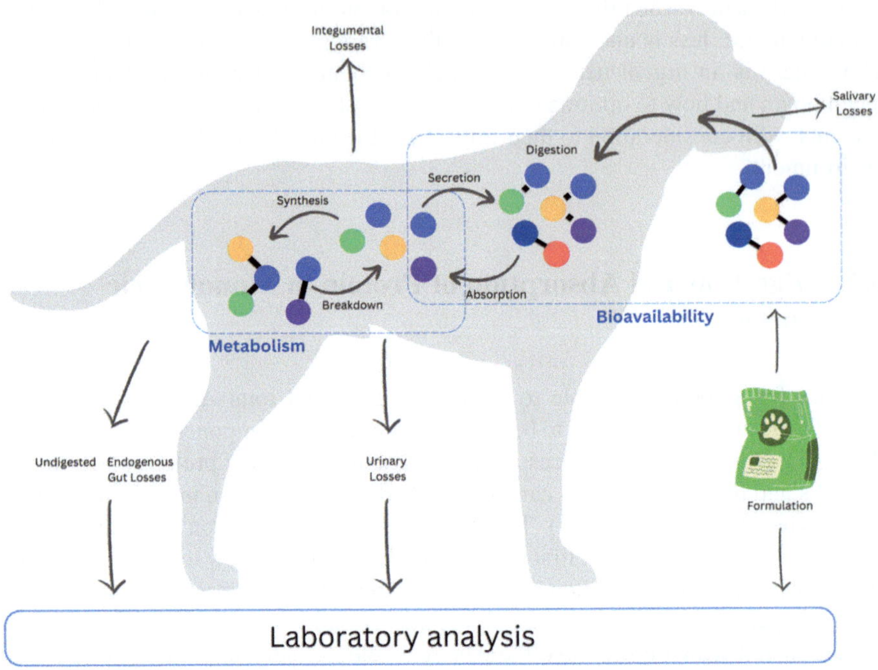

Fig. 5.1 Graphic depicting the different processes from ingestion of food to excretion of waste in the dog. Starting with formulation, dog foods are formulated based on requirements and can then be analyzed to understand AA content of the diet after processing. Ideally, formulation is based on the combination of the animal's AA requirements and expected bioavailability of the dietary AAs. However, as described in the chapter, much work still needs to be done to get to this point. After ingestion comes digestion, which is simply the chemical breakdown of large proteins into smaller polypeptides and AAs whereas a part of the proteins remain undigested. Of those AAs that were liberated by digestion, some are absorbed in a form that is available to the animal to use while others are not (bioavailability). Available AAs can then be utilized for protein synthesis, among other processes. Those proteins can also be broken down into free AAs that can be secreted and transported to other tissues in the body (metabolism). Primary AA losses occur as endogenous gut losses excreted together with the undigested AAs, but AAs can also be lost in saliva, urine, and integumental losses, such as hair loss. We can analyze AA content of said losses in order to better account for the animal's physiological requirements and measure endogenous losses to better account for AA digestibility estimates

5.8 Assessments of in Vivo Digestibility

5.8.1 Ileal Digestibility

There are three different approaches to ileal digestibility measurements: apparent, standardized, and true. In apparent ileal digestibility (AID), there is no distinction between undigested AAs from the diet or endogenous AAs secreted into the gastrointestinal tract (GIT). It is based on the difference between amounts of AA consumed and total AA (i.e., dietary and endogenous) appearing at the end of the ileum, therefore often underestimating digestibility coefficients (Stein et al. 2007b). Standardized ileal digestibility (SID) accounts for basal endogenous losses as determined using a N-free diet. These basal losses are related to sloughed cells, unabsorbed endogenous enzymes from the pancreas, bile acid and brush border enzymes, ingested hair, and unabsorbed mucus (Stein et al. 2007b). As these losses are accounted for, SID values are, in more cases, additive for individual ingredients compared to AID values in a mixed diet and can be better used in diet formulation (Stein et al. 2005). True ileal digestibility (TID) accounts for both basal and specific endogenous losses and therefore, theoretically, it is the gold standard. Specific endogenous losses are those induced by specific characteristics of the ingredient, such as fiber content. However, it is laborious and expensive to measure specific endogenous losses. True ileal digestibility is impractical for diet formulation as the digestibility values are not additive because it does not differentiate between ingredients that might stimulate the secretion of specific endogenous losses (Stein et al. 2007b). For example, some ingredients may stimulate endogenous losses of specific AA more than others. When comparing all three methods, TID will give the highest estimates followed by SID and then AID.

As mentioned above, ATTD does not yield accurate estimates for the absorbed AAs or N from the diet because of the microbial conversions of AAs and N absorption/secretion in the large intestine. In comparison to SID and AID in dogs across 5 canine dry foods, ATTD was significantly greater or tended to be greater for all AAs except Ile, Met, and Ala when AID was calculated and was significantly greater or tended to be greater than all AAs when SID was calculated with greatest difference observed for His and Thr (Hendriks et al. 2013).

5.8.2 Animal Models

There are various in vivo methodologies used in the pet food industry to evaluate digestibility. For a detailed review of these methodologies the reader is directed to de Godoy et al. (2016). Here we focus on several methodologies pertinent to understanding individual digestibility values of the AAs to be used for protein quality calculations. First, ileal cannulated animals are used to measure AID, SID, or TID of ingredients and/or diets (Sauer et al. 1983). A cannula is surgically inserted into the terminal ileum pigs and ingredients are fed as part of a complete diet in a typical feed matrix with a dietary marker to enable determination of nutrient digestibility. This method has also been used to evaluate AA AID and SID in dogs (Walker et al. 1994) and in humans with a small intestinal stoma (Moughan et al. 2005; Hodgkinson et al. 2022). The ileal cannulated dog is no longer considered to be an ethically acceptable model but in swine and human nutrition (ileostomates), the collection of ileal digesta through a cannula and SID is considered the gold standard, despite being invasive (FAO 2013). Similarly, in the euthanasia method, often done in poultry, pigs and rats, the animal is fed the diet of interest for 4–5 days and, on the last day, fed hourly before euthanasia (Kong and Adeola 2014). The terminal ileum is dissected and digesta removed, allowing only one digesta collection/measurement per animal. This method has largely been replaced by ileal cannulated swine and the precision-fed cecectomized rooster, however it was applied to pet food using rats (Hendriks et al. 1999). The cecectomized rooster method involves removing the ceca from the bird, therefore eliminating microbial AA conversions (Sibbald 1979; Parsons 1985). It can be used to determine AID, SID, and TID coefficients of individual ingredients, fed alone, and/or complete diets. The roosters can be reused although the tube-feeding and invasive surgery is not physiologically normal. It has been suggested that the rooster and pig can be used as a model for the dog. Johnson et al. (1998) reported high correlation coefficients (0.89–0.94) for the digestibility of several AAs (Lys, Cys, Thr, Met) between ileal cannulated dogs and cecectomized roosters for several animal by-product meals. Mink have been suggested as a model for the cat, given that both are obligate carnivores. Mink have been used to determine ATTD of ingredients due to their rapid rate of passage, small colon with limited microbial activity (Szymeczko and Skrede 1990). However, the cat has a considerably longer transit time than the mink (Peachey et al. 2000). Only Sundling et al. (2012) compared ATTD coefficients between mink and cats and cats had greater digestibility coefficients for CP for standard cat, mink, and ferret diets.

In North America, many ingredients specific to pet food have been evaluated using the precision-fed cecectomized rooster as a model; single meat-based ingredients (Faber et al. 2010), protein hydrolysates (Folador et al. 2006; Hsu et al. 2023), meat meals (Folador et al. 2006; Deng et al. 2016; Oba et al. 2023), meat meals processed at different temperatures (Johnson et al. 1998), raw chicken versus different forms of processed chicken (Oba et al. 2019), insect meal (Do et al. 2020; Smola et al. 2023), lightly cooked whole diets (Roberts et al. 2023), extruded whole diets (Roberts et al. 2023), and single plant-based ingredients (de Godoy et al. 2009; Oba et al. 2023; Reilly et al. 2021, 2020a,b). Although it has become more common, the ethical demand to move to in vitro methods may phase out this model as done for ileal cannulated dogs.

In human nutrition, the pig is often used as a model. Hodgkinson et al. (2022) compared ileal digestibility coefficients in human ileostomates and pigs for seven protein sources. The only AA that had a significantly different digestibility coefficient was total Lys but not reactive (available) Lys. Pig and human digestibility coefficients were highly ($r > 0.7$) and significantly correlated, indicating that the pig can be used as a model for the human when determining SID coefficients of AAs. Given the pressure to move toward more ethically acceptable models and likely remove the need for animal models entirely, one must accept that no animal model will ever be further validated for the application of pet food protein quality assessment.

5.9 Assessments of in Vitro Digestibility

There are different types of in vitro assays with varying degrees of simulating digestive physiology. The reader is directed to a review by Butts et al. (2012) for an explanation of common methodological approaches. There are two major types of in vitro assays. Static digestion systems are the simplest approach where oral, stomach, and small intestinal digestion are simulated in vitro. It lacks dynamic mechanical actions of the GIT, such as mixing and peristalsis, and digestion conditions cannot be changed over time (Colombo et al. 2021). Dynamic digestion systems mimic motility and changing digestion conditions, however, it is more expensive and complex than static and not readily developed and performed in laboratories.

With in vivo research under increasing scrutiny in many countries nowadays, there is a demand for, ideally validated, in vitro methodologies to study digestion and absorption. In general, in vitro methodologies are cheaper, faster, and easier than in vivo methodologies and they have few, if any, ethical constraints. If conducted well and under standardized conditions, they are also often more reproducible (within and between labs) and have fewer inter-individual variables. However, in many cases in vitro methods have been poorly validated and further, lack simulation of either specific digestive mechanisms or biological factors such as age or physiological state. A major step in the development of harmonized static in vitro methodologies occurred when Minekus et al. (2014) developed the COST Action Infogest protocol that standardized the methodologies for the digestion time for each stage of the GIT (oral, stomach and small intestine), gastrointestinal fluid preparation, addition of enzymes, and sample collection and storage for humans. However, given that it is a static method, it uses constant ratios of meal:digestive fluids and a constant pH for each step of digestion, therefore it is not suitable for simulating digestion kinetics (Minekus et al. 2014). The Infogest method was tested among laboratories using skim milk powder according to either the labs in-house protocol or the harmonized protocol by Minekus et al. (2014). The peptides released during digestion were correlated among labs and had high correlations ($r = 0.84$ vs 0.79) when the in-house versus harmonized methods were used for the gastric phase of digestion, respectively (Egger et al. 2016). Correlation improved by 0.16 (from 0.48 to 0.64) for the intestinal phase of digestion when the harmonized methods were used in contrast to the in-house methods (Egger et al. 2016). Egger et al. (2017) investigated digestion of skim milk powder using Infogest compared to collecting samples along the GIT of pigs consuming the same skim milk powder for 14 h followed by euthanasia. The peptides released during digestion had a high correlation ($r = 0.8$) in the gastric phase of both the pig and Infogest (Egger et al. 2017). The intestinal phase of the Infogest correlated best with a sample taken from the middle jejunum of the pig ($r = 0.6$; Egger et al.

2017). Perhaps in the future with more validation data, the Infogest method can be applied to digestibility in cats and dogs, given that the pig is a model for the human and Infogest is correlated with in vivo pig digestibility. Minor adaptations might still be required, such as omission of the oral phase of incubation, to come closer to the digestive physiology of dogs and cats. Despite the advancements in this area, similar to the in vivo models above, it is still necessary to understand the digestibility of ingredients included in a typical food matrix.

5.10 Evaluating Single Ingredients Versus Ingredients in a Processed Diet

Many of the pet food specific ingredients that have been evaluated in the rooster have been single ingredients not part of a typical pet food matrix or whole diets where the digestibility values apply exclusively to that specific diet. When considering pet food, the different processing methods that are applied to complete diets with many different ingredients must also be considered. There is limited information on effects extrusion or retorting on the protein and AAs in complete and balanced pet food. The effect of retorting on AA TID using the rat as a model was investigated by Hendriks et al. (1999) where there was a decrease in the TID of all AAs and AAN with increasing heat treatment. Properties of an ingredient may change when extruded and ingredient interactions may occur when included in a complete diet resulting in a more or less digestible nutrients in a final diet. Work has been done in other species, for example, a review in swine by Lancheros et al. (2020) reports that extrusion generally improved nutrient digestibility of various ingredients. Ingredient interactions in a matrix (corn and soybean meal) were also evaluated. Extrusion, or heat and pressure, is often used to reduce ANFs. For example, Hugman et al. (2021) compared cold-pelleted, steam-pelleted, and extruded field peas to raw field peas in pigs and reported the lowest trypsin inhibitor activity in the extruded field pea. In addition, the SID coefficient for Leu, Thr, Trp, Val, Ala, Ser, and Tyr were all greater in the extruded field pea compared to the raw field pea. Even if we start to look at extruded ingredients or ingredient interactions in pet food, complete pet foods can combine almost an unlimited number of ingredients and extrusion varies between manufacturers. More evaluations need to be completed to investigate how ingredients perform when in combination with other ingredients as well as understand, in quantitative terms, the impact of processing relative to original ingredients on digestibility in order to have the data be more widely applicable to all pet foods. Furthermore, it is of interest to understand to what degree the chosen extrusion conditions matter in changing the AA digestibility of a food.

5.11 Direct Versus Difference Method of Measuring Digestibility

Understanding the specific and unique digestibility of nutrients for an ingredient when it is included in a complete food containing other ingredients is key for formulation of foods that match the nutrient requirements of the target animals. There are two methods to discern the AA digestibility of an ingredient present in a complex food matrix. First, the direct method uses a diet where the test ingredient is formulated as the only source of AA of interest in the diet (Zhang and Adeola 2017). Second, the difference method uses a basal diet containing a known amount of the test ingredient and test diets containing an additional amount of the test ingredient. Digestibility then, essentially becomes the difference between the test diet and basal diet (Zhang and Adeola 2017). Both of these methods are commonly employed in swine ileal digestibility studies. For example, when comparing the direct versus difference method in pigs for barley and canola meal, AID estimates for Thr, Ala, Asp, Gly, and Ser were less when the direct method was applied compared to the difference method for barley, but there were no differences between methods for canola meal, likely due to the higher protein content of canola meal (Fan and Sauer 1995). These approaches can help in further understanding the nutritional value of pet food ingredients as well as the potential interactions that might occur among ingredients during processing or digestion in the animal.

5.12 Understanding Additivity of Protein Sources in Complete Foods

Whether ingredients might interact or whether the digestibility values of single ingredients are additive when combined in a pet food are important for formulation purposes and achieving the desired performance of the pet food. The idea is that the SID values of individual ingredients should predict the SID values when those ingredients are mixed in a diet. For example, Osho et al. (2019) investigated wheat, canola meal and sorghum distillers dried grains in broilers and collected digesta after 5 days. The predicted SID values were less than measured SID values for Cys and Ser in the wheat-canola meal mixed diet, but only by 2.8 and 1.9%, respectively. All predicted SID values in the wheat-canola meal-sorghum distillers dried grains mixed diet were similar to measured SID values. Fanelli et al. (2021) investigated two types of breakfast cereal and dry milk in ileal cannulated pigs, as a model for humans, fed for 7 days. Predicted SID values were greater than measured SID values for Ile, Glu, and Ser in the cornflakes-dry milk mixed diet, but all AAs were similar in the quick oats-dry milk mixed diet. In order for digestibility estimates to be useful in formulation of pet foods as they are in swine diets, they need to be additive. To the authors' knowledge, the concept of additivity in pet food has not been investigated. Using the cecectomized rooster model to estimate AA SID estimates of complete dog diets and individual ingredients, we can calculate predicted versus measured digestibility values in order to understand if the same holds true for pet food (Banton et al. unpublished). If this is true, then we can apply the Digestible Indispensable Amino Acid Score (DIAAS; discussed below) to complex diets and not just ingredients. This allows a metric to be applied to final products and would significantly improve clear communication to the consumer. If the pet food industry hopes to move toward using digestibility estimates for formulation, resulting in more effective raw material usage and lower environmental impact, work needs to be done, not only in measuring digestibility of ingredients but understanding if those individual ingredient digestibility values may be additive in a complete diet.

5.13 DIAAS as a Tool to Predict Protein Quality

From the above it becomes clear that multiple advancements can be made in quantifying the AA requirements in adult dogs and cats as well as assessing the AA digestibility of ingredients to come to a diet that more accurately supplies the required AAs. The data gathered from the above methodologies to measure AA requirements and AA digestibility coefficients can be used in predicting protein quality of ingredients. Protein quality is defined as the ability of a single or combination of protein ingredients to meet the IAA requirements of an individual (Nosworthy and House 2017). Currently the DIAAS is considered the gold standard for application to human foods by the FAO (2013). Ideally, DIAAS uses SID coefficients from ileal cannulated pigs as a model for humans, however, as discussed above, this method has ethical constraints, and the data does not exist for every ingredient imaginable. A more convenient, ethical, and less expensive approach would be to use in vitro estimates in calculating DIAAS. DIAAS is calculated by dividing the mg of digestible dietary IAA in 1 g of the dietary protein by the mg of the same dietary IAA in 1 g of the reference[1] protein and multiplying by 100%. The values are not truncated, therefore allowing a DIAAS of over 100%. The DIAAS of ingredients can then be directly compared against each other in terms of protein quality. Furthermore, it allows for low-quality protein sources to be paired with high-quality protein sources, like in the example above with breakfast cereal and milk (Fanelli et al. 2021). Breakfast cereal alone has a low DIAAS and is not a good quality protein source to meet the recommended scoring pattern. However, when combined with milk, the DIAAS becomes close to or greater than 100%, suggesting that the low-quality cereal protein compliments the higher quality milk protein to produce a meal that is balanced for IAA (Fanelli et al. 2021). In addition, the predicted DIAAS was significantly greater than the measured DIAAS in a cornflakes-dry milk mixed diet but there was no difference in DIAAS values in a quick oats-dry milk mixed diet when both the recommended AA scoring pattern for a child of 6–36 months or for over 36 months was used (Fanelli et al. 2021). This highlights an important consideration of DIAAS, which

[1] Recommended amino acid scoring pattern.

is that the correct reference pattern needs to be used in order to accurately assess protein quality. Further studies need to be conducted to more accurately quantify the AA requirements of cats and dogs and establish reference patterns for different canine and feline populations. The IAAO method is more valuable in determining the AA requirements that can be used as the reference pattern for DIAAS in pet foods, in combination with in vitro digestibility estimates. In addition, in order to further standardize procedures in the pet food industry, one digestibility model needs to be used to represent the dog and/or cat and ingredients specific to pet food should be used with a DIAAS score calculated for each one. Finally, it is important to point out that there is a difference between digestibility and bioavailability (Fig. 5.1) and DIAAS uses digestibility estimates as a proxy for bioavailability.

5.14 Chemical Characterization of AA

For the evaluation of protein quality, it is evident that accurate analytical techniques are needed to measure AAs in various forms, sample types, and across laboratories. There is currently no official standardized method for AA analysis although the Association of Official Analytical Chemists (AOAC) has a number of validated methods (AOAC 2023). The AOAC procedure for measuring AAs in feed uses acid hydrolysis followed by separation on an ion-exchange chromatograph for determination of all AAs except for Met, Cys, and Trp (AOAC 2023). Given that Met and Cys are often limiting in companion animal diets, they will be discussed here. Traditional analytical methods for Met and Cys involve acid oxidation and hydrolysis to convert Met to methionine sulfone and Cys to cysteic acid (AOAC 2023). Methionine sulfoxide can be utilized for protein synthesis in some animals but at a limited capacity and methionine sulfone and cysteic acid are not bioavailable (Rutherfurd and Moughan 2008). So, if there are significant amounts of methionine sulfoxide and sulfone in the feed, Met will be overestimated. For example, Tansil et al. (2023) reported the SID coefficient for Met in black soldier fly larvae meal as 90.4% when fed to ileal cannulated pigs and the bioavailability as 53.3% when the IAAO technique was used. This discrepancy was likely due to the hydrolysis method used to quantify Met from the digesta of the pigs, whereas with IAAO, chemical analysis of AAs is not required. This discrepancy could also lead to possible deficiencies when formulating if only 53% of the Met is bioavailable to the animal and this should be considered when safety margins in industry guidelines are developed.

There is also a need for more accurate ways to measure modified AAs, like methionine sulfoxide and methionine sulfone present in the diet already before hydrolysis. New methodologies, such as those employed in foodomics, use a non-destructive approach where feed samples do not undergo hydrolysis but instead compounds in the sample are often extracted using a combination of methanol, chloroform, and water (Hu and Xu 2013). Compounds are then analyzed using a combination of liquid or gas chromatography and mass spectrometry, allowing for thousands of molecules to be compared against known standards (Hu and Xu 2013). In this case, different forms of AAs, such as methionine sulfoxide, methionine sulfone, and Met can be distinguished. This has been applied to dog food once where over 600 compounds were identified (Smith et al. 2021) and is currently being used to investigate correlations between food compounds with AAs in the blood of dogs consuming the food (Banton et al. unpublished). This method does not come without its own limitations. There is a lack of standardization of sample extraction that makes results from different laboratories and even results from the classic hydrolysis method incomparable (Hu and Xu 2013). Not only that but given the massive size of the metabolome, it is impossible to extract all compounds using one extraction method and thus compounds at smaller concentrations are often lost (Hu and Xu 2013). Finally, the equipment needed is complex and expensive and, therefore not possible to use in all scenarios. However, this area of research is relatively new and with improvements in methodology and validation, it may be a more viable option in the future.

5.15 Conclusion

From the above, more empirical data are required on the AA requirements of adult dogs and cats, especially because the IAAO method, the current method of choice, shows that the few current minimum requirement estimates for specific AAs of the NRC are too low. In addition, AA requirements for different sizes, ages, and physiological states (such as illness and activity) of dogs and cats need to be urgently generated to better formulate diets to maintain health, especially in dogs where size differences can lead to larger variation in requirements for large versus small breed dogs. Protein quality is a metric of these requirements and bioavailability (digestibility) of an ingredient. Therefore, in order to standardize it in pet food, a better understanding is needed of both. There is a lack of thoroughly validated in vivo digestion models that can represent the dog or cat and with continuous pressure to move away from using animals due to ethical considerations, it is advisable to place greater emphasis on development of in vitro methodologies to measure digestibility. These advancements are required to comprehend the quality of protein we provide daily to our pets and to improve the effective use of the ingredients and lower the environmental impact of pet keeping. Given the current pressure on the food system, there is a pressing need to increase collaboration between industry, academia, and government to facilitate the development and implementation of these advancements in how we make quality foods for the millions of beloved dogs and cats we keep not only as companions, but also as service dogs, therapy dogs/cats, and other services.

References

Association of American Feed Control Officials (2016) AAFCO manual. AAFCO Inc., West Lafayette (IN)

Anderson PA, Baker DH, Corbin JE (1979) Lysine and arginine requirements of the domestic cat. J Nutr 109:1368–1372

Anderson PA, Baker DH, Sherry PA, Corbin JE (1980a) Histidine, phenylalanine-tyrosine and tryptophan requirements for growth of the young kitten. J Anim Sci 50:479–483

Anderson PA, Baker DH, Sherry PA, Teeter RG, Corbin JE (1980b) Threonine, isoleucine, valine and leucine requirements of the young kitten. J Anim Sci 50:266–271

Anderson PJB, Rogers QR, Morris JG (2002) Cats require more dietary phenylalanine or tyrosine for melanin deposition in hair than for maximal growth. J Nutr 132:2037–2042

AOAC (2023) Animal feed—general. In: Official methods of analysis of AOAC INTERNATIONAL (22nd edn). Oxford University Press, New York

Baker DH (1986) Problems and pitfalls in animal experiments designed to establish dietary requirements for essential nutrients. J Nutr 116:2339–2349

Banton S, Baynham A, Pezzali JG, Von Massow M, Shoveller AK (2021) Grains on the brain: a survey of dog owner purchasing habits related to grain-free dry dog foods. PLoS ONE 16:e0250806

Batterham ES (1992) Availability and utilization of amino acids for growing pigs. Nutr Res Rev 5:1–18

Bhat ZF, Morton JD, Bekhit AE-DA, Kumar S, Bhat HF (2021) Thermal processing implications on the digestibility of meat, fish and seafood proteins. Compr Rev Food Sci Food Saf 20:4511–4548

Borman A, Wood TR, Black HC, Anderson EG, Oesterling MJ, Womack M, Rose WC (1946) The role of arginine in growth with some observations on the effects of argininic acid. JBC 166:585–594

Breitkreutz R, Pittack N, Nebe CT, Schuster D, Brust J, Beichert M, Hack V, Daniel V, Edler L, Dröge W (2000) Improvement of immune functions in HIV infection by sulfur supplementation: two randomized trials. J Mol Med (berl) 78:55–62

Burger IH, Smith PM (1987) Amino acid requirements of adult cats. In: Proceedings of an international symposium. Nutrition, Malnutrition and Dietetics in the Dog and Cat, Hanover, pp 49–51

Burger IH, Johnson JV (1991) Dogs large and small: the Allometry of energy requirements within a single species. J Nutr 121:S18–S21

Burns RA, Milner JA (1981) Sulfur amino acid requirements of immature Beagle dogs. J Nutr 111:2117–2124

Burns RA, Milner JA, Corbin JE (1981) Arginine: an indispensable amino acid for mature dogs. J Nutr 111:1020–1024

Burns RA, Milner JA (1982) Threonine, tryptophan and histidine requirements of immature beagle dogs. J Nutr 112:447–452

Burns RA, Garton RL, Milner JA (1984) Leucine, isoleucine and valine requirements of immature beagle dogs. J Nutr 114:204–209

Butts CA, Monro JA, Moughan PJ (2012) In vitro determination of dietary protein and amino acid digestibility for humans. Br J Nutr 108:S282–S287

Campbell RG, Taverner MR, Curic DM (1985) Effects of sex and energy intake between 48 and 90 kg live weight on protein deposition in growing pigs. Anim Sci J 40:497–503

Cargo-Froom C, Shoveller AK, Marinangeli CPF, Columbus DA (2020) Methods for processing pulses to optimize nutritional functionality and maximize amino acid availability in foods and feeds. CFW 65:1–10

Che DS, Nyingwa PS, Ralinala KM, Maswanganye GMT, Wu G (2021) Amino acids in the nutrition, metabolism, and health of domestic cats. Adv Exp Med Biol 1285:217–231

Colombo R, Ferron L, Frosi I, Papetti A (2021) Advances in static in vitro digestion models after the COST action Infogest consensus protocol. Food Funct 12:7619–7636

Cooper L, Ball RO, Pencharz PB, Sakai R, Elango R (2020) Dispensable amino acids, except glutamine and proline, are ideal nitrogen sources for protein synthesis in the presence of adequate indispensable amino acids in adult men. J Nutr 150:2398–2404

Costello MJ, Morris JG, Rogers QR (1980) Effect of dietary arginine level on urinary orotate and citrate excretion in growing kittens. J Nutr 110:1204–1208

Czarnecki GL, Baker DH (1982) Utilization of D- and L-tryptophan by the growing dog. J Anim Sci 55:1405–1410

Czarnecki GL, Baker DH (1984) Urea cycle function in the dog with emphasis on the role of arginine. J Nutr 114:581–590

de Godoy MRC, Bauer LL, Parsons CM, Fahey GC Jr (2009) Select corn coproducts from the ethanol industry and their potential as ingredients in pet foods. J Anim Sci 87:189–199

de Godoy MRC, Hervera M, Swanson KS, Fahey GC (2016) Innovations in canine and feline nutrition: technologies for food and nutrition assessment. Ann Rev Anim Biosci 4:311–333

Delaney SJ, Hill AS, Backus RC, Czarnecki-Maulden GL, Rogers QR (2001) Dietary crude protein concentration does not affect the leucine requirement of growing dogs. J Anim Physiol Anim Nutr (berl) 85:88–100

Deng P, Utterback PL, Parsons CM, Hancock L, Swanson KS (2016) Chemical composition, true nutrient digestibility, and true metabolizable energy of novel pet food protein sources using the precision-fed cecectomized rooster assay. J Anim Sci 94:3335–3342

Di Buono M, Wykes LJ, Ball RO, Pencharz PB (2001) Dietary cysteine reduces the methionine requirement in men. Am J Clin Nutr 74:761–766

Do S, Koutsos L, Utterback PL, Parsons CM, de Godoy MRC, Swanson KS (2020) Nutrient and AA digestibility of black soldier fly larvae differing in age using the precision-fed cecectomized rooster assay1. J Anim Sci 98:skz363

Egger L, Ménard O, Delgado-Andrade C et al (2016) The harmonized INFOGEST in vitro digestion method: from knowledge to action. Int Food Res J 88:217–225

Egger L, Schlegel P, Baumann C, Stoffers H, Guggisberg D, Brügger C, Dürr D, Stoll P, Vergères G, Portmann R (2017) Physiological comparability of the harmonized INFOGEST in vitro digestion method to in vivo pig digestion. Int Food Res J 102:567–574

Elango R, Ball RO, Pencharz PB (2008) Indicator amino acid oxidation: concept and application. J Nutr 138:243–246

Elango R, Humayun MA, Ball RO, Pencharz PB (2009) Indicator amino acid oxidation is not affected by period of adaptation to a wide range of lysine intake in healthy young men. J Nutr 139:1082–1087

Ennis MA, Elango R (2021) A discussion on the “dispensable” amino acids. Curr Opin Clin Nutr Metab Care 24:395–401

Euromonitor (2019) Winning in PetCare. Nestle´ Investor Seminar, Nestle´ Purina Petcare, May 7 2019 London

European Pet Food Industry Federation (2018) Nutritional guidelines for complete and complementary pet food for cats and dogs. FEDIAF Inc., Bruxelles (Belgium)

Erickson RH, Kim YS (1990) Digestion and absorption of dietary protein. Ann Rev Med 41:133–139

Faber TA, Bechtel PJ, Hernot DC, Parsons CM, Swanson KS, Smiley S, Fahey GC Jr (2010) Protein digestibility evaluations of meat and fish substrates using laboratory, avian, and ileally cannulated dog assays. J Anim Sci 88:1421–1432

Falany CN, Johnson MR, Barnes S, Diasio RB (1994) Glycine and taurine conjugation of bile acids by a single enzyme. Molecular cloning and expression of human liver bile acid CoA: amino acid N-acyltransferase. J Biol Chem 269:19375–19379

Fan MZ, Sauer WC (1995) Determination of apparent ileal amino acid digestibility in barley and canola meal for pigs with the direct, difference, and regression methods. J Anim Sci 73:2364–2374

FAO (2013) Dietary protein quality evaluation in human nutrition: report of an FAO expert consultation, 31 March-2 April, 2011, Auckland, New Zealand. Food and Agriculture Organization of the United Nations, Rome

Fanelli NS, Bailey HM, Guardiola LV, Stein HH (2021) Values for Digestible Indispensable Amino Acid Score (DIAAS) determined in pigs are greater for milk than for breakfast cereals, but DIAAS values for individual ingredients are additive in combined meals. J Nutr 151:540–547
Finkelstein JD, Mudd SH (1967) Trans-sulfuration in mammals. J Biol Chem 242:873–880
Folador JF, Karr-Lilienthal LK, Parsons CM, Bauer LL, Utterback PL, Schasteen CS, Bechtel PJ, Fahey GC Jr (2006) Fish meals, fish components, and fish protein hydrolysates as potential ingredients in pet foods. J Anim Sci 84:2752–2765
Fuller MF, Reeds PJ, Cadenhead A, Seve B, Preston T (1987) Effects of the amount and quality of dietary protein on nitrogen metabolism and protein turnover of pigs. Br J Nutr 58:287–300
Fuller MF, Chamberlain AG (1985) Protein requirements for pigs. Recent developments in pig nutrition. Butterworths, Toronto, pp 85–96
Fuller MF, Reeds PJ (1998) Nitrogen cycling in the gut. Annu Rev Nutr 18:385–411
Gatel F, Fekete J (1989) Lysine and threonine balance and requirements for weaned piglets 10–25 kg liveweight fed cereal-based diets. Livest Prod Sci 23:195–206
Gaylord L, Remillard R, Saker K (2018) Risk of nutritional deficiencies for dogs on a weight loss plan. J Small Anim Pract 59:695–703
German AJ (2016) Weight management in obese pets: the tailoring concept and how it can improve results. Acta Vet Scand 58:57
Grant CE, Shoveller AK, Blois S, Bakovic M, Monteith G, Verbrugghe A (2020) Dietary intake of amino acids and vitamins compared to NRC requirements in obese cats undergoing energy restriction for weight loss. BMC Vet Res 16:426
Hardy AJ, Morris JG, Rogers QR (1977) Valine requirement of the growing kitten. J Nutr 107:1308–1312
Hargrove DM, Rogers QR, Morris JG (1983) The tryptophan requirement of the kitten. Br J Nutr 50:487–493
Hargrove DM, Rogers QR, Morris JG (1984) Leucine and isoleucine requirements of the kitten. Br J Nutr 52:595–605
Hargrove DM, Morris JG, Rogers QR (1994) Kittens choose a high leucine diet even when isoleucine and valine are the limiting amino acids. J Nutr 124:689–693
Harper AE (1965) Effect of variations in protein intake on enzymes of amino acid metabolism. Can J Biochem 43:1589–1603
Harper AE, Benevenga NJ, Wohlhueter RM (1970) Effects of ingestion of disproportionate amounts of amino acids. Physiol Rev 50:428–558
Harrison M, Thomas G, Gilham M, Gray K, Colyer A, Allaway D (2020) Short-term determination and long-term evaluation of the dietary methionine requirement in adult dogs. Br J Nutr 123:1333–1344
Heger J, Mengesha S, Vodehnal D (1998) Effect of essential: total nitrogen ratio on protein utilization in the growing pig. Br J Nutr 80:537–544
Heger J (2003) Essential to non-essential amino acid ratios. In: D'Mello JPF (ed) Amino acids in Animal Nutrition, 2nd edn. CABI Publishing, UK, pp 103–124
Hendriks WH, Tarttelin MF, Moughan PJ (1998) The amino acid composition of cat (Felis catus) hair. Anim Sci 67:165–170
Hendriks WH, Emmens MM, Trass B, Pluske JR (1999) Heat processing changes the protein quality of canned cat foods as measured with a rat bioassay. J Anim Sci 77:669
Hendriks WH (2003) Canine and feline amino acid requirements for different physiological functions. Amino acids in animal nutrition, 2nd edn. CABI Publishing, Cambridge, pp 411–426
Hendriks WH, Van Baal J, Bosch G (2012) Ileal and faecal protein digestibility measurement in humans and other non-ruminants—a comparative species view. Br J Nutr 108:S247–S257
Hendriks WH, Thomas DG, Bosch G, Fahey GC Jr (2013) Comparison of ileal and total tract nutrient digestibility of dry dog foods1. J Anim Sci 91:3807–3814
Hendriks WH, Bakker EJ, Bosch G (2015) Protein and amino acid bioavailability estimates for canine foods. J Anim Sci 93:4788–4795

Hirakawa DA, Baker DH (1985) Sulfur amino acid nutrition of the growing puppy: determination of dietary requirements for methionine and cystine. Nutr Res 5:631–642

Hirakawa DA, Baker DH (1986) Lysine requirement of growing puppies fed practical and purified diets. Nutr Res 6:527–538

Hodgkinson SM, Stroebinger N, Van der Wielen N, Mensink M, Montoya C, Hendriks WH, De Vries S, Stein HH, Moughan PJ (2022) Comparison of true ileal amino acid digestibility between adult humans and growing pigs. J Nutr 152:1635–1646

Hoenig M, Laflamme PD, Klaser DA, Singer MJ, Ferguson DC (2001) Glucose tolerance and lipid profiles in dogs fed different fiber diets. Vet Ther 2:160–169

Huh Y, Son KY (2022) Association between total protein intake and low muscle mass in Korean adults. BMC Geriatr 22:319

Hsu C, Utterback PL, Parsons CM, Marx F, Guldenpfennig R, de Godoy MRC (2023) Standardized amino acid digestibility and protein quality in extruded canine diets containing hydrolyzed protein using a precision fed rooster assay. J Anim Sci skad289

Hu C, Xu G (2013) Mass-spectrometry-based metabolomics analysis for foodomics. TrAC Trends Anal Chem 52:36–46

Hugman J, Wang LF, Beltranena E, Htoo JK, Zijlstra RT (2021) Nutrient digestibility of heat-processed field pea in weaned pigs. Anim Feed Sci Technol 274:114891

Humbert B, Bleis P, Martin L, Dumon H, Darmaun D, Nguyen P (2001) Effects of dietary protein restriction and amino acids deficiency on protein metabolism in dogs. J Anim Physiol Anim Nutr 85:255–262

Johnson ML, Parsons CM, Fahey GC Jr, Merchen NR, Aldrich CG (1998) Effects of species raw material source, ash content, and processing temperature on amino acid digestibility of animal by-product meals by cecectomized roosters and ileally cannulated dogs. J Anim Sci 76:1112–1122

Kaplan JL, Stern JA, Fascetti AJ et al (2018) Taurine deficiency and dilated cardiomyopathy in golden retrievers fed commercial diets. PLoS ONE 13:e0209112

Keene JC, Austic RE (2001) Dietary supplements of mixtures of indispensable amino acids lacking threonine, phenylalanine or histidine increase the activity of hepatic threonine dehydrogenase, phenylalanine hydroxylase or histidase, respectively, and prevent growth depressions in chicks caused by dietary excesses of threonine, phenylalanine, or histidine. J Nutr Biochem 12:274–284

Kim K-I (1993) Requirement for phenylalanine and replacement value of tyrosine for phenylalanine in rainbow trout (Oncorhynchus mykiss). Aquacul 113:243–250

Ko KS, Backus RC, Berg JR, Lame MW, Rogers QR (2007) Differences in taurine synthesis rate among dogs relate to differences in their maintenance energy requirement. J Nutr 137:1171–1175

Kong C, Adeola O (2014) Evaluation of amino acid and energy utilization in feedstuff for swine and poultry diets. Asian-Australas J Anim Sci 27:917–925

Laflamme DP, Hannah SS (2013) Discrepancy between use of lean body mass or nitrogen balance to determine protein requirements for adult cats. J Feline Med Surg 15:691–697

Lancheros JP, Espinosa CD, Stein HH (2020) Effects of particle size reduction, pelleting, and extrusion on the nutritional value of ingredients and diets fed to pigs: a review. Anim Feed Sci Technol 268:114603

Lenis NP, van Diepen HTM, Bikker P, Jongbloed AW, van der Meulen J (1999) Effect of the ratio between essential and nonessential amino acids in the diet on utilization of nitrogen and amino acids by growing pigs. J Anim Sci 77:1777–1787

Li P, Wu G (2023) Amino acid nutrition and metabolism in domestic cats and dogs. J Anim Sci Biotechnol 14:19

Li P, Yin Y-L, Li D, Kim SW, Wu G (2007) Amino acids and immune function. Br J Nutr 98:237–252

Lopez MJ, Mohiuddin SS (2023) Biochemistry, essential amino acids. In: StatPearls. StatPearls Publishing, Treasure Island

Mansilla WD, Gorman A, Fortener L, Shoveller AK (2018) Dietary phenylalanine requirements are similar in small, medium, and large breed adult dogs using the direct amino acid oxidation technique. J Anim Sci 96:3112–3120

Mansilla WD, Templeman JR, Fortener L, Shoveller AK (2020a) Minimum dietary methionine requirements in Miniature Dachshund, Beagle, and Labrador Retriever adult dogs using the indicator amino acid oxidation technique. J Anim Sci 98:skaa324

Mansilla WD, Fortener L, Templeman JR, Shoveller AK (2020b) Adult dogs of different breed sizes have similar threonine requirements as determined by the indicator amino acid oxidation technique. J Anim Sci 98:skaa066

Meguid MM, Matthews DE, Bier DM, Meredith CN, Young VR (1986a) Valine kinetics at graded valine intakes in young men. Am J Clin Nutr 43:781–786

Meguid M, Matthews D, Bier D, Meredith C, Soeldner J, Young V (1986b) Leucine kinetics at graded leucine intakes in young men. Anim J Clin Nutr 43:770–780

Millward DJ (2012) Identifying recommended dietary allowances for protein and amino acids: a critique of the 2007 WHO/FAO/UNU report. Br J Nutr 108:S3–S21

Milner JA (1979a) Assessment of indispensable and dispensable amino acids for the immature dog. J Nutr 109:1161–1167

Milner JA (1979b) Assessment of the essentiality of methionine, threonine, tryptophan, histidine and isoleucine in immature dogs. J Nutr 109:1351–1357

Milner JA (1981) Lysine requirements of the immature dog. J Nutr 111:40–45

Milner JA, Garton RL, Burns RA (1984) Phenylalanine and tyrosine requirements of immature beagle dogs. J Nutr 114:2212–2216

Minekus M, Alminger M, Alvito P et al (2014) A standardised static in vitro digestion method suitable for food—an international consensus. Food Funct 5:1113–1124

Moehn S, Bertolo RFP, Pencharz PB, Ball RO (2004) Indicator amino acid oxidation responds rapidly to changes in lysine or protein intake in growing and adult pigs. J Nutr 134:836–841

Moehn S, Bertolo RFP, Pencharz PB, Ball RO (2005) Development of the indicator amino acid oxidation technique to determine the availability of amino acids from dietary protein in pigs. J Nutr 135:2866–2870

Morris JG, Rogers QR, Pacioretty LM (1990) Taurine: an essential nutrient for cats. J Small Anim Pract 31:502–509

Morris JG, Rogers QR (1994) Assessment of the nutritional adequacy of pet foods through the life cycle. J Nutr 124:2520S-2534S

Morris JG, Rogers QR, O'Donnell JA (2004) Lysine requirement of kittens given purified diets for maximal growth. J Anim Physiol Anim Nutr (berl) 88:113–116

Motil KJ, Matthews DE, Bier DM, Burke JF, Munro HN, Young VR (1981) Whole-body leucine and lysine metabolism: response to dietary protein intake in young men. Am J Physiol 240:E712-721

Motil KJ, Opekun AR, Montandon CM, Berthold HK, Davis TA, Klein PD, Reeds PJ (1994) Leucine oxidation changes rapidly after dietary protein intake is altered in adult women but lysine flux is unchanged as is lysine incorporation into VLDL-apolipoprotein B-100. J Nutr 124:41–51

Moughan PJ (2003) Simulating the partitioning of dietary amino acids: new directions. J Anim Sci 81:E60–E67

Moughan PJ, Butts CA, van Wijk H, Rowan AM, Reynolds GW (2005) An acute ileal amino acid digestibility assay is a valid procedure for use in human ileostomates. J Nutr 135:404–409

Moughan PJ, Ravindran V, Sorbara JOB (2014) Dietary protein and amino acids—Consideration of the undigestible fraction. Poult Sci 93:2400–2410

Nosworthy MG, House JD (2017) Factors influencing the quality of dietary proteins: implications for pulses. Cereal Chem 94:49–57

Novotny MJ, Hogan PM, Paley DM, Adams HR (1991) Systolic and diastolic dysfunction of the left ventricle induced by dietary taurine deficiency in cats. Am J Physiol Heart Circ Physiol 261:H121–H127

NRC (1989) Recommended dietary allowances. national research council (10th edn). National Academies Press, Washington

NRC (2006) Nutrient requirements of dogs and cats. national research council (10th edn). National Academies Press, Washington

Oba PM, Utterback PL, Parsons CM, de Godoy MRC, Swanson KS (2019) Chemical composition, true nutrient digestibility, and true metabolizable energy of chicken-based ingredients differing by processing method using the precision-fed cecectomized rooster assay1. J Anim Sci 97:998–1009

Oba PM, Utterback PL, Longshaw M, Parsons CM, Swanson KS (2023) Comparing the standardized amino acid digestibility of an alternative protein source with commercially available protein-based ingredients using the precision-fed cecectomized rooster assay. J Anim Sci 101:skad236

Oberbauer AM, Larsen JA (2021) Amino acids in dog nutrition and health. Adv Exp Med Biol 1285:199–216

Osho SO, Babatunde OO, Adeola O (2019) Additivity of apparent and standardized ileal digestibility of amino acids in wheat, canola meal, and sorghum distillers dried grains with solubles in mixed diets fed to broiler chickens. Poult Sci 98:1333–1340

Parsons CM (1985) Influence of caecectomy on digestibility of amino acids by roosters fed distillers' dried grains with solubles. J Agric Sci 104:469–472

Peachey SE, Dawson JM, Harper EJ (2000) Gastrointestinal transit times in young and old cats. Comp Biochem Physiol A Mol Integr Physiol 126:85–90

Pencharz PB, Ball RO (2003) Different approaches to define individual amino acid requirements. Ann Rev Nutr 23:101–116

Quam DD, Morris JG, Rogers QR (1987) Histidine requirement of kittens for growth, haematopoiesis and prevention of cataracts. Br J Nutr 58:521–532

Rabin B, Nicolosi RJ, Hayes KC (1976) Dietary influence on bile acid conjugation in the cat. J Nutr 106:1241–1246

Reeds PJ (2000) Dispensable and indispensable amino acids for humans. J Nutr 130:1835S-1840S

Rehman Z, Shah WH (2005) Thermal heat processing effects on antinutrients, protein and starch digestibility of food legumes. Food Chem 91:327–331

Reilly LM, von Schaumburg PC, Hoke JM, Davenport GM, Utterback PL, Parsons CM, de Godoy MRC (2020a) Macronutrient composition, true metabolizable energy and amino acid digestibility, and indispensable amino acid scoring of pulse ingredients for use in canine and feline diets. J Anim Sci 98:skaa149

Reilly LM, von Schaumburg PC, Hoke JM, Davenport GM, Utterback PL, Parsons CM, de Godoy MRC (2020b) Use of precision-fed cecectomized rooster assay and digestible indispensable amino acid scores to characterize plant- and yeast-concentrated proteins for inclusion in canine and feline diets. Transl Anim Sci 4:txaa133

Reilly LM, von Schaumburg PC, Hoke JM, Davenport GM, Utterback PL, Parsons CM, de Godoy MRC (2021) Use of the precision-fed cecectomized rooster assay to determine standardized amino acid digestibility, true metabolizable energy content, and digestible indispensable amino acid scores of plant-based protein by-products used in canine and feline diets. Transl Anim Sci 5:txab025

Roberts LJ, Oba PM, Utterback PL, Parsons CM, Swanson KS (2023) Amino acid digestibility and nitrogen-corrected true metabolizable energy of mildly cooked human-grade vegan dog foods using the precision-fed cecectomized and conventional rooster assays. Transl Anim Sci 7:txad020

Robinson JL, Bartlett RK, Harding SV, Randell EW, Brunton JA, Bertolo RF (2016) Dietary methyl donors affect in vivo methionine partitioning between transmethylation and protein synthesis in the neonatal piglet. Amino Acids 48:2821–2830

Romano C, Corsetti G, Flati V, Pasini E, Picca A, Calvani R, Marzetti E, Dioguardi FS (2019) Influence of diets with varying essential/nonessential amino acid ratios on mouse lifespan. Nutrients 11:1367

Rutherfurd SM, Moughan PJ (2008) Determination of sulfur amino acids in foods as related to bioavailability. J AOAC Int 91:907–913

Sanderson SL, Gross KL, Ogburn PN, Calvert C, Jacobs G, Lowry SR, Bird KA, Koehler LA, Swanson LL (2001) Effects of dietary fat and L-carnitine on plasma and whole blood taurine concentrations and cardiac function in healthy dogs fed protein-restricted diets. Am J Vet Res 62:1616–1623

Sasse CE, Baker DH (1972) The phenylalanine and tyrosine requirements and their interrelationship for the young chick. Poult Sci 51:1531–1536

Sauer WC, Jørgensen H, Berzins R (1983) A modified nylon bag technique for determining apparent digestibilities of protein in feedstuffs for pigs. Can J Anim Sci 63:233–237

Schwesinger WH, Kurtin WE, Page CP, Stewart RM, Johnson R (1999) Soluble dietary fiber protects against cholesterol gallstone formation. Am J Surg 177:307–310

Shoveller AK, Brunton JA, House JD, Pencharz PB, Ball RO (2003) Dietary cysteine reduces the methionine requirement by an equal proportion in both parenterally and enterally fed piglets. J Nutr 133:4215–4224

Sibbald IR (1979) A bioassay for bvailable amino acids and true metabolizable energy in feedingstuffs. Poult Sci 58:668–673

Smalley KA, Rogers QR, Morris JG (1983) Methionine requirement of kittens given amino acid diets containing adequate cystine. Br J Nutr 49:411–417

Smith CE, Parnell LD, Lai CQ, Rush JE, Freeman LM (2021) Investigation of diets associated with dilated cardiomyopathy in dogs using foodomics analysis. Sci Rep 11:15881

Smola MA, Oba PM, Utterback PL, Sánchez-Sánchez L, Parsons CM, Swanson KS (2023) Amino acid digestibility and protein quality of mealworm-based ingredients using the precision-fed cecectomized rooster assay. J Anim Sci 101:skad012

Stein HH, Pedersen C, Wirt AR, Bohlke RA (2005) Additivity of values for apparent and standardized ileal digestibility of amino acids in mixed diets fed to growing pigs. J Anim Sci 83:2387–2395

Stein HH, Sève B, Fuller MF, Moughan PJ, De Lange CFM (2007a) Invited review: amino acid bioavailability and digestibility in pig feed ingredients: terminology and application. J Anim Sci 85:172–180

Stein HH, Fuller MF, Moughan PJ, Sève B, Mosenthin R, Jansman AJM, Fernández JA, de Lange CFM (2007b) Definition of apparent, true, and standardized ileal digestibility of amino acids in pigs. Livest Sci 109:282–285

Stoll B, Henry J, Reeds PJ, Yu H, Jahoor F, Burrin DG (1998) Catabolism dominates the first-pass intestinal metabolism of dietary essential amino acids in milk protein-fed piglets. J Nutr 128:606–614

Sundling L, Ahlstrøm Ø, Tauson A-H (2012) Comparative digestibility of nutrients and energy in ferrets (Mustela putorius furo), mink (Neovison vison) and cats (Felis catus). In: Larsen PF, Møller SH, Clausen T, et al (eds) Proceedings of the Xth International Scientific Congress in fur animal production. Academic Publishers, Wageningen, pp 112–120

Sutherland KAK, Mansilla WD, Fortener L, Shoveller AK (2020) Lysine requirements in small, medium, and large breed adult dogs using the indicator amino acid oxidation technique. Transl Anim Sci 4:txaa082

Szymeczko R, Skrede A (1990) Protein digestion in mink. Acta Agri Scand 40:189–200

Tansil F, Pezzali JG, Cargo-Froom C, Huber L-A, Kiarie EG, Courtney-Martin G, Levesque CL, Shoveller AK (2023) Evaluation of standardized ileal digestibility of amino acids and metabolic availability of methionine, using the indicator amino acid oxidation method, in black soldier fly larvae (Hermetia illucens) meal fed to growing pigs. J Anim Sci 101:skac420

Taylor TP, Morris JG, Kass PH, Rogers QR (1997) Increasing dispensable amino acids in diets of kittens fed essential amino acids at or below their requirement increases the requirement for arginine. Amino Acids 13:257–272

Teeter RG, Baker DH, Corbin JE (1978) Methionine and cystine requirements of the cat. J Nutr 108:291–295

Templeman JR, Mansilla WD, Fortener L, Shoveller AK (2019) Tryptophan requirements in small, medium, and large breed adult dogs using the indicator amino acid oxidation technique1. J Anim Sci 97:3274–3285

Titchenal CA, Rogers QR, Indrieri RJ, Morris JG (1980) Threonine imbalance, deficiency and neurologic dysfunction in the kitten. J Nutr 110:2444–2459

Tôrres CL, Biourge VC, Backus RC (2022) Plasma and whole blood taurine concentrations in dogs may not be sensitive indicators of taurine deficiency when dietary sulfur amino acid content is reduced. Front Vet Sci 9:873460

Viviano KR, Lavergne SN, Goodman L, Vanderwielen B, Grundahl L, Padilla M, Trepanier LA (2009) Glutathione, cysteine, and ascorbate concentrations in clinically ill dogs and cats. J Vet Intern Med 23:250–257

Walker JA, Harmon DL, Gross KL, Collings GF (1994) Evaluation of nutrient utilization in the canine using the ileal cannulation technique. J Nutr 124:2672S-2676S

Ward J (1976) The amino acid requirements of the adult dog. University of Cambridge, Ph.D.

Waterlow JC (1986) Metabolic adaptation to low intakes of energy and protein. Annu Rev Nutr 6:495–526

WHO/FAO/UNU Expert Consultation (2007) Protein and amino acid requirements in human nutrition. World Health Organ Tech Rep Ser 1–265

Williams JM, Morris JG, Rogers QR (1987) Phenylalanine requirement of kittens and the sparing effect of tyrosine. J Nutr 117:1102–1107

Williamson E, Kato H, Volterman KA, Suzuki K, Moore DR (2019) The effect of dietary protein on protein metabolism and performance in endurance-trained males. Med Sci Sports Exer 51:352–360

Wolfe RR, Rutherfurd SM, Kim I-Y, Moughan PJ (2016) Protein quality as determined by the digestible indispensable amino acid score: evaluation of factors underlying the calculation. Nutr Rev 74:584–599

Wolfe RR, Kim I-Y, Park S, Ferrando A (2022) Tracing metabolic flux to assess optimal dietary protein and amino acid consumption. Exp Mol Med 54:1323–1331

Womack M, Rose WC (1947) The role of proline, hydroxyproline, and glutamic acid in growth. J Biol Chem 171:37–50

Wu G, Fang Y-Z, Yang S, Lupton JR, Turner ND (2004) Glutathione metabolism and its implications for health. J Nutr 134:489–492

Wu G (2009) Amino acids: metabolism, functions, and nutrition. Amino Acids 37:1–17

Wu G (2014) Dietary requirements of synthesizable amino acids by animals: a paradigm shift in protein nutrition. J Anim Sci Biotech 5:34

Yen JT (2001) Anatomy of the digestive system and nutritional physiology. Swine Nutrition, 2nd edn. CRC Press, New York, pp 46–78

Young VR, Marchini JS (1990) Mechanisms and nutritional significance of metabolic responses to altered intakes of protein and amino acids, with reference to nutritional adaptation in humans. Am J Clin Nutr 51:270–289

Zello GA, Pencharz PB, Ball RO (1993) Dietary lysine requirement of young adult males determined by oxidation of L-[1-13C]phenylalanine. Am J Physiol 264:E677-685

Zello GA, Wykes LJ, Ball RO, Pencharz PB (1995) Recent advances in methods of assessing dietary amino acid requirements for adult humans. J Nutr 125:2907–2915

Zhang F, Adeola O (2017) Techniques for evaluating digestibility of energy, amino acids, phosphorus, and calcium in feed ingredients for pigs. Anim Nutr 3:344–435

Chapter 6
Functions and Metabolism of Amino Acids in the Hair and Skin of Dogs and Cats

Erin D. Connolly and Guoyao Wu

Abstract The hair and skin of domestic cats or dogs account for 2% and 12–24% of their body weight, respectively, depending on breed and age. These connective tissues contain protein as the major constituent and provide the first line of defense against external pathogens and toxins. Maintenance of the skin and hair in smooth and elastic states requires special nutritional support, particularly an adequate provision of amino acids (AAs). Keratin (rich in cysteine, serine and glycine) is the major protein both in the epidermis of the skin and in the hair. Filaggrin [rich in some AAs (e.g., serine, glutamate, glutamine, glycine, arginine, and histidine)] is another physiologically important protein in the epidermis of the skin. Collagen and elastin (rich in glycine and proline plus 4-hydroxyproline) are the predominant proteins in the dermis and hypodermis of the skin. Taurine and 4-hydroxyproline are abundant free AAs in the skin of dogs and cats, and 4-hydroxyproline is also an abundant free AA in their hair. The epidermis of the skin synthesizes melanin (the pigment in the skin and hair) from tyrosine and produces *trans*-urocanate from histidine. Qualitative requirements for proteinogenic AAs are similar between cats and dogs but not identical. Both animal species require the same AAs to nourish the hair and skin but the amounts differ. Other factors (e.g., breeds, coat color, and age) may affect the requirements of cats or dogs for nutrients. The development of a healthy coat, especially a black coat, as well as healthy skin critically depends on AAs [particularly arginine, glycine, histidine, proline, 4-hydroxyproline, and serine, sulfur AAs (methionine, cysteine, and taurine), phenylalanine, and tyrosine] and creatine. Although there are a myriad of studies on AA nutrition in cats and dogs, there is still much to learn about how each AA affects the growth, development and maintenance of the hair and skin. Animal-sourced foodstuffs (e.g., feather meal and poultry by-product meal) are excellent sources of the AAs that are crucial to maintain the normal structure and health of the skin and hair in dogs and cats.

Keywords Amino acids · Animal · Hair · Metabolism · Nutrition · Skin

---

E. D. Connolly · G. Wu (✉)
Department of Animal Science, Texas A&M University, College Station, TX 77843, USA
e-mail: g-wu@tamu.edu

G. Wu (ed.), *Nutrition and Metabolism of Dogs and Cats*, Advances in Experimental Medicine and Biology 1446, https://doi.org/10.1007/978-3-031-54192-6_6

Abbreviations

AA	Amino acid
SAA	Sulfur-containing amino acid
EAA	Nutritionally essential amino acid
CEAA	Conditionally essential amino acid
NEAA	Nutritionally non-essential amino acid
BW	Body weight
SND	Superficial necrolytic dermatitis
HPS	Hepatocutaneous syndrome
NRC	National research council

6.1 Introduction

The skin and hair are large connective tissues in dogs and cats, and together account for approximately 14–26% of their total body weight, depending on breeds and ages (Hendriks et al. 1997; Moriello 2022a, b; Mundt and Stafforst 1987). These tissues protect the internal organs from external threats, sense environmental changes (e.g., touch and pressure), participate in immune responses against pathogens, contribute to thermoregulation, and aid in maintaining proper hydration in the body. Healthy skin and hair should be smooth and elastic while maintaining their normal color, whereas unhealthy skin and hair result in infections, itchiness, irritation, discomfort, and alterations of an animal's behavior. The health and well-being of dogs and cats critically depends on the integrity and health of their skin and hair.

Protein is the most abundant component of the skin and hair. Catabolism of AAs occurs in these two connective tissues (Wu 2020). Thus, up to 35% of daily protein intake is used to maintain both tissues in adult dogs and cats. Dogs and cats have similar qualitative requirements for dietary amino acids (AAs) with the exception of arginine and taurine (Li and Wu 2023). For the proper functioning of the skin and hair, both animals require 20 proteinogenic AAs and, in the case of cats and some breeds of dogs, also taurine; these AAs need to be obtained from diets (Li and Wu 2023). The dietary requirements of dogs and cats are not identical, but the deficiency of one or more AAs in their diets important for the function of the skin and hair results in the abnormal pigmentation, dehydration, peeling, toughness, lesions, rashes, and itchiness of the skin, as well as abnormally colored and brittle hair (Watson et al. 2017). However, cats have shown a need for a higher dietary intake for sulfur-containing AAs (SAAs) than dogs (Baker and Czarnecki-Maulden 1991). In cats and dogs, the major AAs in the hair are cysteine, serine, glutamate, glycine, methionine and histidine (Hendriks et al. 1998). The AA composition for hair development in dogs was determined to be alike compared to the cat's (Hendriks et al. 1998; Marshall 1986).

Typically, it is not common for household dogs and cats to develop any AA deficiency when fed commercial pet foods containing adequate amounts of proteinogenic AAs and, in the case of cats and most breeds of dogs, also taurine (Sato et al. 2020). However, with the rising popularity of homemade diets for the companion animals, AA deficiencies may be more of a concern for pet owners than before. AAs are an important part of any dog or cat's diet (Oberbauer and Larsen 2021). According to pet owners, they often view hair quality as a fundamental way to determine if their pet's food is nutritionally adequate (Li and Wu 2023; Case et al. 2011). Determining the ideal intake of dietary AAs by companion animals, therefore, is important for commercial pet-food companies to provide high-quality products and relieve consumers' concerns.

This review highlights the literature currently available on the function and metabolism of AAs in the hair and skin of dogs and cats. The metabolism of traditionally classified "nutritionally essential AAs (EAAs)", "nutritionally nonessential AAs (NEAAs)", and "conditionally essential AAs (CEAAs)" that play major roles in the growth, development, and maintenance of the hair and skin (Wu and Li 2022). These AAs will be further discussed to determine how their function in cats or dogs can be impacted by reducing or increasing dietary intakes. In particular, tyrosine and phenylalanine are crucial for maintaining hair color in the animals, whereas cysteine, glycine, serine, arginine, and proline play key roles in the synthesis and structure of hair proteins.

6.2 Structure of the Hair

The hair is a unique tissue in mammals (including dogs and cats) and accounts for 2.0% of their adult body weight (Hendriks et al. 1997; Mundt and Stafforst 1987). The hair consists of two distinct parts: the hair follicle (embedded within the skin) and the hair shaft (the visible part of the hair above the skin) (Moriello 2022a, b). The hair shaft is made of highly keratinized cells and is derived from hair bulb stem cells (the germinal matrix cells) in the hair follicles through a series of steps involving cell differentiation, proliferation, differentiation, and migration (Fig. 6.1) (Ma et al. 2024; Méchin and Simon 2023). There are three layers in the hair shaft: the medulla, the cortex, and the cuticle, with their major cell types being medullar, cortical, and cuticle keratinocytes, respectively. The sebaceous glands localized above the hair bulb produce sebum through enzyme-catalyzed reactions that is excreted into the hair follicle through the sebaceous duct and then appears on the surface of the skin through the hair's growth. The sebum is a mixture of oily and waxy substances, including triglycerides, squalene, free cholesterol, and fatty acids, as well as esters of glycerol, cholesterol, and wax.

At the base of the hair root, there is the germinal matrix (also known as hair matrix). The germinal matrix consists of primarily trichocytes and, to a lesser extent, other epithelial cells, and melanocytes (the cells that synthesize and release melanin).

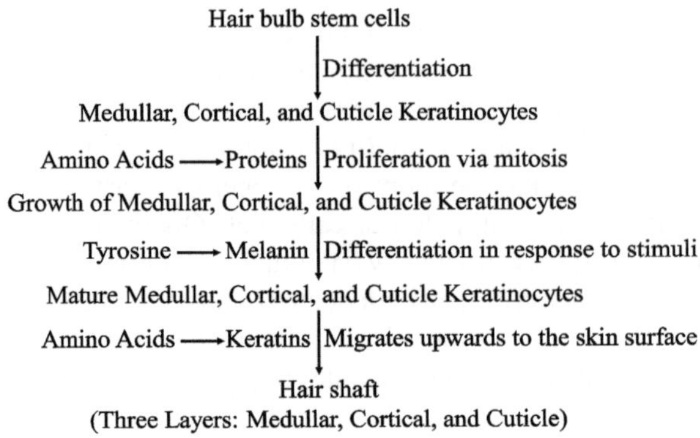

Fig. 6.1 The formation of the hair in mammals including dogs and cats. The hair bulb stem cells differentiate into medullar, cortical, and cuticle keratinocytes. These cells then proliferate via mitosis; further differentiate in response to factors from the neighboring cells such as the hair keratinocytes, fibroblasts, and endothelial cells. The differentiated keratinocytes take up melanin granules released by melanocytes and migrate upwards to the skin surface. These processes require the synthesis of proteins (including keratins) from amino acids as well as the conversion of tyrosine into melanin. Thus, the hair grows by forming new cells at the base of the hair root. This physiological process depends on sufficient amino acids and other nutrients

Trichocytes are also known as hair keratinocytes, and they are the specialized epithelial cells that primarily form the hair and produce the keratin protein in the hair. There are three types of trichocytes (medullar, cortical, and cuticle keratinocytes) that are derived from the follicle bulb stem cells. The ratio of hair keratinocytes to melanocytes in the hair bulb is approximately 5 to 1, and this value is greater than the ratio of keratinocytes to melanocytes in the epidermis of the skin. In adult cats, the rate of hair growth varies with breed, season, and hair type (i.e., primary or guard hair versus secondary or undercoat), with the mean value being 0.33 mm/day for the primary hair and 0.25 mm/day for the secondary hair (Baker 1974; Hendriks et al. 1997). Likewise, in adult dogs, the rate of hair growth may range from 0.31 to 0.71 mm/day, depending on breed, anatomical site, and season (Gunaratnam and Wilkinson 1983).

The hair is made up of approximately 90% keratin protein (e.g., 86% keratin in cats and 92% keratin in dogs; Hendriks et al. 1998). The hair also contains keratin-associated proteins, such as elastin (produced by fibroblasts) and filaggrin. Elastin helps to anchor hair follicles and allows the hair to remain flexible and avoid breakage, whereas filaggrin is produced by the inner root sheath cells of the hair follicles. Cysteine is the most abundant AA in the hair of dogs and cats. Lesser abundant AAs include serine, glycine, proline, arginine, threonine, glutamate, leucine, valine, glutamine, lysine, aspartate, tyrosine, isoleucine, and asparagine (Table 6.2). Collin et al. (1992) reported that human cytokeratin 2 protein consisted of 645 AA residues, including 39 glutamate, 31 glutamine, 25 aspartate, and 24 asparagine residues. Thus, the ratios (mol/mol) of glutamate to glutamine and of aspartate to asparagine in the hair are 0.56:0.44 [similar to 0.57:0.43 for sheep wool (Holt et al. (1971)]

and 0.51:0.49, respectively. Note that, in contrast to dogs, the content of proline in the cat's hair appears to be much lower than that of glycine (Hendriks et al. 1998). Future studies are needed to confirm this finding. The hair has many physiological and social functions, including serving as an insulating layer between the skin and the external environment; protection against external factors; pulling up sebum to the skin surface; impact on social interactions; and thermoregulation.

6.3 Structure of the Skin

As a large and complex organ in the body, the skin accounts for 12–24% of the body weight in dogs and cats, depending on species, breed, and age (Moriello 2022a, b). The skin consists of three main parts: the epidermis, the dermis, and the hypodermis (Solano et al. 2020). The skin provides the first line of physical barriers from the external environment (including water, water-soluble toxic chemicals, UV–VIS irradiation, pathogens, and mechanical pressure); is a major component of the innate immunity in animals; serves as a sensory organ; helps to regulate body temperature; and prevents dehydration (Wu 2020). The abundant AAs in skin proteins are shown in Table 6.1, with collagen being the most abundant protein (Li and Wu 2018; Devlin 2006). Taurine (an important osmolyte and an antioxidant) is an abundant free AA in the epidermis (5–18 mM; mainly in keratinocytes) and the dermis (3 to 17 mM), whereas concentrations of 4-hydroxyproline (an important antioxidant) in all the layers of the skin are about 0.4–0.8 mM, depending on species, breeds, age, and nutritional status (Lobo et al. 2001; Yoshimura et al. 2021). For comparison, the concentrations of taurine and 4-hydroxyproline in the plasma of dogs and cats are approximately 0.1 and 0.04 mM, respectively (Li and Wu 2023). In dogs and cats, the epidermis is the epithelial tissue of the skin and is composed of 4 to 5 sub-layers of epithelial cells, depending on its location in the body. Within the epidermis, there is a layer called the stratum corneum, which is rich in keratinocytes (Solano et al. 2020). The epidermis is made up of 95% keratinocytes and 5% of other cells (i.e., melanocytes, Langerhans cells, Merkel cells, and inflammatory cells) (Lai-Cheong and McGrath 2021). In both dogs and cats, the epidermis contains sweat glands (Moriello 2022a, b). The hair follicles in the skin's epidermis project further into the dermis. A layer of water-insoluble lipids covers the skin surface and consists of mainly triglycerides, wax esters, squalene, and long-chain fatty acids and, to a lesser extent, cholesterol, cholesterol esters, and diglycerides. In the epidermis of the skin, keratinocytes produce keratins (cysteine-rich proteins) and filaggrin (a serine-rich protein). Keratins are the main component of the epidermis and account for about 95% of epidermal proteins to maintain the physical strength of the skin. Filaggrin (produced by the granular cells of the skin's epidermis) promotes keratin aggregation in the skin. Epidermal keratinocytes also produce *trans*-urocanate (UA) from histidine for protecting the skin from ultraviolet irradiation. Furthermore, epidermal melanocytes convert tyrosine (and phenylalanine) into melanin. The type and amount of melanin (e.g., eumelanin, pheomelanin, and mixed melanin) determine the color of the skin and the hair.

The dermis, which accounts for ~ 90% of the skin's weight, contains fibroblasts (the primary cell type), collagen, elastin, proteoglycan; hair follicles, sweat glands, sebaceous glands; mast cells, lymphocytes, macrophages; white adipocytes; nerves; lymphatic vessels; and blood capillaries (Moriello 2022a, b). The blood vessels in the dermis extend to the hypodermis. The artery provides nutrients and oxygen to the skin, and the vein removes metabolites from the skin. The lymphatics help to drain excess fluid from tissues in the skin back into the blood circulation. Collagen accounts for 70% of dry matter in the skin's dermis, 90% of total proteins in the skin, and 75% of the skin's dry matter, and helps to maintain the strength of the skin (Lai-Cheong and McGrath 2021). Proteoglycans (proteins that are heavily glycosylated) confer both viscosity and hydration to the dermis. Elastin is a protein in the extracellular matrix of connective tissue. Elastin (accounting for 5% of dry matter in the skin's dermis and 2–4% of the skin's dry matter) provides normal elasticity and flexibility in the skin. In other words, the elastin protein allows the skin to retain its firmness and return to its original position after being stretched. The dermis contributes nutritional support to the skin, immune responses, thermoregulation, and the sensation of physical contact.

The hypodermis (the innermost layer of the skin) consists of white adipose tissue (the primary tissue of the hypodermis) and connective tissue. The connective tissue contains fibroblasts, sweat glands, nerves, macrophages, lymphatic vessels, and

Table 6.1 Abundant amino acids in major proteins of the skin of mammals including dogs and cats[a]

Protein	Constituent in the skin and hair	Major amino acids (%, mol/mol)
Collagen[b]	About 70% of dry matter in the dermis (this layer is 90% of the skin's weight); 90% of total proteins in the skin; 75% of the skin's dry matter	Gly, 33; and Pro + Hydroxyproline, 23
Elastin	A physiologically important protein in the skin and hair; 10% of total proteins in the skin, 5% of dry matter in the skin's dermis, and 2–4% of the skin's dry matter; 5% of total proteins in the hair	Gly, 36; and Pro + Hydroxyproline, 15
Keratins[c]	About 90% of dry matter in the hair; 95% of epidermal proteins in the skin of dogs and cats; 80–90% of dry matter in the epidermis of the skin	Cys, 16; Ser, 11; and Gly 9
Filaggrin	4–5% of dry matter in the epidermis of the skin; 3–4% of dry matter in the hair	Ser, 20; Glu + Gln, 20; Gly, 17; Arg 12; and His, 8

[a]Hendriks et al. (1998) and Steinert et al. (1981)
[b]The most abundant protein in the skin and the whole body of mammals including dogs and cats
[c]The most abundant protein in the hair of mammals including dogs and cats
Note: Collagen fibers make up 90% of the skin, and elastin makes up 10%

Table 6.2 Composition of amino acids (mol/100 mol of residues) in the hair of mammals[a]

Amino acid	Cat	Dog	Human	Horse	Sheep
Alanine	5.1	5.1	4.6	5.5	5.2
Arginine	6.1	6.3	5.8	7.9	6.2
Asparagine	3.6	3.4	2.2	3.9	3.8
Aspartate	2.0	1.9	1.7	2.1	2.1
Asp + Asn[b]	5.6	5.3	4.9	6.0	5.9
Cysteine	15.8	16.7	17.8	14.4	13.1
Glutamate	6.5	6.3	6.5	6.4	6.3
Glutamine	4.9	4.8	4.9	4.9	4.8
Glu + Gln[c]	11.4	11.1	11.4	11.3	11.1
Glycine	9.5	7.8	6.4	6.4	8.6
Histidine	1.2	0.9	0.9	1.1	0.8
Isoleucine	2.5	2.5	2.6	3.6	3.0
Leucine	6.7	6.1	5.8	7.5	7.2
Lysine	2.9	3.9	2.7	2.9	2.7
Methionine	0.9	0.9	0.6	0.2	0.5
Phenylalanine	2.3	1.7	1.6	2.5	2.5
Proline	4.9	7.3	8.4	7.8	6.6
Serine	10.6	10.5	11.7	9.6	10.8
Threonine	6.4	6.2	6.8	5.5	6.5
Tyrosine	3.0	2.7	2.0	1.9	3.8
Valine	4.9	4.9	5.8	5.9	5.7

[a]Hendriks et al. (1998)
[b]The ratio (mol/mol) of aspartate to asparagine is 0.35:0.65 (Holt et al. 1971)
[c]The ratio (mol/mol) of glutamate to glutamine is 0.57:0.43 (Holt et al. 1971)

blood vessels. Collagen is the major protein in the hypodermis and is produced by fibroblasts. The hypodermis helps to store energy and connects the dermis layer of the skin to muscles. The hypodermis provides insulation, thermoregulation, and protection to the body.

6.4 Metabolism of AAs in the Skin and Hair of Dogs and Cats

Hair and skin tissues of mammals including cats and dogs cannot form de novo the carbon skeletons of citrulline, cysteine, histidine, isoleucine, leucine, lysine, methionine, phenylalanine, taurine, threonine, tryptophan, tyrosine, and valine (Wu 2022). Thus, all of these AAs are not synthesized de novo in the hair or skin (Table 6.3).

Note that the α-ketoacids of branched-chain AAs (BCAAs), methionine, phenylalanine, tryptophan, and tyrosine can be transaminated with glutamate to yield the corresponding α-AAs. In addition, cats have a very limited ability to synthesize (a) citrulline and arginine from glutamine/glutamate and proline; and (b) taurine from methionine and cysteine; and cannot form niacin from tryptophan (Wu 2022). Likewise, all breeds of dogs cannot synthesize adequately arginine to meet physiological needs due to a low activity of pyrroline-5-carboxylate synthase (Li and Wu 2023). Interestingly, a few breeds of dogs (e.g., the Newfoundland breed, golden retrievers, and American Cocker Spaniels) have a limited ability to synthesize taurine even when fed methionine- and cysteine-adequate diets (Backus et al. 2003; Kittleson et al. 1997) possibly due to lower hepatic activities of cysteine dioxygenase and cysteine sulfinate decarboxylase. To maintain nitrogen balance and survive, both cats and dogs must ingest the following AAs: arginine, histidine, isoleucine, leucine, lysine, methionine, phenylalanine, threonine, tryptophan, and valine (Wu 2022). Sufficient provision of other proteinogenic AAs is crucial for the optimal growth and health of both dogs and cats (Li and Wu 2023). At present, little is known about the metabolism of free AAs in the hair of animals including dogs and cats. Interestingly, protein citrullination, an irreversible post-translational enzymatic conversion of arginine residues to citrulline residues by peptidylarginine deiminases (also known as protein arginine deiminases), plays an important role in the growth, development, and function of the hair follicles (Méchin and Simon 2023). Thus, the diets of dogs and cats must contain adequate arginine for the integrity and health of the skin and hair, as well as for the growth of the hair.

The skin tissues of animals (including dogs and cats) can synthesize alanine, aspartate, asparagine, glutamate, glutamine, glycine, ornithine, proline, and serine via multiple metabolic pathways (Wu 2020). These synthetic pathways may also occur in the hair follicles. In addition, the skin tissues may form glycine from 4-hydroxyproline via the hydroxyproline oxidase pathway as reported for other tissues such as skeletal muscle, liver, and kidneys (Hu et al. 2023a, b), thereby recycling 4-hydroxyproline (a product of collagen degradation) into glycine for the production of new collagen molecules (Hu et al. 2022).

6.4.1 Metabolism of Traditionally Classified EAAs

Skin tissues of animals including dogs and cats can degrade the following AAs: arginine, histidine, isoleucine, leucine, phenylalanine, tryptophan, tyrosine, and valine (Wu 2020). BCAAs are degraded via BCAA transaminase and branched-chain α-ketoacid dehydrogenase. Because of the large mass of skeletal muscle, it is the major tissue to initiate BCAA catabolism in mammals. Phenylalanine is classified as a nutritionally essential AA (Dickinson et al. 2003; Rogers and Morris 1979) and plays an important role in biosynthetic processes, leading to melanin production when dietary tyrosine is limited (Watson et al. 2018). The breakdown of phenylalanine occurs via phenylalanine hydroxylase and either tyrosine transaminase or tyrosine

Table 6.3 De novo synthesis and degradation of amino acids in the skin

AA	De novo synthesis	Degradation
Alanine	Yes	Yes
β-Alanine	No	No
Arginine[a]	Yes	Yes
Aspartate	Yes	Yes
Asparagine	Yes	?
BCAAs	No	Yes
Citrulline	No	Yes
Cysteine	No	?
GABA	No	?
Glutamate	Yes	Yes
Glutamine	Yes	Yes
Glycine	Yes	Yes
Histidine	No	Yes
Lysine	No	Yes
Methionine[b]	No	Yes
Phenylalanine	No	Yes
Proline	Yes	Yes
OH-Pro	Yes	Yes
Ornithine	Yes	Yes
Serine	Yes	Yes
Taurine	No	No
Threonine	No	?
Tryptophan	No	Yes
Tyrosine[c]	Yes	Yes

[a]Synthesis from citrulline
[b]Conversion of methionine into S-adenosylmethionine
[c]Formation of tyrosine from phenylalanine by phenylalanine hydroxylase (a tetrahydrobiopterin-dependent enzyme)
BCAAs, branched-chainamino acids (leucine,isoleucine and valine); GABA, γ-aminobutyrate; OH-Pro, 4-hydroxyproline plus 3-hydroxyproline; ?, unknown.

hydroxylase (Wu 2022). Following its breakdown, phenylalanine is utilized for the maintenance of melanin in the hair (Watson et al. 2018; Morris et al. 2002; Anderson et al. 2002). Overall, the majority of phenylalanine functions in the skin and hair of dogs and cats relate to its conversion to tyrosine via the phenylalanine hydroxylase system (Dickinson et al. 2004). Lysine is another EAA important in the skin (Solano et al. 2020). Its protein-bound derivatives (desmosine and isodesmosine), which are formed by lysyl oxidase, are vital for cross-links in collagen and elastin in the skin (Solano et al. 2020). In addition to these AAs, histidine is important for skin health due to its part in maintaining stratum corneum moisture (Solano et al. 2020). Through the process of deamination, histidine is converted into urocanic acid which plays a role in maintaining the acidic pH in the stratum corneum and preventing fungal and bacterial growth in the skin (Solano et al. 2020). Lastly, arginine (an EAA for both cats and dogs) helps the skin heal its wounds (Solano et al. 2020; Stechmiller et al. 2005) through the tetrahydrobiopterin- and NADPH-dependent production of nitric oxide (NO) by NO synthase; NO plays a central role in modulating vascular homeostasis, inflammation, and antimicrobial response that are the three major components of the wound healing process (Solano et al. 2020; Childress and Stechmiller 2002; Kim et al. 2012; Alexander and Supp 2014). In addition, NO regulates blood flow and nutrient supplies to the skin and hair follicles (Li et al. 2009). Cats and dogs have a limited ability to synthesize de novo arginine as noted previously, so this AA is required to be part of their dietary intake, or their health will suffer (Wu 2022). In contrast, skin and hair tissues can readily convert L-citrulline into L-arginine in the presence of aspartate via argininosuccinate synthase and lyase (Wu 2020). Thus, dietary citrulline can effectively replace dietary arginine in both dogs and cats.

Two important sulfur-containing AAs (SAAs) in the development and maintenance of the hair and skin are methionine and cysteine (Harrison et al. 2020). Methionine is the precursor for cysteine in the mammalian liver (Harrison et al. 2020). Thus, cysteine is capable of fulfilling ~ 50% of a mammal's dietary requirement for methionine (Wu 2022). However, there is no conversion of methionine into cysteine in the skin or hair. In animals including cats and dogs, methionine and cysteine play important roles in structural proteins of the hair shaft and are essential for the development and maintenance of the hair and skin (Sato et al. 2019). In addition, cysteine is capable of facilitating cysteinyl-DOPA formation for pheomelanin production (Watson et al. 2018). Because of cysteine's important role in melanin biosynthesis, the larger the quantity of cysteine, the more the hair will show red/yellow pigmentation instead of black (Watson et al. 2018; Ito 1993). This observation highlights the importance of feeding cats and dogs with black coats the correct ratio of cysteine to other AAs, thereby affecting melanin biosynthesis; otherwise, their black coats will be tainted with reddish hairs. At present, there is not much information about the dietary requirements of methionine plus cysteine during the life cycle of companion animals mainly due to dependency on the provision of other AAs (Harrison et al. 2020). Moreover, taurine is another SAA that is important as an antioxidant, osmolyte, and a regulator of gene expression and metabolism within the mammalian body. Notably, taurine synthesis is much more limited in cats than in dogs due to a lower activity of cysteine dioxygenase in the former (Wu 2022). In this case, taurine should be supplemented to make up for the insufficiency of de novo synthesis. Taurine is essential to the function and health of the skin and hair, and is formed from methionine plus serine, crucial AAs for the growth and development of the skin and hair. In addition, taurine is a physiological regulator of gene expression and cell signaling to optimize tissue protein synthesis and modulate hepatic methionine oxidation (Harrison et al. 2020). Because of this, taurine can influence methionine availability, hence impacting skin and hair maintenance.

6.4.2 Metabolism of Traditionally Classified NEAAs

The skin of animals including dogs and cats can degrade glutamine via the phosphate-activated glutaminase (a mitochondrial enzyme) and glutamine:fructose-6-phosphate transaminase (a cytosolic enzyme) pathways; alanine, glutamate, and aspartate via transaminases (both cytosolic and mitochondrial); proline via proline oxidase (a mitochondrial enzyme); and ornithine via ornithine aminotransferase (a mitochondrial enzyme) and ornithine decarboxylase (a cytosolic enzyme) (Wu 2022). In addition, 4-hydroxyproline is a derivative of the AA proline (Hu et al. 2022). Proline hydroxylation within collagen generates 4-hydroxyproline and 3-hydroxyproline (Hu et al. 2022). 4-Hydroxyproline is important for the structure and function of collagen proteins in mammals (Wu 2022). In addition to its role in collagen, elastin contains a large amount of 4-hydroxyproline (Hu et al. 2022; Li and Wu 2018; Solano 2020). Note that the proline derivatives, 4-hydroxyproline and 3-hydroxyproline, are synthesized with proline residues found in the collagen and elastin by a vitamin C and iron-dependent post-translational hydroxylation (Hu et al. 2022). It is now known that animals including dogs and cats have dietary requirements for NEAAs (Wu and Li 2022; Li and Wu 2023).

6.4.3 Conditionally Essential AAs

CEAAs are AAs that need to be supplemented for optimal conditions in animals such as growth, development, reproduction, lactation, and health under circumstances when the normal rate of synthesis is not enough to keep up with the rate of utilization (Wu 2022). Under this definition, tyrosine, cysteine, serine, glutamine, proline, and glycine are classified as CEAAs in dogs and cats that are essential for the production of major proteins in the skin and hair (Wu 2022). Tyrosine is an aromatic AA with special significance in the companion animals. Without a sufficient dietary intake of tyrosine, there is a limited amount of dopaquinone available for melanin synthesis (Yu et al. 2001). Eumelanin and phaeomelanin, which are two pigments of melanin, are synthesized from dopaquinone (Yu et al. 2001). The tyrosine used in melanin synthesis may derive from the hydroxylation of phenylalanine or from tyrosine in the blood (Anderson et al. 2002). In total, tyrosine can account for up to 50% of growing kitten and puppies' aromatic AA requirements (Baker and Czarnecki-Maulden 1991; Anderson et al. 1980; Milner et al. 1984). Other CEAAs are glycine and proline that are crucial for the structure and integrity of the skin of mammals such as dogs and cats, and they are the main AAs present in collagen and elastin proteins of the skin (Wu 2022). In addition, cysteine and serine play important roles in the synthesis of keratins and the development of the hair as well (Li and Wu 2023). Furthermore, the main AAs in the feline hair also include glutamate, proline and glycine; the AA composition of dog hair is similar to cats with the exception of proline possibly being lower in the cat's hair (Hendriks et al. 1998). Glycine, which is one of three AAs that constitute the majority of collagen (Li and Wu 2023), can be formed from both serine and 4-hydroxyproline (Wu 2022).

6.4.4 *Creatine*

Another important substance in the skin is creatine. As a metabolically active organ, the skin has high rates of protein turnover (including collagen synthesis and degradation) and energy metabolism (Solano 2020; Wu 2020). Similarly, hair growth requires energy. In the skin and hair keratinocytes, creatine kinase is expressed for storing energy as phosphocreatine, and phosphocreatine is converted into ATP and creatine in response to physiological needs of animals for energy. In mammals (including dogs and cats), creatine is synthesized from arginine, glycine, and methionine through the cooperation of the kidneys, liver, and pancreas (Wu 2022). However, as, compared with young adults, creatine synthesis is reduced in aging animals, including cats and dogs. Thus, adequate availability of creatine in the body is crucial for skin health (including protection from oxidative and ultraviolet damage) and hair quality (Lenz et al. 2005; Peirano et al. 2011). Note that, like taurine, creatine is present in animal-sourced foodstuffs but is absent from plant-sourced foodstuffs.

6.5 Functions of AAs in Skin and Hair of Dogs and Cats

6.5.1 *SAAs*

In addition to SAA's crucial roles in skin maintenance, methionine and cysteine are important for the development of hair (Oberbauer and Larsen 2021). It was reported that a deficiency or excess of methionine and cysteine available for hair protein synthesis led to the cat's hair taking on a brittle appearance (Sato et al. 2019; Zhang et al. 2013). Due to the greater number of hairs on a cat or dog, they require more cysteine in their diets for hair uptake, as compared to mammals with less hair (Li and Wu 2023). Both dogs and cats have basically the same AA requirements for the hair (Hendriks et al. 1998). Since cysteine is the most prevalent AA in the hair of dogs and cats, it is essential for the structure, growth, and development of their hair (Li and Wu 2023). One study reported that the methionine and cysteine requirement in kittens is between 301 and 664 mg/kg BW, dependent on the kitten's age (Sato et al. 2020).

The exact functional role of taurine in the skin is unknown, but taurine has been identified in the keratinocytes of the granular layer of skin in dogs and rats (Lobo et al. 2001). In fact, taurine was found in high concentrations in the epidermis (including the upper spinous layer) of the skin (Lobo et al. 2001). Because there are high levels of cysteine present in the skin, it has been hypothesized that taurine may play a role in regulating cutaneous cysteine metabolism, as taurine is known to be synthesized fromcysteine in the liver and to function as an osmolyte, an antioxidant, and a protector of tissue integrity (Lobo et al. 2001; Blackburn 1961).

6.5.2 Aromatic AAs

The aromatic AAs, tyrosine and phenylalanine, have an important impact on synthesis of hair pigment (Oberbauer and Larsen 2021). Melanin is a special product of AA metabolism in the skin. In the epidermis and dermis of the skin, melanocytes oxidize phenylalanine and tyrosine into Dopa. Dopa is then converted into dopaquinone. Dopaquinone is further oxidized to eumelanin (dark-brown pigment). Dopaquinone can also combine with cysteine to produce cysteinyl-dopa, which is subsequently oxidized to benzothiazine and then further oxidized to pheomelanin (a brown–red pigment). Mixed melanins are formed from 5,6-dihydroxyindole and benzothiazine. The type and amount of melanin determine the color of the skin and hair in dogs and cats.

There are recommendations that the diet of a kitten should contain at least 8.5 g/kg of phenylalanine and tyrosine and at least half of this requirement should be fulfilled by phenylalanine (Dickinson et al. 2004; NRC 2006). MacDonald et al. (1984) had determined that a kitten requires about 0.8–0.9% phenylalanine and tyrosine in diets to maintain healthy. Another study demonstrated the importance of dietary tyrosine and phenylalanine in black-coated cats by observation of their coats (Anderson et al. 2002). Sufficient amounts of dietary tyrosine and phenylalanine ($\geq$ 18 g Phe + Tyr per kg diet) are needed to prevent the appearance of visually discernible red hair in black-coated cats. This feline study also revealed that a deficiency in dietary tyrosine and phenylalanine (< 16 g Phe + Tyr per kg diet) resulted in the change of hair color from black to reddish brown, indicating limited melanogenesis (Anderson et al. 2002). Even though a deficiency in tyrosine induces this coat discoloration, it was found that these effects could easily be reversed by re-implementing higher levels of tyrosine and phenylalanine back in the diet (Morris et al. 2002). Dogs with black hair also exhibited these symptoms when fed a phenylalanine- and tyrosine-deficient diet (Watson et al. 2015), confirming that dietary intakes of both aromatic AAs affect the proportions of eumelanin (brown–black) to pheomelanin (reddish-brown) that are synthesized in follicular melanocytes. In further support of this notion, the dogs fed diets supplemented with phenylalanine and tyrosine visually had more black pigmentation in their hair than those fed the deficient diets (Watson et al. 2018). It has also been noted that not only coat color, but also coat patterning, is affected by the amount of eumelanin and pheomelanin produced by melanocytes (Watson et al. 2017; Ortonne and Prota 1993). This directly ties in with AA intake since it has been determined that tyrosine is a key component in the process of melanogenesis (Watson et al. 2017). An increase in tyrosine intake stimulates tyrosinase activity, therefore, leading to more eumelanin production in the hair (Watson et al. 2018; Slominski 1989). Phenylalanine and tyrosine deficiencies are much more apparent in darker colored animals, so studies have primarily used black dogs and cats as models. Nonetheless, there has been a note of similar results for orange-coated cats as well (Yu et al. 2001). Unfortunately, these results were not presented, because the primary focus of the study was on black kittens. It was observed that the change was not as noticeable in orange cats as in black cats, but the orange cats did turn a paler orange when given an AA deficient diet (Yu et al. 2001). This result indicates the need for studies in future to also focus on if lighter-colored kittens and puppies would benefit from supplemental tyrosine and phenylalanine to support optimal melanin production.

6.5.3 Roles of Dietary Glycine, Proline, Serine, and SAAs in Health of the Skin and Hair

Collagen has high concentrations of glycine, proline, and 4-hydroxyproline (Li and Wu 2018). Both glycine and proline are CEAAs for dogs and cats (Li and Wu 2023), with glycine, proline, and 4-hydroxyproline being potent antioxidants and cell signaling molecules (Hu et al. 2022; Wang et al. 2014; Wu et al. 2011) to maintain the health of bone and joints. Due to inadequate syntheses in the body, these AAs must be provided in diets to support the deposition of collagen and keratin in the skin and hair of growing animals including dogs and cats and to maintain the integrity and health of their connective tissues during adulthood. In addition, serine is the second most abundant AA in hair proteins, and may not be adequately synthesized by dogs or cats. Notably, proline, hydroxyproline, cysteine, serine, glutamate, and glutamine had traditionally been classified as NEAAs for dogs and cats (NRC 2006). However, this term is now recognized as a misnomer in nutritional sciences (Hou and Wu 2017). Dogs and cats have dietary requirements for these NEAAs for optimum growth and health (Li and Wu 2023). Furthermore, the skin of animals including dogs and cats needs a plethora of nutrients (including AAs, polyunsaturated fatty acids, vitamins, and minerals) for maintaining whole-body homeostasis and health (Outbridge and Owens 2023). For example, methionine not only participates in one-carbon metabolism but is also required for the synthesis of collagen and keratin in the body (Sato et al. 2020). Keratin is another important protein in the skin which is comprised of mainly cysteine and serine, as noted previously. The keratinization of the skin and the normal growth of hair may account for 25–30% of the daily protein requirements by cats and dogs (Scott et al. 1995; Watson 1998). The aromatic AA, tyrosine, is influential in the skin as well as the hair due to its function in melanogenesis. Melanin is also found in the skin. Because of this, tyrosine has an important role in skin pigment (Zoran 2002), and many studies focus on the effects of tyrosine on hair and possibly skin color. Collectively, AAs in diets, particularly arginine, cysteine, glutamate, glutamine, glycine, lysine, proline, hydroxyproline, serine, taurine, and tyrosine, are crucial for the growth and health of the skin and hair.

A deficiency in SAAs, particularly methionine and cysteine, can lead to dermatitis of the mucocutaneous skin surrounding the mouth and nose of cats (Che et al. 2021). It has also been reported in dogs that a deficiency of AAs can lead to dermatological lesions (Oberbauer and Larsen 2021). Based on this data, it is crucial that cats and dogs receive adequate AA intake in their diets for the maintenance of healthy skin. Interestingly, Hirakawa and Baker discovered that AA imbalances due to inappropriate ratios of SAAs in diets can lead to skin lesions as well (1985). When cysteine accounted for 66–75% of the total SAAs in the puppy's diet, necrosis of the skin began to appear, indicating that proper AA balance is essential (Hirakawa and Baker 1985). Watson (1998) reported that a nutritionally inadequate diet in dogs and cats led to symptoms correlated with dermatosis, including greasy skin and excessive scaling of the skin. In addition to this observation, Watson stated that cats and dogs fed diets deficient in methionine and cysteine manifest symptoms such as the thin, inelastic, and hyperpigmented skin (Watson 1998). These studies came to similar conclusions—that a deficiency or even an excess of one AA can negatively impact the skin of dogs and cats. Thus, a deficiency of dietary protein in both dogs and cats can result in pale, dry, thin, inelastic, and flaking skin; skin atrophy; the loss and impaired growth of hair; brittle and depigmented hair; and the appearance of gray hair color (Scott et al. 1995; Che et al. 2021).

Growing dogs and cats gain protein in the body and therefore require more dietary AAs than adult dogs. Adult dogs and cats need dietary AAs for the maintenance of nitrogen balance, as well as hair growth. It is likely that elderly adults require more dietary AAs than young adults. However, to date, there are no data on the dietary requirements of elderly dogs or cats for dietary AAs. There is evidence that elderly adult cats require more arginine, tryptophan, glutamate, and glycine than young adult cats (Li and Wu 2023). In both dogs and cats, adequate intakes of dietary AAs can help to mitigate the loss of muscle protein and alleviate skin atrophy during the aging process. AAs that are synthesized de novo must also be included in the diets of dogs and cats to support the maximum growth and optimum health of their skin and hair.

6.6 Provision of AAs from Animal-Sourced Foodstuffs

Animal-sourced foodstuffs (e.g., feather meal and poultry by-product meal) are excellent sources of the AAs that are crucial to maintain the normal structure and health of the skin and hair in dogs and cats. (Li and Wu 2020, 2022; Li et al. 2021; Wu et al. 2011). For example, feather meal contains the highest content of the following AAs among all common animal protein feedstuffs (% of foodstuff): proline, 12.2; glycine, 9.36; serine, 9.31; arginine, 6.09; 4-hydroxyproline, 5.19; and cysteine, 4.35. In addition, blood meal, spray-dried poultry plasma, and egg products are rich in arginine, cysteine, proline, and serine. Furthermore, meat-bone meal, chicken visceral digest, and poultry by-product meal (pet-food grade) are rich in glycine, 4-hydroxyproline, and taurine.

Animal-sourced foodstuffs, particularly feather meal and poultry are useful for balancing AAs (e.g., glycine, proline, methionine, cysteine, lysine, threonine, and tryptophan) and provide other functional nutrients (e.g., taurine, creatine, and 4-hydroxyproline) in the diets of dogs and cats to maintain the normal structure and health of their skin and hair (Wu 2022; Yao et al. 2011). Inclusion levels of individual animal-sourced foodstuffs in the diets of dogs and cats are 10–30%, depending on the sources, AA composition, and palatability of ingredients, as well as the physiological needs of the companion animals. A mixture of animal-sourced foodstuffs can be used for the diets of dogs and cats to bring about complementary effects in their nutrition.

6.7 Future Research Direction

Dogs and cats have the potential to develop a disorder called superficial necrolytic dermatitis (SND). SND is an uncommon metabolic condition in dogs and even less common in cats that manifests as skin lesions (Loftus et al. 2022). Canine patients with SND exhibit hypoaminoacidemia and aminoaciduria, and severe deficiencies of cysteine, proline, and 4-hydroxyproline have been thought to cause severe skin lesions (Loftus et al. 2022). Dogs given IV infusions consistently of AAs including proline ultimately survived longer after their SND diagnosis than those not given the infusions (Loftus et al. 2022). In addition, dogs with hepatocutaneous syndrome (HPS), a condition similar to SND, displayed generalized hypoaminoacidemia (Loftus et al. 2017). In dogs with HPS, plasma AA analysis revealed a severe depletion in cysteine, proline and hydroxyproline: three AAs crucial to the development and maintenance of skin (Loftus et al. 2017). Both HPS and SND patients showcase the importance of AA balance in the skin (Loftus et al. 2017). Although these are not commonly seen disorders of the skin, plasma AA concentrations can help determine what AAs are responsible for different aspects of skin health. Dogs and cats with these kinds of metabolic disorders would benefit from additional AA treatments to lessen their symptoms. Research focused on what cocktail of AAs would be most beneficial could make SND and HPS patients more comfortable.

In addition to the foregoing, a newfound concern of dog and cat owners within the last decade about pet food is how to improve their pet's nutrition. Increasing skin and hair quality through optimizing the intake of dietary AAs has great potential to satisfy consumers by providing a visual result to indicate the nutritional status of dogs and cats. Present studies on AA deficiencies in the skin and hair of dogs and cats are focused primarily on growing animals. To expand this area, new research on more age groups of animals such as adult and senior dogs and cats would greatly increase knowledge on how different age groups respond to a specific AA deficiency. Overall, there are many studies focused on symptoms manifesting in the skin and hair based on phenylalanine, tyrosine, methionine, or cysteine deficiencies. Studies on other AAs relevant to the skin and hair such as glycine, proline, and histidine have not received much attention, as compared to the AAs mentioned previously. More research would allow for insight into how symptoms correlated with AA deficiency can be prevented with proper AA nutrition.

6.8 Conclusion

The skin and hair are specialized connective tissues in mammals including dogs and cats. The hair grows from the hair follicle embedded in the dermis layer of the skin. Proper AA nutrition is essential for both dogs and cats. It is critical for the health, growth, and development of hair and skin that diets include complete EAAs and NEAAs to avoid detrimental effects. AA deficiencies can become apparent in a dog or cat's appearance. An increase in hair discoloration, skin fragility, or skin hyperpigmentation can all indicate a deficiency in AA intake. Without the proper ratios of AAs in the diets of dogs and cats, their health will also begin to deteriorate. This indicates the importance of AAs in the integrity and function of skin and hair. Thus, diets must provide all proteinogenic AAs and taurine in adequate amounts and proper ratios to meet the requirements of dogs and cats for their maintenance, as well as their optimum growth, reproduction, lactation, and health (including the health of the skin and hair). Because proteins in the skin and hair contain large amounts of arginine, cysteine, glutamate, glutamine, glycine, proline, 4-hydroxyproline, serine, and tyrosine, a sufficient provision of all these AAs is crucial for the normal structure and health of the skin and hair. Animal-sourced foodstuffs (e.g., feather meal and poultry by-product meal) are abundant sources of these AAs in canine and feline diets. Maintaining healthy skin and hair is crucial for the health and well-being of both dogs and cats.

Acknowledgements This work was supported by Texas A&M AgriLife Research (H-8200). We thank Dr. Peng Li (North American Renderers Association, Alexandria, VA, USA) for helpful discussions.

Conflict of Interest The authors have no conflicts of interest to declare.

References

Alexander JW, Supp DM (2014) Role of arginine and omega-3 fatty acids in wound healing and infection. Adv Wound Care 3:682–690

Anderson PJB, Rogers QR, Morris JG (2002) Cats require more dietary phenylalanine or tyrosine for melanin deposition in hair than for maximal growth. J Nutr 132:2037–2042

Anderson PA, Baker DH, Sherry PA, Corbin JE (1980) Histidine, phenylalanine-tyrosine and tryptophan requirements for growth of the young kitten. J Anim Sci 50:479–483

Baker KP (1974) Hair growth and replacement in the cat. Br Vet J 130:327–335

Baker DH, Czarnecki-Maulden GL (1991) Comparative nutrition of cats and dogs. Annu Rev Nutr 11:239–263

Backus RC, Cohen G, Pion PD, Good KL, Rogers QR, Fascetti AJ (2003) Taurine deficiency in Newfoundlands fed commercially available complete and balanced diets. J Am Vet Med Assoc 223:1130–1136

Blackburn S (1961) The chemical composition of keratins and skin proteins. In: Biochemists Handbook (Long C, King EJ, Sperry WM eds). D. Van Nostrand Co, Princeton, NJ, 701–705

Case LP, Daristotle L, Hayek MG, Raasch MF (2011) Canine and feline nutrition, 3rd edn. Mosby, Maryland Heights, Missouri

Che D, Nyingwa PS, Ralinala KM, Maswanganye GM, Wu G (2021) Amino acids in the nutrition, metabolism, and health of domestic cats. Adv Exp Med Biol 1285:217–231

Childress BB, Stechmiller JK (2002) Role of nitric oxide in wound healing. Biol Res Nurs 4:5–15

Collin C, Moll R, Kubicka S, Ouhayoun JP, Franke WW (1992) Characterization of human cytokeratin 2, an epidermal cytoskeletal protein synthesized late during differentiation. Exp Cell Res 202:132–141

Devlin TM (2006) Textbook of biochemistry with clinical correlations. Wiley-Liss Press, Hoboken, New Jersey

Dickinson PJ, Anderson PJ, Williams DC, Powell HC, Shelton GD, Morris JG, LeCouteur RA (2004) Assessment of the neurologic effects of dietary deficiencies of phenylalanine and tyrosine in cats. Am J Vet Res 65:671–680

Gunaratnam P, Wilkinson GT (1983) A study of normal hair growth in dogs. J Small Anim Pract 24:445–453

Harrison M, Thomas G, Gilham M, Gray K, Colyer A, Allaway D (2020) Short-term determination and long-term evaluation of the dietary methionine requirement in adult dogs. Br J Nutr 123:1333–1344

Hendriks WH, Tarttelin MF, Moughan PJ (1997) Seasonal hair growth in the adult domestic cat (*Felis catus*). Comp Biochem Physiol A 116:29–35

Hendriks WH, Tarttelin MF, Moughan PJ (1998) The amino acid composition of cat (*Felis catus*) hair. Anim Sci 67:165–170

Hirakawa DA, Baker DH (1985) Sulfur amino acid nutrition of the growing puppy: determination of dietary requirements for methionine and cystine. Nutr Res 5:631–642

Holt LA, Milligan B, Roxburgh CM (1971) Aspartic acid, asparagine, glutamic acid, and glutamine contents of wool and two derived protein fractions. Aust J Biol Sci 24:509–514

Hou YQ, Wu G (2017) Nutritionally nonessential amino acids: A misnomer in nutritional sciences. Adv Nutr 8:137–139

Hu SD, He WL, Wu G (2022) Hydroxyproline in animal metabolism, nutrition, and cell signaling. Amino Acids 54:513–528

Hu SD, He WL, Bazer FW, Johnson GA, Wu G (2023a) Synthesis of glycine from 4-hydroxyproline in tissues of neonatal pigs. Exp Biol Med 248:1206–1220

Hu SD, He WL, Bazer FW, Johnson GA, Wu G (2023b) Synthesis of glycine from 4-hydroxyproline in tissues of neonatal pigs with intrauterine growth restriction. Exp Biol Med 248:1446–1458

Ito S (1993) High-performance liquid chromatography (HPLC) analysis of eu-and pheomelanin in melanogenesis control. J Invest Dermatol 100:S166–S171

Kim H, Lim YJ, Park JH, Cho Y (2012) Dietary silk protein, sericin, improves epidermal hydration with increased levels of filaggrins and free amino acids in NC/Nga mice. Br J Nutr 108:1726–1735

Kittleson MD, Keene B, Pion PD, Loyer CG (1997) Results of the multicenter spaniel trial (MUST): taurine- and carnitine-responsive dilated cardiomyopathy in American cocker spaniels with decreased plasma taurine concentration. J Vet Intern Med 11:204–211

Lai-Cheong JE, McGrath JA (2021) Structure and function of skin, hair and nails. Medicine 49:337–342

Lenz H, Schmidt M, Welge V, Schlattner U, Wallimann T, Elsasser HP, Wittern KP, Wenck H, Stab F, Blatt T (2005) The creatine kinase system in human skin: protective effects of creatine against oxidative and UV damage in vitro and in vivo. J Invest Dermatol 124:443–452
Li P, Wu G (2018) Roles of dietary glycine, proline, and hydroxyproline in collagen synthesis and animal growth. Amino Acids 50:29–38
Li P, Wu G (2020) Composition of amino acids and related nitrogenous nutrients in feedstuffs for animal diets. Amino Acids 52:523–542
Li P, Wu G (2022) Functional molecules of intestinal mucosal products in animal nutrition and health. Adv Exp Med Biol 1354:263–277
Li P, Wu G (2023) Amino acid nutrition and metabolism in domestic cats and dogs. J Anim Sci Biotechnol 14:19
Li P, He WL, Wu G (2021) Composition of amino acids in foodstuffs for humans and animals. Adv Exp Med Biol 1332:189–210
Li XL, Bazer FW, Gao H, Jobgen W, Johnson GA, Li P, McKnight JR, Satterfield MC, Spencer TE, Wu G (2009) Amino acids and gaseous signaling. Amino Acids 37:65–78
Lobo MV, Alonso FJ, Latorre A, Martin del Rio R (2001) Taurine levels and localization in the stratified squamous epithelia. Histochem Cell Biol 115:341–347
Loftus JP, Center SA, Lucy JM, Stanton JA, McDonough SP, Peters-Kennedy J et al (2017) Characterization of aminoaciduria and hypoaminoacidemia in dogs with hepatocutaneous syndrome. Am J Vet Res 78:735–744
Loftus JP, Miller AJ, Center SA, Peters-Kennedy J, Astor M (2022) Treatment and outcomes of dogs with hepatocutaneous syndrome or hepatocutaneous-associated hepatopathy. J Vet Intern Med 36:106–115
Ma C, Cheng M, Wu Y, Xu X (2024) The role of mesenchymal stem cells in hair regeneration and hair cycle. Stem Cells Dev 33:1–10
MacDonald ML, Rogers QR, Morris JG (1984) Nutrition of the domestic cat, a mammalian carnivore. Annu Rev Nutr 4:521–562
Marshall RC (1986) Nail, claw, hoof and horn keratin. In: Biology of the integument: 2 Vertebrates. Berlin, Heidelberg: Springer Berlin Heidelberg, pp 722–738
Méchin MC, Simon M (2023) Deimination in epidermal barrier and hair formation. Philos Trans R Soc Lond B Biol Sci 378(1890):20220245
Milner JA, Garton RL, Burns RA (1984) Phenylalanine and tyrosine requirements of immature beagle dogs. J Nutr 114:2212–2216
Moriello KA (2022a) Structure of the skin in dogs. Merck Manual. https://www.merckvetmanual.com/dog-owners/skin-disorders-of-dogs/structure-of-the-skin-in-dogs. Accessed 15 Nov 2023
Moriello KA (2022b) Structure of the skin in cats. Merck Manual. https://www.merckvetmanual.com/cat-owners/skin-disorders-of-cats/structure-of-the-skin-in-cats. Accessed 15 Nov 2023
Morris JG, Yu S, Rogers QR (2002) Red hair in black cats is reversed by addition of tyrosine to the diet. J Nutr 132:1646S-1648S
Mundt H-C, Stafforst C (1987) Production and composition of dog hair. In: Nutrition, malnutrition and dietetics in the dog and cat (Edney ATB ed). British Veterinary Association, London, UK, pp 62–65
National Research Council (NRC, 2006) Nutrient requirements of dogs and cats. National Academies Press, Washington DC
Oberbauer AM, Larsen JA (2021) Amino acids in dog nutrition and health. Adv Exp Med Biol 1285:199–216
Ortonne JP, Prota G (1993) Hair melanins and hair color: ultrastructural and biochemical aspects. J Invest Dermatol 101:82S-89S
Outerbridge CA, Owens TJ (2023) Nutritional management of skin diseases. In: Fascetti AJ, Delaney SJ, Larsen JA, Villaverde C (eds) Applied veterinary clinical nutrition. Wiley, Hoboken, New Jersey
Peirano RI, Achterberg V, Dusing HJ, Akhiani M, Koop U, Jaspers S, Kruger A, Schwengler H, Hamann T, Wenck H et al (2011) Dermal penetration of creatine from a face-care formulation

containing creatine, guarana and glycerol is linked to effective antiwrinkle and antisagging efficacy in male subjects. J Cosmet Dermatol 10:273–281
Rogers QR, Morris JG (1979) Essentiality of amino acids for the growing kitten. J Nutr 109:718–723
Sato J, Merenda MEZ, Uemoto AT, Dos Santos MP, Barion MRL, Carciofi AC, de Paula Dorigam JC, Ribeiro LB, Vasconcellos RS (2020) Modelling of sulphur amino acid requirements and nitrogen endogenous losses in kittens. Arch Anim Nutr 74:414–427
Sato J, Ogawa CYL, Sandrini M, Medina AN, Sato F, Vasconcellos RS (2019) Study of keratin hair of domestic cat under methionine and cystine experimental diet using FT-Raman spectroscopy. Vib Spectrosc 100:1–5
Scott DW, Miller WH, Griffin CE (1995) Nutritional skin diseases. Muller and Kirk's small animal dermatology. W.B. Saunders Co., Philadelphia, PA, pp 891–901
Slominski A (1989) L-tyrosine induces synthesis of melanogenesis related proteins. Life Sci 45:1799–1803
Solano F (2020) Metabolism and functions of amino acids in the skin. Adv Exp Med Biol 1265:187–199
Stechmiller JK, Childress B, Cowan L (2005) Arginine supplementation and wound healing. Nutr Clin Pract 20:52–61
Steinert P, Cantieri J, Teller D, Lonsdale-Eccles J, Dale B (1981) Characterization of a class of cationic proteins that specifically interact with intermediate filaments. Proc Natl Acad Sci USA 78:4097–4101
Wang WW, Wu ZL, Lin G, Hu SD, Wang B, Dai ZL, Wu G (2014) Glycine stimulates protein synthesis and inhibits oxidative stress in pig small-intestinal epithelial cells. J Nutr 144:1540–1548
Watson A, Le Verger L, Guiot AL, Feugier A, Biourge V (2017) Nutritional components can influence hair coat colouration in white dogs. J Appl Anim Nutr 5:e5
Watson A, Servet E, Hervera M, Biourge VC (2015) Tyrosine supplementation and hair coat pigmentation in puppies with black coats—a pilot study. J Appl Anim Nutr 3:e10
Watson A, Wayman J, Kelley R, Feugier A, Biourge V (2018) Increased dietary intake of tyrosine upregulates melanin deposition in the hair of adult black-coated dogs. Anim Nutr 4:422–428
Watson TDG (1998) Nutrition and skin disease in dogs and cats. J Nutr 128(Suppl 12):2783S-2789S
Wu G (2022) Amino acids: biochemistry and nutrition. CRC Press, Boca Raton, Florida
Wu G (2020) Metabolism and functions of amino acids in sense organs. Adv Exp Med Biol 1265:201–217
Wu G, Li P (2022) The "ideal protein" concept is not ideal in animal nutrition. Exp Biol Med 247:1191–1201
Wu G, Bazer FW, Burghardt RC, Johnson GA, Kim SW, Knabe DA, Li P, Li XL, McKnight JR, Satterfield MC, Spencer TE (2011) Proline and hydroxyproline metabolism: implications for animal and human nutrition. Amino Acids 40:1053–1063
Yao K, Yin YL, Feng ZM, Tang ZR, Fang J, Wu G (2011) Tryptophan metabolism in animals: important roles in nutrition and health. Front Biosci S3:286–297
Yoshimura T, Inokuchi Y, Mutou C, Sakurai T, Nagahama T, Murakami S (2021) Age-related decline in the taurine content of the skin in rodents. Amino Acids 53:429–434
Yu S, Rogers QR, Morris JG (2001) Effect of low levels of dietary tyrosine on the hair colour of cats. J Small Anim Pract 42:176–180
Zhang HH, Jiang QK, Sun WL, Xu C, Cong B, Yang FH, Li GY (2013) Effects of different dietary protein levels and DL-methionine supplementation on hair growth and pelt quality in mink (Neovision vision). J Anim Physiol Anim Nutr 97:1036–1042
Zoran DL (2002) The carnivore connection to nutrition in cats. J Am Vet Med Assoc 221:1559–1567

Chapter 7
Functions and Metabolism of Amino Acids in Bones and Joints of Cats and Dogs

Arianna N. Lopez, Fuller W. Bazer, and Guoyao Wu

Abstract The bone is a large and complex organ (12–15% of body weight) consisting of specialized connective tissues (bone matrix and bone marrow), whereas joints are composed of cartilage, tendons, ligaments, synovial joint capsules and membranes, and a synovial joint cavity filled with synovial fluid. Maintaining healthy bones and joints is a dynamic and complex process, as bone deposition (formation of new bone materials) and resorption (breakdown of the bone matrix to release calcium and phosphorus) are the continuous processes to determine bone balance. Bones are required for locomotion, protection of internal organs, and have endocrine functions to maintain mineral homeostasis. Joints are responsible for resisting mechanical stress/trauma, aiding in locomotion, and supporting the overall musculoskeletal system. Amino acids have multiple regulatory, compositional, metabolic, and functional roles in maintaining the health of bones and joints. Their disorders are prevalent in mammals and significantly reduce the quality of life. These abnormalities in companion animals, specifically cats and dogs, commonly lead to elective euthanasia due to the poor quality of life. Multiple disorders of bones and joints result from genetic predisposition and are heritable, but other factors such as nutrition, growth rate, trauma, and physical activity affect how the disorder manifests. Treatments for cats and dogs are primarily to slow the progression of these disorders and assist in pain management. Therapeutic supplements such as Cosequin and formulated diets rich in amino acids are used commonly as treatments for companion animals to reduce pain and slow the progression of those diseases. Also, amino acids (e.g., taurine, arginine, glycine, proline, and 4-hydroxyproline), and glucosamine reduce inflammation and pain in animals with bone and joint disorders. Gaining insight into how amino acids function in maintaining bone and joint health can aid in developing preventative diets and therapeutic supplementations of amino acids to improve the quality of life in companion animals.

Keywords Amin acid · Bone · Companion animal · Joint · Metabolism · Nutrition

---

A. N. Lopez · F. W. Bazer · G. Wu (✉)
Department of Animal Science, Texas A&M University, College Station, TX 77843, USA
e-mail: g-wu@tamu.edu

G. Wu (ed.), *Nutrition and Metabolism of Dogs and Cats*, Advances in Experimental Medicine and Biology 1446, https://doi.org/10.1007/978-3-031-54192-6_7

Abbreviations

BMP	Bone morphogenetic proteins
CKD–MBD	Chronic kidney disease–mineral and bone disorder
DJD	Degenerative joint disease
ECM	Extracellular matrix
FGF23	Fibroblast growth factor-23
GAGs	Glycosaminoglycans
IGF1	Insulin-growth factor 1
MAR	Mineral apposition rate
MSFs	Marrow stromal fibroblasts
MT1-MMP	Membrane-type matrix metalloproteinase 1
NO	Nitric oxide
NOS	Nitric oxide synthase
OA	Osteoarthritis
NSAIDs	Non-steroidal anti-inflammatory drugs
PTH	Parathyroid hormone

7.1 Introduction

Bones and joints aid in locomotion and have structural, protective, endocrine, and other dynamic roles in the body. The endocrine role of these structures maintains mineral homeostasis through continuous bone remodeling and cellular turnover in response to biochemical stimuli (Quarles 2008). The stability and integrity of bones and joints is critical for their function in resisting mechanical stress and trauma. Mechanical stress and biochemical stimuli both elicit the continuous processes of bone deposition (formation of new bone materials) and resorption (breakdown of the bone matrix to release calcium and phosphorus), as well as cellular turnover to maintain bone and joint health (Hughes-Fulford 2004; Quarles 2008). The major protein in bones and joints is collagen (Fig. 7.1). Amino acids (AAs) have important compositional, functional, regulatory, and therapeutic roles which heavily impact the health of bones and joints. AAs have anti-inflammatory and antioxidant roles that reduce oxidative stress and inflammation (Wu 2022). Disorders in maintaining bone and joint health occur in both humans and animals and can significantly reduce the quality of life. These disorders do not manifest the same in all individuals, as environmental and other factors determine the severity and frequency. Those factors include nutrition, physical activity, trauma, age, and growth rate (Fig. 7.2). Most bone and joint disorders in animals (including cats and dogs) are without a cure and are treated through pain management and to slow disease progression. The standard care for these disorders is non-steroidal anti-inflammatory drugs (NSAIDs) for pain management; however, NSAIDs have high levels of toxicity and can affect kidney function. Therapeutic AA supplements and diets have low toxicity and may be

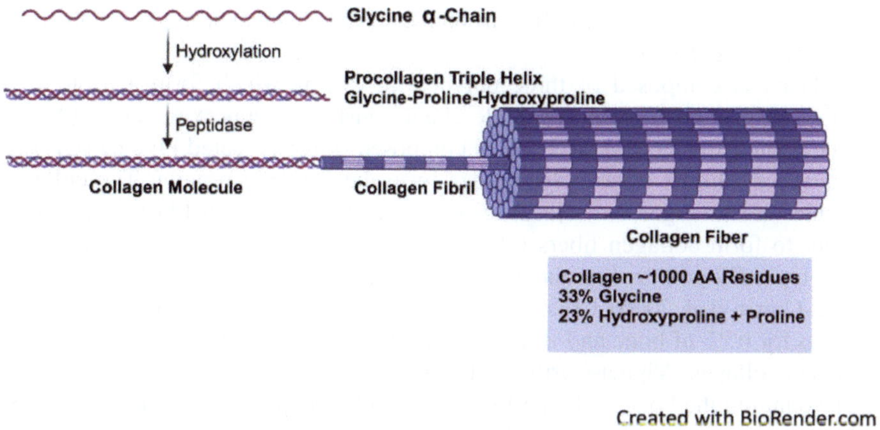

Fig. 7.1 Collagen formation in mammals including dogs and cats. Each of the three α-chains (polypeptides) is synthesized from amino acids, mainly glycine and proline in the rough endoplasmic reticulum. Hydroxylation of some proline and lysine residues in each polypeptide results in the formation of a procollagen triple helix. Under the enzymatic action of peptidase, the triple helical procollagen molecule becomes a triple-helical polypeptide collagen molecule. The collagen molecules arrange in concentric weaves to form collagen fibrils, which bundle together to form collagen fibers. The primary amino acid sequence of collagen is glycine-proline-X or glycine-X-hydroxyproline, where X may be any of other proteinogenic amino acids. Depending on collagen type, the three polypeptides may be identical (homotrimers, as in collagens II and III) or may differ in at least one chain (heterotrimers, as in collagens I, V and XI). Each α-chain consists of approximately 1000 amino acid residues (Feng 2009; Li and Wu 2018; Viguet-Carrin et al. 2006).

better suited for long term pain management (Lascelles et al. 2010; McAlindon et al. 2000). The roles of AAs (e.g., arginine, glycine, proline, and 4-hydroxyproline) and glucosamine in bones and joints heavily impact the overall health and quality of life in individuals with disorders in these structures. This review serves to examine the current literature regarding AA metabolism and functions in maintaining the health of bones and joints in felines and canines.

7.2 Bone Structure and Cellular Composition

In mammals including dogs and cats, bone accounts for 12–15% of the body weight and consists of specialized connective tissues: bone matrix (cortex and trabeculae) and bone marrow (red and yellow marrows). Bone cortex (osteoblasts, osteoclasts, and cylinders of matrix) and trabeculae (lacunae, osteocytes, red bone marrow, and matrix) contribute to 80% and 20% of the total bone matrix, respectively. In chemical composition, bone is composed of 30% organic matter, 60% inorganic matter, and 10% water. The organic portion of bone is comprised of collagen (primarily Type I collagen), non-collagenous proteins, and proteoglycans (Feng 2009). Collagen and elastin represent 90 and 1% of the organic component in the bone matrix, respectively.

Therefore, if properties of the collagen change, the mechanical properties of the bone are altered (Feng 2009).

Collagen is composed of three polypeptide chains which form a triple-helix structure (Fig. 7.1). Each polypeptide chain consists of approximately 1000 AA residues. The triple helical structure is composed of the repeated glycine (Gly)-X-Y sequences, and its three polypeptides form the collagen protein. The collagen molecules are arranged in concentric weaves to form collagen fibrils that bundle together to form collagen fibers (Viguet-Carrin et al. 2006). The AA referred to as X is typically proline, but may be other AAs, and Y is typically hydroxyproline (Viguet-Carrin et al. 2006). Crystalline hydroxyapatite is the mineral component that accounts for 60% of bone and is arranged in the same orientation of the concentric weaves in collagen (Viguet-Carrin et al. 2006).

Surrounding the bone is the periosteum, which contains osteogenic cells, dense collagenous connective tissue, and fibroblasts (Allen et al. 2004). The endosteum is a specialized connective tissue that lines the walls of spongy bone and developing bone and contain osteoblasts and osteoclasts for remodeling (Allen et al. 2004). The interactions of stem cells within the bones in coordinating responses to biochemical stimuli and mechanical stress were first evident in cells with osteogenic potential in bone marrow (Bianco 2015). During early studies of osteogenic stem cells in the 1960s, cell culture was used for the isolation of hematopoietic cells from non-hematopoietic stromal cells (Bianco 2015). Transplanted non-hematopoietic stromal cells generated heterotopic bone, thus defining them as the non-hematopoietic stromal cells with osteogenic potential (Bianco 2015). These findings, first shown in rodents and later in humans, formed the concept of osteogenic stem cells (Bianco 2015). It was demonstrated that osteogenic stem cells can differentiate into osteoblasts (Kuznetsov et al. 1997). Experiments completed in that study revealed that marrow stromal fibroblasts (MSFs) have osteogenic potential, but not all portions of MSFs have equal osteogenic capacity (Kuznetsov et al. 1997).

In earlier work, Parfitt (1988)studied bone histomorphometry, including the terminology to describe cell types, tissues, and units, as well as the dynamics of bone and mineral metabolism. The cell types of the bone consist of osteoclasts and those of osteoblast lineage: osteoprogenitor stem cells and osteocytes (Datta et al. 2008). Cells from the osteoblast linage are polarized and deposit osteoid along the osteoblast–bone interface, leading to bone formation (Datta et al. 2008). Osteoblasts and osteoclasts work in a coordinated manner to carry out continuous remodeling of bone (Datta et al. 2008). Osteoprogenitor stem cells arise from mesenchymal stem cells and are activated postnatally during repair and remodeling of bone to stimulate osteoblast activity (Datta et al. 2008). Osteocytes communicate within the extracellular matrix (ECM) through canaliculi (osteocyte dendritic processes) and may act as mechanosensory cells (Datta et al. 2008). Osteocytes are multifunctional cells as they express membrane-associated proteins and genes which mediate bone remodeling by mediating osteoblast differentiation (Zhang et al. 2006). Osteoclasts are responsible for bone resorption and are located within cavities of the matrix called

Howship's lacuna. Endothelial cells within bones are responsible for multiple functions such as the transport of nutrients and metabolites in response to multifunctional cell signaling molecules (Yin et al. 2021).

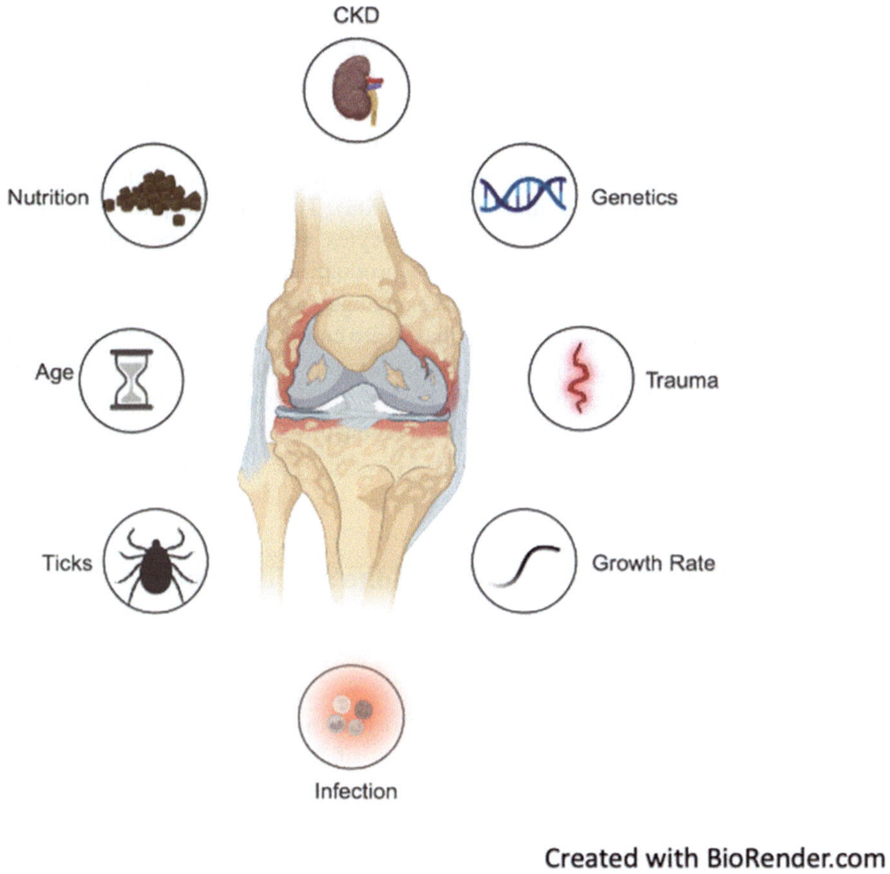

Fig. 7.2 Risk factors impacting bone and joint health in dogs and cats. Chronic kidney disease is common in companion animals and leads to chronic kidney disease-mineral and bone disorder (CKD-MBD) that causes osteodystrophy in the bones and joints. It is common for dogs and cats to be genetically predisposed to many bone and joint disorders, and environmental factors determine how they manifest in these animals. Factors such as nutrition, infection, trauma, growth rate, age, and tick-borne diseases determine severity and frequency of multiple bone and joint diseases (Chen et al. 2018; Foster 2016; Geddes et al. 2015; Maniaki et al. 2021; Parker et al. 2020; Perry 2016; Rychel 2010; Schachner and Lopez 2015; Smith 1998).

7.3 Joint Structure and Cellular Composition

Joints consist of cartilage, tendons, ligaments, synovial joint capsules and membranes, as well as a synovial joint cavity filled with synovial fluid. The articular surface of long bones that form the joint are covered in hyaline cartilage for protection (Birch et al. 2013). The cells of hyaline cartilage are chondrocytes, which produce a large amount of ECM composed of collagen fibers (mainly types II, IX and XI collagens), glycosaminoglycans, proteoglycans, elastin fibers, and other proteins. Collagens account for ~ 70% of the dry matter in the cartilage.

Tendons have specialized fibroblast cells called tenocytes, which produce ECM, collagen, and elastin (Benjamin et al. 2008). The synovial joint membrane has specialized fibroblast cells called synoviocytes which produce collagen and hyaluronic acid, and are responsible for the production of synovial fluid components, absorption of fluids, and blood/fluid exchanges. Approximately 75 and 1.5% of tendons and ligaments (on a dry matter basis) are composed of collagen (primarily type I collagen) and elastin, respectively. As collagen is formed through the hydroxylation of proline and lysine, the extent of the hydroxylation determines stability of that collagen. The ligaments and tendons are responsible for resisting movement and for support of the joints (Birch et al. 2013).

7.4 Formation and Resorption of Bone as Continuous Remodeling Processes

Bone remodeling is controlled by mechanical forces, cell turnover, hormones, cytokines, and transcription factors (Fig. 7.3) (Datta et al. 2008). Parfitt (1988) studied the dynamics of the continuous process of bone remodeling and outlined methods for determining mineralization activity and mineral apposition rate (MAR) over the osteoid surface. Osteocytes act as mechanosensory cells through the selective expression of membrane associated proteins and genes (Zhang et al. 2006). Selective proteins are expressed in response to mechanical stress at sites of bone remodeling (Zhang et al. 2006). The mechanosensory function of these cells is due to the canaliculi dendritic processes (Holmbeck et al. 2005). The expression of membrane-type matrix metalloproteinase 1 (MT1-MMP) is required for the formation and maintenance of the canaliculi which is an early event in osteoblast differentiation (Holmbeck et al. 2005). Mature osteocytes secrete sclerostin, which inhibits bone formation by inhibition of bone morphogenic proteins (BMPs) (Poole et al. 2005).

Parathyroid hormone (PTH) produced by the parathyroid glands maintains calcium homeostasis through endocrine cell signaling for the remodeling of bone (Raggatt and Partridge 2010). Increased secretion of PTH down-regulates the expression of osteocyte-derived sclerostin (Bellido et al. 2005). Through down-regulation of osteocyte derived sclerostin, pro-osteoblastogenic transcription factors and growth hormones initiate osteoblast differentiation (Bellido et al. 2005). When calcium concentrations are sufficient in plasma, PTH is secreted to induce a cell signaling cascade which mediates bone remodeling (Raggatt and Partridge 2010). Earlier work revealed that, in response to PTH secretion, osteoclast formation was induced (Takahashi et al. 1988).

Osteoclasts secrete organic acids, proteases, and chemokines, which collectively degrade the ECM and expose binding sites for osteoclast attachment (Raggatt and Partridge 2010). Osteoclast attachment creates a sealed microenvironment for remodeling whereas resorption of the mineralized matrix occurs in what are known as Howship's lacunae (Raggatt and Partridge 2010). The resorption of mineralized matrix allows for transport of calcium, phosphate, and other proteins to the basolateral membrane that increases concentrations of calcium for maintaining calcium homeostasis (Datta et al. 2008). Once resorption is complete (Fig. 7.3), osteoblasts and osteogenic stem cells then occupy the Howship's lacunae (Fig. 7.4) (Raggatt and Partridge 2010). These cells undergo differentiation and secrete type I collagen, proteoglycans, glycosylated proteins, and lipids that makeup the composition of new osteoid (Raggatt and Partridge 2010). When calcium concentrations are sufficient in circulation, calcitonin is secreted from the thyroid gland, which inhibits resorption and increases osteoblastic bone formation and calcium recruitment (Hsiao et al. 2020). In the formation of new osteoid, calcium is recruited thus maintaining calcium homeostasis. Bone mineralization is achieved by incorporation of the inorganic component hydroxylapatite, and osteoblasts undergo apoptosis, thus completing this continuous remodeling process (Raggatt and Partridge 2010). Major mediators of bone formation and resorption are summarized in Fig. 7.5.

7.5 AA Metabolism in Bone and Joints

Bone and joint tissues of mammals including cats and dogs cannot form de novo the carbon skeletons of citrulline, cysteine, histidine, isoleucine, leucine, lysine, methionine, phenylalanine, taurine, threonine, tryptophan, tyrosine, and valine and, therefore, cannot synthesize these AAs (Wu 2022). In contrast, bone and joint tissues can convert L-citrulline into L-arginine in the presence of aspartate via argininosuccinate synthase and lyase, whereas the α-ketoacids of branched-chain AAs (BCAAs), methionine, phenylalanine, tryptophan, and tyrosine can be transaminated with glutamate to yield the corresponding α-AAs. In addition, the bone and joint tissues are capable of synthesizing alanine, aspartate, asparagine, glutamate, glutamine, glycine, ornithine, proline, and serine (Wu 2022). Like other tissues (e.g., skeletal muscle, liver, and kidney) of animals (Hu et al. 2023a, b), bone and joint tissues may generate glycine from 4-hydroxyproline, thereby recycling 4-hydroxyproline (a product of collagen degradation) into glycine for the production of new collagen molecules (Hu et al. 2022).

Bone and joint tissues of mammals including cats and dogs can degrade the following: (1) L-arginine via both arginase and nitric oxide (NO) synthase (NOS)

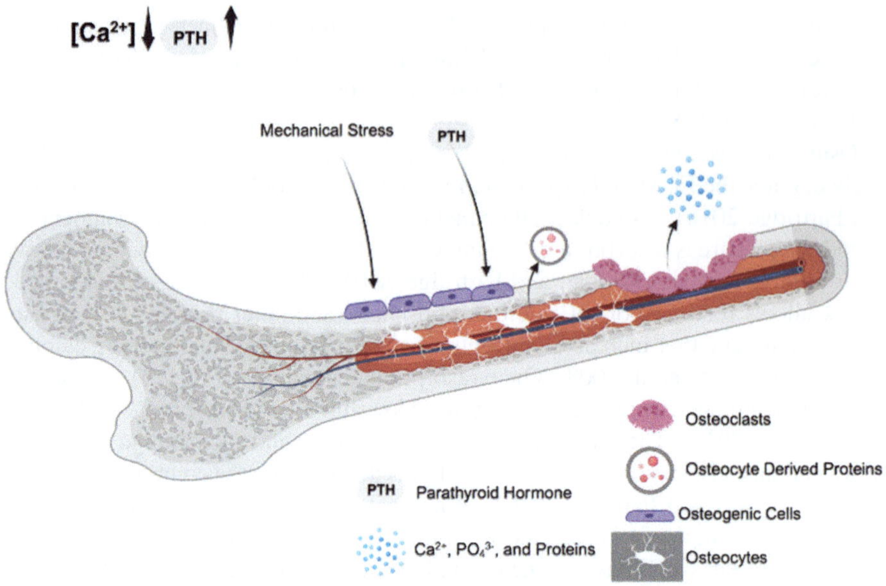

Fig. 7.3 A summary of key events in bone resorption. Due to low concentrations of calcium, parathyroid hormone (PTH) is secreted which promotes bone resorption. Bone resorption is also promoted by mechanical stress and/or trauma. Mechanosensory osteocytes then secrete proteins, which promote bone resorption through cell signaling cascades and initiation of osteoclast formation. By osteoclast resorption of bone within the Howship's lacunae, it allows for the transport of calcium, phosphorus, and proteins to maintain calcium and phosphorus homeostasis (Bellido et al. 2005; Raggatt and Partridge 2010; Poole et al. 2005; Takahashi et al. 1988; Zhang et al. 2006)

pathways; (2) branched-chain AAs (BCAA) via BCAA transaminase and branched-chain α-ketoacid dehydrogenase; (3) glutamine via phosphate-activated glutaminase and glutamine:fructose-6-phosphate transaminase; (4) alanine, glutamate, and aspartate via transaminases; (5) proline via proline oxidase; and (6) ornithine via ornithine aminotransferase and ornithine decarboxylase (Wu 2022). Bone marrow expresses phenylalanine hydroxylase to convert phenylalanine into tyrosine but not vice versa (Harding et al. 2003), but it is unknown whether this enzyme is present in joints. Both bone and joint tissues of mammals can interconvert serine and glycine, but have little or no ability to degrade cysteine, histidine, lysine, methionine, phenylalanine, taurine, threonine, tryptophan, and tyrosine into CO_2 (Wu 2022; Yao et al. 2011).

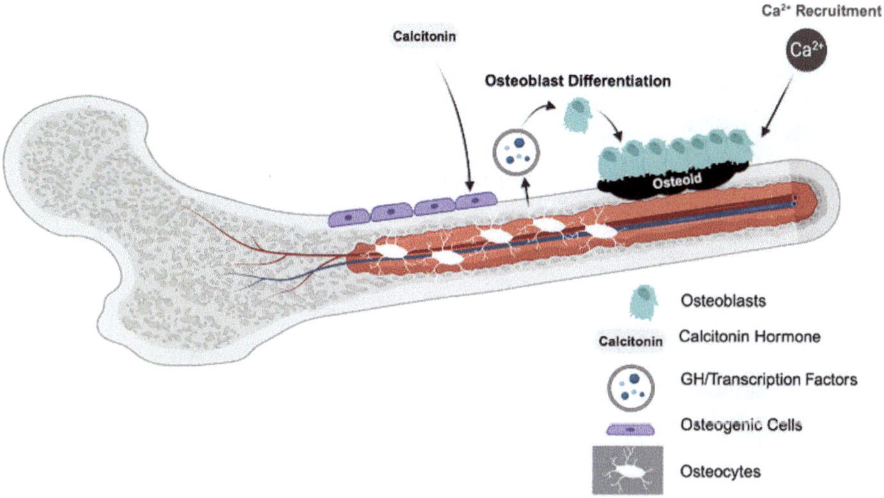

Fig. 7.4 A summary of events in bone formation. Once bone resorption is complete, bone formation takes place subsequently. When calcium concentrations are sufficient in plasma, calcitonin is then secreted for promotion of bone formation and recruitment of calcium to maintain calcium homeostasis. Calcitonin secretion leads to cell signaling cascades, and secretion of growth hormones and transcription factors. These growth hormones and transcription factors initiate osteoblast differentiation and activation. Osteoblasts then deposit osteoid within the Howship's lacuna and calcium is recruited, which maintains calcium homeostasis (Datta et al. 2008; Holmbeck et al. 2005; Hsiao et al. 2020; Raggatt and Partridge 2010; Zhang et al. 2006)

7.6 Roles of AAs in Bone Remodeling

AAs have multiple roles in mediating bone remodeling and cellular turnover that are compositional, nutritional, and physiological. A master transcription factor known as runt-related transcription factor 2 (Runx2) and other factors [such as osterix proteins and activating transcription factor 4 (ATF4)] induce osteogenic stem cell differentiation (Wagley et al. 2020). These transcription factors are stimulated by bone morphogenetic proteins (BMP) and growth hormones (Wagley et al. 2020). These growth hormones and transcription factors are synthesized from AAs, which emphasizes the importance of AAs in bone remodeling (Wagley et al. 2020). Metabolism of AAs, specifically arginine, directly impact bone remodeling as it is a substrate for NO synthesis through the enzymatic activity of NOS (Evans and Ralston 1996). Catabolism of arginine leads to NO/NOS effects on bone remodeling and joint health. As a vasodilator, NO controls blood flow and, thus, the amounts of nutrients to be utilized for biological responses in bone and joints (Li et al. 2009; Wu et al. 2021). It has been suggested that NO and NOS also function in the inhibition of osteoclast formation and promotion of osteoid deposition although its role as a mediator for cytokine-induced bone resorption (Evans and Ralston 1996). Administration of L-arginine prevents bone loss through the synthesis of NO because low concentrations of NO promote bone resorption (Fiore et al. 2000). Glucosamine is

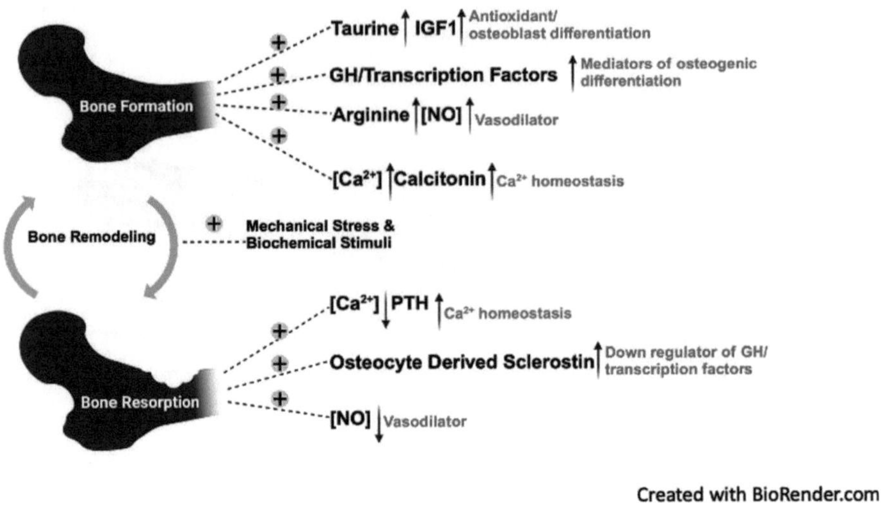

Fig. 7.5 A depiction of the major mediators of bone formation and resorption. If calcium concentration is sufficient in plasma, it leads to calcitonin secretion which promotes bone formation to recruit calcium and maintain homeostasis. However, when calcium is low, then PTH is secreted to promote bone resorption and increase calcium concentrations. An amino acid, specifically arginine, is metabolized into NO, which is a vasodilator and controls blood flow. When concentrations of NO are increased, bone formation is promoted and bone resorption is inhibited. However, low NO concentrations support/promote bone resorption. Growth hormones such as BMPs and FGF, and transcription factors such as osterix promote bone formation in response to biochemical stimuli. Osteocytes secrete sclerostin protein which inhibits these growth hormones, such as BMPs, and thus promotes bone resorption. In mammals, taurine increases IGF1 concentrations, which promotes bone formation. Mechanical stress and biochemical stimuli initiate the bone remodeling process (Evans and Ralston 1996; Fiore et al. 2000; Hsiao et al. 2020; Pérez-Hernández et al. 2022; Stegen et al. 2022; Wagley et al. 2020; Yang et al. 2023)

an amino monosaccharide formed from fructose-6-phosphate (a product of glucose metabolism) and the amino group donor glutamine (Yang et al. 2023). Glucosamine, an inhibitor of NO synthesis by both constitutive (Wu et al. 2001) and inducible (Meininger et al. 2000) NOS, elicits an anti-inflammatory effect to provide relief from osteoarthritis (OA) (Bhathal et al. 2017). Animal-sourced foodstuffs, which contain glycosaminoglycans, provide relatively large amounts of glucosamine and related substances, including N-acetyl-glucosamine, N-acetyl-galactosamine, hyaluronic acid, heparin, heparan sulfate, dermatan sulfate, chondroitin sulfates, and keratan sulfate (Wu 2022) and, therefore, could be beneficial for the health of bones and joints. In addition, serine promotes chondrocyte proliferation during the continuous bone remodeling process (Stegen et al. 2022). Furthermore, through the generation of melanin, which promotes the proliferation and differentiation of osteoblasts, dietary tyrosine enhances bone mineral density and bone strength in animals (Yoo et al. 2017).

Chondrocytes depend on transcription factors and growth hormones for activation (Stegen et al. 2022). Gene expression by chondrocytes is controlled by glutamine-derived metabolites through epigenetic mechanisms (Stegen et al. 2022). AAs, specifically serine, mediate nucleotide production to support the epigenetic mechanisms controlling the expression of genes by chondrocytes (Stegen et al. 2022). Taurine has metabolic roles in bone remodeling as an upstream regulator of insulin-growth factor 1 (IGF1) production, which regulates osteoblast differentiation (Pérez-Hernández et al. 2022). Taurine supplementation increases IGF1 levels and thus supports bone mineralization and formation (Pérez-Hernández et al. 2022). This nutritional measure is more important for cats as they have a very limited ability to synthesize taurine (Li and Wu 2023). Rather, dietary supplementation with a combination of AAs (leucine, lysine, isoleucine, phenylalanine, valine, methionine, arginine, threonine, histidine, and tryptophane) increases bone mineral density, concentrations of IGF1, bone formation, and inhibits bone resorption in rats (Ammann et al. 2002). These findings emphasize the importance of AA metabolism and functions in maintaining the health of bones.

The health of joints, which are composed of tendons, ligaments, and synovial membranes, depends on AAs as well (Wu 2018). As collagen is the primary protein of tendons in the joint, the stability of the collagen determines the stability of the tendons, and thus the stability of the overall joint (Birch et al. 2013). Glycine, an AA, is required for formation of the polypeptide chains of collagen (Viguet-Carrin et al. 2006). Proline and hydroxyproline are also components of the collagen triple helix configuration (Viguet-Carrin et al. 2006). The importance of glycine, proline, and 4-hydroxyproline is emphasized as their contribution is 57% of total AA residues in collagen, which accounts for approximately a third of total proteins in animals (Li and Wu 2018). In addition, glycine, proline, and 4-hydroxyproline are potent antioxidants and cell signaling molecules (Hu et al. 2022; Wang et al. 2014; Wu et al. 2011) to maintain the health of bone and joints. Ligaments have a lesser amount of collagen and a greater amount of glycosaminoglycans which interact with collagen fibrils through interfibrillar proteoglycans (Franchi et al. 2010).

AAs also have roles in eliciting anti-inflammatory and antioxidant responses which support joint health. Taurine and its derivatives inhibit excess production of inflammatory mediators and reduce oxidative stress (Pérez-Hernández et al. 2022). Taurine exerts beneficial effects on bone and joints via multiple mechanisms. First, taurine stimulates the differentiation of osteoblasts and chondrocytes to promote bone formation and healing. Second, taurine increases the activity of osteocytes to enhance the uptake of Ca and P and the deposition of the Ca–P salt into the bone matrix. Third, taurine inhibits osteoclast activity, thereby alleviating osteoporosis, bone disfigurement, osteoarthritis, and rheutmatoid arthritis. In addition to taurine, it has also been reported that collagen supplements rich in glycine, proline, and hydroxyproline elicit anti-inflammatory and antioxidant effects (Mobasheri et al. 2021). Furthermore, a mixture of arginine, glycine, and methionine, substrates for the synthesis of creatine via the interorgan metabolism involving primarily the kidneys and liver (Wu 2022), plays an important role in activating osteoblasts (i.e., cells involved in bone formation) (Gualano et al. 2016).

Adequate intakes of dietary AAs, along with Ca, P, Cu and Zn, are essential for the growth and health of tissues (including bone and Joints) in animals (Wu 2018). Thus, a deficiency of dietary protein and minerals in companion and farm animals can promote bone resorption, increase the risk of bone fractures, reduce bone mass and strength, induce osteoporosis, decrease the strength of tendons, and impair recovery of injured bone and joints (Darling et al. 2021). Animal-sourced foodstuffs are abundant sources of arginine, glutamine, glycine, lysine, proline, and 4-hydroxyproline, as well as all other proteinogenic AAs for companion and farm animals (Li and Wu 2020, 2022; Li et al. 2021). For comparison, plant-sourced feedstuffs contain only a negligible amount of 4-hydroxyproline, and corn grain, soybean meal, wheat flour, and rice contain much less AAs than animal-sourced foodstuffs (Wu 2022). Because different animal-sourced foodstuffs have their unique composition of functional AAs and related metabolites (such as creatine and 4-hydroxyproline), a mixture of animal-sourced foodstuffs can be included in the diets of animals to bring about complementary effects in nutrition. For example, feather meal can be used along with one or more another animal protein foodstuffs, such as poultry by-product meal (pet-food grade), spray-dried porcine mucosal peptones, meat–bone meal, blood meal, and black soldier fly larvae meal (Li and Wu 2020, 2022; Li et al. 2021).

7.7 Bone and Joint Disorders

Bone and joint disorders in companion animals are a common health concern in felines and canines, which reduce the quality of life and commonly lead to elective euthanasia (Harari 2022a, b). For example, almost all types of dogs encounter bone and joint problems, and large breeds are more susceptible than small and medium breeds. Growth spurt during the early stage of postnatal development may lead to bone and joint problems in adults, including discomfort and bone disfigurement. In addition, feline arthritis (also known as degenerative joint disease) is very common (Harari 2022b). This disease occurs when the cartilage within a joint is injured. This lack of cartilage in the joint can cause chronic unbearable pain and can lead to poor quality of life. To better understand these disorders, it is important to know how they occur, how they are diagnosed, and the susceptibility of animals to these disorders. Only then can advances in preventative management techniques be realized.

One common disorder is chronic kidney disease–mineral and bone disorder (CKD–MBD), which is diagnosed by determining circulating concentrations of PTH using a second generation intact parathyroid hormone (iPTH) assay (Chen et al. 2018). This disorder occurs as a complication of CKD, which leads to an imbalance of calcium and phosphorus, and thus an abnormal PTH activity (Chen et al. 2018). As PTH mediates bone remodeling to maintain calcium and phosphorus homeostasis, CKD–MBD leads to bone and mineral malformations (Chen et al. 2018). Another common disorder is arthritis, which can be in different forms such as immune-mediated, infectious, tick-borne, and osteoarthritis (OA). Diagnosis is primarily assessed through radiography and obtaining synovial fluid samples from affected joints (May 2005). Clinical characteristics for infectious arthritis in joints are lameness, heat production, and fluid-induced distension (May 2005). Another form of arthritis is OA, which is diagnosed through physical examinations and radiographs of the affected joints (Rychel 2010). Osteochondrosis is another disease characterized by a malformation of cartilage within bones and joints (Kippenes and Johnston 1998). Diagnosis of osteochondrosis involves the radiography of osteochondritis lesions and magnetic resonance to provide information on fragment stability and changes in cartilage (Kippenes and Johnston 1998). Hip dysplasia affects both humans and companion animals and is characterized by a disorder in the development of the hip joints. Diagnosis and screening for hip dysplasia require radiography and physical examinations (Butler and Gambino 2017). These bone and joint disorders, along with others, can be due to poor nutrition (particularly inadequate and imbalanced AA intakes), injury, genetics, growth rate, exercise, hormones, age, and other less well-defined conditions.

7.7.1 *Feline Bone and Joint Health*

Multiple bone and joint disorders that occur in humans, also manifest in felines. One disorder found in both is CKD–MBD characterized by disturbances in calcium, phosphorus, and magnesium homeostasis (Foster 2016). These disturbances caused by CKD in cats lead to phosphorus retention that stimulates the production of FGF23 and the subsequent secretion of PTH (Foster 2016). Renal osteodystrophy in cats is the associated bone and joint malformations caused by CKD–MBD (Geddes et al. 2015). Elevated concentrations of FGF23 in plasma is a marker for subclinical CKD–MBD, and an independent predictor of mortality of CKD–MBD in cats (Geddes et al. 2015). Treatment of CKD–MBD is aimed at slowing the progression of the disorder. Studies in rats revealed that low phosphorus diets may slow the progression of CKD–MBD (Moe et al. 2009).

One of the most common bone and joint disorders affecting felines is degenerative joint disease (DJD) characterized by the progressive destruction of cartilage, bone, ligaments, and/or the synovial capsule (Lascelles and Duncan 2010). Diagnosis of DJD requires radiographs (Lascelles and Duncan 2010). As DJD is associated with OA, the typical treatment in humans is NSAIDs for pain management. However, in cats, there is concern for the use of NSAIDs as these drugs can affect kidney function and CKD is common in cats (Lascelles et al. 2010). Therefore, advances in developing therapeutic diets and dietary supplementation have gained significant research interests (Lascelles et al. 2010). A therapeutic dietary treatment of glucosamine hydrochloride and collagen peptides in cats with joint degeneration effectively elicit anti-inflammatory effects with no adverse side effects (Minami et al. 2011). As supportive treatment of DJD using glucosamine and collagen supplementation has no toxic effects, it is better suited for long-term supportive treatment than NSAIDs in cats (Minami et al. 2011). Causes and risk factors that may predispose cats to

DJD are obesity and previous trauma (Maniaki et al. 2021). Thus, prevention of obesity through improving nutrition such as adequate AA and energy intakes plays an important role in reducing risk for DJD.

Hip dysplasia is a joint disorder that reduces quality of life in cats, and diagnosis requires analysis of radiographs (Perry 2016). This abnormality is known to be an inheritable disorder which may manifest differently dependent on environmental factors (Perry 2016). Large breeds of cats such as Maine Coons and Persians are predisposed to hip dysplasia due to a narrow genetic-pool, greater body weight, and larger frame-size (Perry 2016). Factors such as nutrition, growth rate, trauma, and age may influence how hip dysplasia manifests in cats.

Another disorder in felines is synovial osteochondromatosis, which is characterized by ossified nodules that form along the joints and arise from cartilage (Allan 2000). In addition, mucopolysaccharidosis is a disorder characterized by the accumulation of glycosaminoglycans (GAGs) within the connective tissues such as chondrocytes, fibroblasts, smooth muscle cells, and Kupffer cells (Allan 2000). The accumulation of GAGs causes skeletal malformations and decreased bone opacity (Allan 2000). With insight into how these disorders occur, the factors which cause them, and interactions among them will allow for prevention and pain management of these disorders in cats.

The use of therapeutic dietary supplementation in treatment and management of feline degenerative joint diseases has been reported, based on results of multiple studies, to be preferrable for long-term support due to a lack of toxic side effects (Lascelles et al. 2010). As AA functions and metabolism have significant roles in maintaining bone and joint health in felines, preventative therapeutics utilizing AAs may benefit the quality of life in these animals (Table 7.1).

7.7.2 Canine Bone and Joint Health

In canines, the disorder of CKD–MBD has been associated with a poor nutritional status, increases in PTH in blood, and increased concentrations of FGF23 (Parker et al. 2020). Like humans and cats, CKD–MBD is typically the consequence of CKD. In dogs, increases in concentrations of phosphorus, calcium, metabolites of phosphorus, and FGF23 in blood are negatively correlated with survival (Parker et al. 2020). As PTH acts to maintain calcium and phosphorus homeostasis by meditating bone remodeling, increases in PTH in the circulation lead to malformations in bone known as renal osteodystrophy (Parker et al. 2020). In dogs, the maxilla and mandible bones are preferentially affected by renal osteodystrophy (Parker et al. 2020). As CKD causes irreversible renal damage, the treatment of this disease is to manage the progression of CKD by correcting imbalances in calcium, phosphorus, and FGF23 (Bartges 2012). In 2015, about 1 in 4 dogs were diagnosed with some forms of arthritis in the United States (Bland 2015). The inflammatory disease, known as OA, is characterized by degeneration of the cartilage, changes in the synovial membrane, and bone malformations, and is most reported for large breeds of dogs (Rychel 2010).

Table 7.1 Roles of amino acids in the health of bone and joints[1]

Amino acid	Pharmacological/ physiological purpose	Cat application	Dog application	References
Glycine	Elicits an anti-inflammatory and antioxidant response and is required for collagen synthesis	Yes	Yes	Mobasheri et al. (2021); Li and Wu (2018)
Taurine	Increases [IGF1], promotes bone formation, and reduces oxidative stress	Yes; e.g., via dietary supplementation	Yes	Pérez-Hernández et al. (2022)
Serine	Promotes chondrocyte proliferation and mediates chondrocyte gene expression	Yes	Yes	Stegen et al. (2022)
Tyrosine	Promotes chondrocyte proliferation and differentiation	Yes	Yes	Yoo et al. (2017)
Arginine	Increases [NO] and prevents bone loss	Yes	Yes	Fiore et al. (2000)
Glutamine	Synthesizes glucosamine which elicits an anti-inflammatory response	Yes	Yes	Bhathal et al. (2017)
Proline	Elicits an anti-inflammatory and antioxidant response, and an amino acid residue in collagen	Yes	Yes	Mobasheri et al. (2021); Li and Wu (2018)
4-Hydroxyproline	Elicits an anti-inflammatory and antioxidant response and an amino acid residue in collagen	Yes	Yes	Mobasheri et al. (2021); Li and Wu (2018)
A mixture of arginine, glycine and methionine[2]	Activates osteoblasts (i.e., cells involved in bone formation)	Yes	Yes	Gualano et al. (2016)
Amino acid complex[3]	Increases bone mineral density, concentrations of IGF1, bone formation, and inhibits bone resorption	Yes	Yes	Ammann et al. (2002)

[1]Except for glycine and taurine, all amino acids are L-isomers
[2]Through the synthesis of creatine
[3]A mixture of leucine, lysine, isoleucine, phenylalanine, valine, methionine, arginine, threonine, histidine, and tryptophane
IGF1, insulin-like growth factor 1; NO, nitric oxide

However, this disorder may affect all breeds, regardless of size, as they age. Diagnosis involves pain assessment during orthopedic and neurologic examinations (Rychel 2010). Once the source of pain has been localized, radiographs are used to confirm the diagnosis. As OA is a lifelong disorder with no cure, the treatment is designed to provide pain management, joint protection, nutritional support, and physical therapy (Rychel 2010). This disorder manifests as cells within these affected joints release pro-inflammatory cytokines that induce an inflammatory response, which leads to inflammation of the synovial membrane, bone thickening, and cartilage deconstruction (Bland 2015). As in humans, the use of NSAID's is the standard pharmacological therapy in dogs for pain management (Bhathal et al. 2017). Also, advances in the use of nutritional supplements in managing OA disease include the use of glucosamine, an amino monosaccharide formed through glucose metabolism with glutamine as the amino-group donor. Supplementation of palmitoyl-glucosamine and curcumin was reported to relieve inflammation and reduce neuropathic pain through anti-inflammatory actions (Gugliandolo et al. 2020). There are various glucosamine and chondroitin supplements currently recommended such as Dasuquin and Cosequin DS for treatment of OA in dogs (Gugliandolo et al. 2020). Dietary supplementation with D-glucosamine and scale collagen protein elicited anti-inflammatory effects in dogs due to increases in proteoglycan production and the anti-inflammatory actions of glucosamine (Minami et al. 2011).

Hip dysplasia is another common developmental problem facing many large breeds of dogs. Hip dysplasia is characterized by subluxation of the femoral head within the hip joint and typically leads to DJD (Smith 1998). Radiographs and physical examinations are the primary method of diagnosis in dogs (Butler and Gambino 2017). Multiple environmental factors, including nutrition, may affect how this disorder manifests in a dog. For example, excess energy consumption increases the frequency and severity of hip dysplasia in canines (Smith 1998). Therefore, maintenance of body condition is a method of prevention of hip dysplasia (Schachner and Lopez 2015). Because arginine reduces obesity in rats, pigs, and humans (Wu 2022) and because dogs have a limited ability to synthesize arginine (Li and Wu 2023), an adequate provision of this dietary AA may be crucial for the prevention of hip dysplasia. As genetics play a role in predisposition for hip dysplasia in dogs, selective breeding is also a means for reducing the occurrence and frequency of this disorder (Schachner and Lopez 2015). These disorders, as well as the secondary associated bone and joint diseases, such as DJD and renal osteodystrophy, all reduce the quality of life in dogs and commonly lead to elective euthanasia.

7.8 Conclusions and Perspectives

Bone and joints provide the skeleton of mammals including dogs and cats. Collagen (rich in glycine, proline, and 4-hydroxyproline) and glycosaminoglycans are major organic substances in the bone matrix and joints. Maintaining bone and joint health is a continuous process involving cellular turnover and bone remodeling, events in which AAs and minerals (e.g., calcium and phosphorus) have functional and metabolic roles. The transcription factors and growth hormones that induce the differentiation of osteogenic stem cells are all synthesized from AAs that play key roles in mediating this continuous process. The AA L-arginine is required for the synthesis of metabolically important molecules that mediate bone remodeling, particularly NO that influences bone resorption. Taurine, creatine, glycine, serine, methionine, tyrosine, and creatine also promote the formation of bone tissues.

AAs are also key to maintaining joint health as they are required for the synthesis of glucosamine, proteoglycans and GAGs that have anti-inflammatory effects to reduce inflammation in joints. Joint health and stability depend on the integrity of the ligaments and tendons which support them. The stability of the tendons and ligaments depends on the stability of collagens that composes them. AAs (particularly glycine and proline) are required for the synthesis of collagen and therefore impact the stability of the overall joint.

Bone and joint disorders common to humans and companion animals such as dogs and cats reduce the quality of life. In dogs and cats, many of these disorders lead to elective euthanasia due to the detrimental effects on the animal's quality of life. The disorders of CKB-MBD, DJD, OA, hip dysplasia, and other common disorders also affect the health of both cats and dogs. Many of these disorders have no true cure; therefore, the therapies are for pain management and to slow the progression of the disorders. The common method of pain management relief is the use of NSAIDs; however, they can have toxic effects. There are multiple factors responsible for bone and joint disorders such as trauma, genetic predisposition, growth rate, and nutrition. Nutrition and genetics heavily impact how these disorders manifest in cats and dogs, and therapeutic diets and supplementation of AAs have been found to reduce pain and relieve inflammation in affected animals. These therapeutic diets and dietary supplements have been found to be better suited for long-term support as they have lower risks for toxic effects. AA supplementation in preventative care may also improve how a disorder manifests in companion animals. Arginine, taurine, and serine are a few AAs which may be utilized in building a preventative nutritional plan and therapeutic treatments. In addition, animal-sourced foodstuffs are useful for balancing AAs and provide functional organic nutrients (e.g., taurine, creatine, and 4-hydroxyproline) as well as essential minerals (e.g., calcium, phosphorus, copper and zinc) in the diets of companion animals to maintain the normal structure and health of their bone and joints. A better understanding of the functions and metabolism of AAs for maintaining bone and joint health can inform strategies to use dietary supplements of select AAs for the prevention and management of bone and joint disorders in felines and canines, thus improving the quality of life.

Acknowledgements This work was supported by Texas A&M AgriLife Research (H-8200). We thank Dr. Peng Li (North American Renderers Association, Alexandria, VA, USA) for helpful discussions.

Conflicts of Interest The authors have no conflicts of interest to declare.

References

Allan GS (2000) Radiographic features of feline joint disease. Vet Clin Am Small Anim Pract 30:281–302

Allen MR, Hock JM, Burr DB (2004) Periosteum: biology, regulation, and response to osteoporosis therapies. Bone 35:1003–1012

Ammann P, Laib A, Bonjour JP, Meyer JM, Rüegsegger P, Rizzoli R (2002) Dietary essential amino acid supplements increase bone strength by influencing bone mass and bone microarchitecture in ovariectomized adult rats fed an isocaloric low-protein diet. J Bone Miner Res 17:1264–1272

Bartges JW (2012) Chronic kidney disease in dogs and cats. Vet Clin Am Small Anim Pract 42:669–692

Bellido T, Ali AA, Gubrij I, Plotkin LI, Fu Q, O'brien CA, Manolagas SC, Jilka RL (2005) Chronic elevation of parathyroid hormone in mice reduces expression of sclerostin by osteocytes: a novel mechanism for hormonal control of osteoblastogenesis. Endocrinology 146:4577–4583

Benjamin M, Kaiser E, Milz S (2008) Structure-function relationships in tendons: a review. J Anat 212:211–228

Bhathal A, Spryszak M, Louizos C, Frankel G (2017) Glucosamine and chondroitin use in canines for osteoarthritis: a review. Open Vet J 7:36–49

Bianco P (2015) Stem cells and bone: a historical perspective. Bone 70:2–9

Birch HL, Thorpe CT, Rumian AP (2013) Specialization of extracellular matrix for function in tendons and ligaments. Muscles Ligaments Tendons J 3:12–22

Bland SD (2015) Canine osteoarthritis and treatments: a review. Vet Sci Dev 5:84–89

Butler JR, Gambino J (2017) Canine hip dysplasia diagnostic imaging. Vet Clin Am Small Anim Pract 47:77–793

Chen H, Han X, Cui Y, Ye Y, Purrunsing Y, Wang N (2018) Parathyroid hormone fragments: new targets for the diagnosis and treatment of chronic kidney disease-mineral and bone disorder. Biomed Res Int 2018:9619253

Darling AL, Millward DJ, Lanham-New SA (2021) Dietary protein and bone health: towards a synthesised view. Proc Nutr Soc 80:165–172

Datta HK, Ng WF, Walker JA, Tuck SP, Varanasi SS (2008) The cell biology of bone metabolism. J Clin Pathol 61:577–587

Evans DM, Ralston SH (1996) Nitric oxide and bone. J Bone and Miner Res 11:300–305

Feng X (2009) Chemical and biochemical basis of cell-bone matrix interaction in health and disease. Curr Chem Biol 3:189–196

Fiore C, Pennisi P, Cutuli P, Prato A, Messine R, Clementi G (2000) L-Arginine prevents bone loss and bone collagen breakdown in cyclosporin A-treated rats. Eur J Pharmacol 408:323–326

Foster JD (2016) Update on mineral and bone disorders in chronic kidney disease. Vet Clin Am Small Anim Pract 46:1131–1149

Franchi M, De Pasquale V, Martini D, Quaranta M, Macciocca M, Dionisi A, Ottani V (2010) Contribution of glycosaminoglycans to the microstructural integrity of fibrillar and fiber crimps in tendons and ligaments. Sci World J 10:1932–1940

Geddes RF, Elliott J, Syme HM (2015) Relationship between plasma fibroblast growth factor-23 concentration and survival time in cats with chronic kidney disease. J Vet Intern Med 29:1494–1501

Gualano B, Rawson ES, Candow DG, Chilibeck PD (2016) Creatine supplementation in the aging population: effects on skeletal muscle, bone and brain. Amino Acids 48:1793–1805

Gugliandolo E, Peritore AF, Impellizzeri D, Cordaro M, Siracusa R, Fusco R, D'amico R, Di Paola R, Schievano C, Cuzzocrea S, Crupi R (2020) Dietary supplementation with palmitoyl-glucosamine co-micronized with curcumin relieves osteoarthritis pain and benefits joint mobility. Animals 10:1827

Harari J (2022a) Bone disorders in dogs. Merck Manual. https://www.merckvetmanual.com/dog-owners/bone,-joint,-and-muscle-disorders-of-dogs/bone-disorders-in-dogs

Harari J (2022b) Bone disorders in cats. Merck Manual. https://www.merckvetmanual.com/cat-owners/bone,-joint,-and-muscle-disorders-of-cats/joint-disorders-in-cats
Harding CO, Neff M, Jones K, Wild K, Wolff JA (2003) Expression of phenylalanine hydroxylase (PAH) in erythrogenic bone marrow does not correct hyperphenylalaninemia in Pah(enu2) mice. J Gene Med 5:984–993
Hsiao CY, Chen TH, Chu TH, Ting YN, Tsai PJ, Shyu JF (2020) Calcitonin induces bone formation by increasing expression of wnt10b in osteoclasts in ovariectomy-induced osteoporotic rats. Front Endocrinol 8:613
Holmbeck K, Bianco P, Pidoux I, Inoue S, Billinghurst RC, Wu W, Chrysovergis K, Yamada S, Birkedal-Hansen H, Poole AR (2005) The metalloproteinase mt1-mmp is required for normal development and maintenance of osteocyte processes in bone. J Cell Sci 118:147–156
Hu SD, He WL, Wu (2022) Hydroxyproline in animal metabolism, nutrition, and cell signaling. Amino Acids 54:513–528
Hu SD, He WL, Bazer FW, Johnson GA, Wu G (2023a) Synthesis of glycine from 4-hydroxyproline in tissues of neonatal pigs. Exp Biol Med 248:1206–1220
Hu SD, He WL, Bazer FW, Johnson GA, Wu G (2023b) Synthesis of glycine from 4-hydroxyproline in tissues of neonatal pigs with intrauterine growth restriction. Exp Biol Med 248:1446–1458
Hughes-Fulford M (2004) Signal transduction and mechanical stress. Sci STKE 2004:249
Kippenes H, Johnston G (1998) Diagnostic imaging of osteochondrosis. Vet Clin Am Small Anim Pract 28:137–160
Kuznetsov SA, Krebsbach PH, Satomura K, Kerr J, Riminucci M, Benayahu D, Robey PG (1997) Single-colony derived strains of human marrow stromal fibroblasts form bone after transplantation in vivo. J Bone Miner Res 12:1335–1347
Lascelles BDX, DePuy V, Thomson A, Hansen B, Marcellin-Little DJ, Biourge V, Bauer JE (2010) Evaluation of a therapeutic diet for feline degenerative joint disease. J Vet Intern Med 24:487–495
Lascelles B, Duncan X (2010) Feline degenerative joint disease. Vet Surg 39:2–13
Li P, Wu G (2018) Roles of dietary glycine, proline, and hydroxyproline in collagen synthesis and animal growth. Amino Acids 50:29–38
Li P, Wu G (2020) Composition of amino acids and related nitrogenous nutrients in feedstuffs for animal diets. Amino Acids 52:523–542
Li P, Wu G (2022) Functional molecules of intestinal mucosal products in animal nutrition and health. Adv Exp Med Biol 1354:263–277
Li P, Wu G (2023) Amino acid nutrition and metabolism in domestic cats and dogs. J Anim Sci Biotechnol 14:19
Li P, He WL, Wu G (2021) Composition of amino acids in foodstuffs for humans and animals. Adv Exp Med Biol 1332:189–210
Li XL, Bazer FW, Gao H, Jobgen W, Johnson GA, Li P, McKnight JR, Satterfield MC, Spencer TE, Wu G (2009) Amino acids and gaseous signaling. Amino Acids 37:65–78
Maniaki E, Murrell J, Langley-Hobbs SJ, Blackwell EJ (2021) Associations between early neutering, obesity, outdoor access, trauma and feline degenerative joint disease. J Feline Med Surg 23:965–975
May C (2005) Diagnosis and management of bacterial infective arthritis in dogs and cats. In Pract 27:316–321
McAlindon TE, La Valley MP, Gulin JP, Felson DT (2000) Glucosamine and chondroitin for treatment of osteoarthritis: a systematic quality assessment and meta-analysis. JAMA 283:1469–1475
Meininger CJ, Kelly KA, Li H, Haynes TE, Wu G (2000) Glucosamine inhibits inducible nitric oxide synthesis. Biochem Biophys Res Commun 279:234–239
Minami S, Hata M, Tamai Y, Hashida M, Takayama T, Yamamoto S, Okada M, Funatsu T, Tsuka T, Imagawa T, Okamoto Y (2011) Clinical application of D-glucosamine and scale collagen peptide on canine and feline orthopedic diseases and spondylitis deformans. Carbohyd Polym 84:831–834

Mobasheri A, Mahmoudian A, Kalvaityte U, Uzieliene I, Larder CE, Iskandar MM, Kubow S, Cesar Hamdan P, Scala de Almeida Jr C, Favazzo LJ, van Loon LJ, Emans PJ, Plapler PG, Zuscik MJ, Mobasheri alimobasheri A, Larder christina CE, Scala de Almeida C, van Loon lvanloon LJ, Pieter Emans maastrichtuniversitynl J (2021) A white paper on collagen hydrolyzates and ultrahydrolyzates: potential supplements to support joint health in osteoarthritis. Curr Rheumatol Rep 23:78

Moe SM, Chen NX, Seifert MF, Sinders RM, Duan D, Chen X, Liang Y, Radcliff JS, White KE, Gattone VH (2009) A rat model of chronic kidney disease-mineral bone disorder. Kidney Int 75:176–184

Parfitt AM (1988) Bone histomorphometry: standardization of nomenclature, symbols and units (summary of proposed system). Bone 9:67–69

Parker VJ, Rudinsky AJ, Benedict JA, Beizaei A, Chew DJ (2020) Effects of calcifediol supplementation on markers of chronic kidney disease-mineral and bone disorder in dogs with chronic kidney disease. J Vet Intern Med 34:2497–2506

Pérez-Hernández E, Pastrana-Carballo JJ, Gómez-Chávez F, Gupta RC, Pérez-Hernández N (2022) A key metabolic regulator of bone and cartilage health. Endocrinol Metab 37:559–574

Perry K (2016) Feline hip dysplasia: a challenge to recognize and treat. J Feline Med Surg 18:203–218

Poole KES, Van Bezooijen RL, Loveridge N, Hamersma H, Papapoulos SE, Löwik CW, Reeve J, Poole K (2005) Sclerostin is a delayed secreted product of osteocytes that inhibits bone formation. FASEB J 19:1842–1844

Quarles LD (2008) Endocrine functions of bone in mineral metabolism regulation. J Clin Invest 118:3820–3828

Raggatt LJ, Partridge NC (2010) Cellular and molecular mechanisms of bone remodeling. J Biol Chem 285:25103–25108

Rychel JK (2010) Diagnosis and treatment of osteoarthritis. Top Companion Anim Med 25:20–25

Schachner ER, Lopez MJ (2015) Diagnosis, prevention, and management of canine hip dysplasia: a review. Vet Med 181:1

Smith GK (1998) Canine hip dysplasia: pathogenesis, diagnosis, and genetic control. Vet Q 20:S22–S24

Stegen S, Loopmans S, Stockmans I, Moermans K, Carmeliet P, Carmeliet G (2022) De novo serine synthesis regulates chondrocyte proliferation during bone development and repair. Bone Res 10:14

Takahashi N, Yamana H, Yoshiki S, Roodman GD, Mundy GR, Jones SJ, Boyde A, Suda T (1988) Osteoclast-like cell formation and its regulation by osteotropic hormones in mouse bone marrow cultures. Endocrinology 122:1373–1382

Viguet-Carrin S, Garnero P, Delmas PD (2006) The role of collagen in bone strength. Osteoporos Int 17:319–336

Wagley Y, Chesi A, Acevedo PK, Lu S, Wells AD, Johnson ME, Grant SFA, Hankenson KD (2020) Canonical notch signaling is required for bone morphogenetic protein-mediated human osteoblast differentiation. Stem Cells 38:1332–1347

Wang WW, Wu ZL, Lin G, Hu SD, Wang B, Dai ZL, Wu G (2014) Glycine stimulates protein synthesis and inhibits oxidative stress in pig small-intestinal epithelial cells. J Nutr 144:1540–1548

Wu G (2018) Principals of animal nutrition. CRC Press, Boca Raton, Floria

Wu G (2022) Amino acids: biochemistry and nutrition. CRC Press, Boca Raton, Floria

Wu G, Haynes TE, Li H, Yan W, Meininger CJ (2001) Glutamine metabolism to glucosamine is necessary for glutamine inhibition of endothelial nitric oxide synthesis. Biochem J 353:245–252

Wu G, Bazer FW, Burghardt RC, Johnson GA, Kim SW, Knabe DA, Li P, Li XL, McKnight JR, Satterfield MC, Spencer TE (2011) Proline and hydroxyproline metabolism: implications for animal and human nutrition. Amino Acids 40:1053–1063

Wu G, Meininger CJ, McNeal CJ, Bazer FW, Rhoads JM (2021) Role of L-arginine in nitric oxide synthesis and health in humans. Adv Exp Med Biol 1332:167–187

Yang Y, Wang J, Yao M, Li X, Lu X, He J, Zhang H, Tian B, Zhou J (2023) An update on the review of microbial synthesis of glucosamine and N-acetylglucosamine. World J Microbiol Biotechnol 39:1–11

Yao K, Yin YL, Feng ZM, Tang ZR, Fang J, Wu G (2011) Tryptophan metabolism in animals: important roles in nutrition and health. Front Biosci S3:286–297

Yin Y, Tang Q, Xie M, Hu L, Chen L (2021) Insights into the mechanism of vascular endothelial cells on bone biology. Biosci Rep 41:1

Yoo HS, Chung KH, Lee KJ, Kim DH, An JH (2017) Melanin extract from Gallus gallus domesticus promotes proliferation and differentiation of osteoblastic MG-63 cells via bone morphogenetic protein-2 signaling. Nutr Res Pract 11:190–197

Zhang K, Barragan-Adjemian C, Ye L, Kotha S, Dallas M, Lu Y, Zhao S, Harris M, Harris SE, Feng JQ, Bonewald LF (2006) E11/gp38 Selective expression in osteocytes: regulation by mechanical strain and role in dendrite elongation. Mol Cell Biol 26:4539–4552

Chapter 8
Roles of Nutrients in the Brain Development, Cognitive Function, and Mood of Dogs and Cats

Guoyao Wu

Abstract The brain is the central commander of all physical activities and the expression of emotions in animals. Its development and cognitive health critically depend on the neural network that consists of neurons, glial cells (namely, non-neuronal cells), and neurotransmitters (communicators between neurons). The latter include proteinogenic amino acids (e.g., L-glutamate, L-aspartate, and glycine) and their metabolites [e.g., γ-aminobutyrate, D-aspartate, D-serine, nitric oxide, carbon monoxide, hydrogen sulfide, and monoamines (e.g., dopamine, norepinephrine, epinephrine, and serotonin)]. In addition, some non-neurotransmitter metabolites of amino acids, such as taurine, creatine, and carnosine, also play important roles in brain development, cognitive health, behavior, and mood of dogs and cats. Much evidence shows that cats require dietary ω3 (α-linolenic acid, eicosapentaenoic acid, and docosahexaenoic acid) and ω6 (linoleic acid and arachidonic acid) polyunsaturated fatty acids for the development of the central nervous system. As an essential component of membranes of neurons and glial cells, cholesterol is also crucial for cognitive development and function. In addition, vitamins and minerals are required for the metabolism of AAs, lipids, and glucose in the nervous system, and also act as antioxidants. Thus, inadequate nutrition will lead to mood disorders. Some amino acids (e.g., arginine, glycine, methionine, serine, taurine, tryptophan, and tyrosine) can help to alleviate behavioral and mood disorders (e.g., depression, anxiety and aggression). As abundant providers of all these functional amino acids and lipids, animal-sourced foods (e.g., liver, intestinal mucosa, and meat) play important roles in brain development, cognitive function, and mood of dogs and cats. This may explain, in part, why dogs and cats prefer to eat visceral organs of their prey. Adequate provision of nutrients in all phases of the life cycle (pregnancy, lactation, postnatal growth, and adulthood) is essential for optimizing neurological health, while preventing cognitive dysfunction and abnormal behavior.

Keywords Arginine · Brain · Cats · Cognition · Dogs · Fatty acids · Mood · Taurine · Tryptophan · Vitamins

---

G. Wu (✉)
Department of Animal Science, Texas A&M University, College Station, TX 77843, USA
e-mail: g-wu@tamu.edu

G. Wu (ed.), *Nutrition and Metabolism of Dogs and Cats*, Advances in Experimental Medicine and Biology 1446, https://doi.org/10.1007/978-3-031-54192-6_8

Abbreviations

AA	Amino acid
BH4	Tetrahydrobiopterin
BPB	Brain protection blend
BW	Body weight
DHA	Docosahexaenoic acid
EPA	Eicosapentaenoic acid
EAA	Nutritionally essential amino acid
GABA	γ-Aminobutyrate
NO	Nitric oxide
PUFAs	Polyunsaturated fatty acids

8.1 Introduction

As the central commander of the body, the brain is a highly complex organ responsible for cognitive function, behavior, and mood in mammals (Stiles and Jernigan 2010). In dogs and cats, the growth and development of this organ begins from neural progenitor cells during early (approximately day 15) gestation when these cells rapidly divide and differentiate into neurons and glia as the major constituents of the central nervous system (de Morais-Pintoet al. 2021; Fox 1971). In both dogs and cats that have a similar gestation length (63–65 days; Fontaine 2012; Nelson and Cooper 1975), the telencephalic vesicles (the first cerebral vesicles leading to the establishment of forebrain) develop rapidly from approximately day 20 of gestation and are well formed by day 28 of gestation. In these two carnivorous species, the fetal brain grows more rapidly than the fetus but are very sensitive to nutritional insults, and brain structures are readily recognizable by day 55 of gestation (Packard 2021). At birth, the brains of dog and cat puppies account for 2.74 and 3.61% of the body weights (BW), respectively, reaching approximately 20% of its adult size (Latimer 1967). Maternal nutrition [e.g., availabilities of amino acids (AAs; particularly taurine, glycine, arginine, cysteine, and tryptophan), glucose, polyunsaturated fatty acids (PUFAs), monounsaturated fatty acids, vitamins, minerals, and water] plays crucial roles in the prenatal development of canine and feline brains (Bosch et al. 2007; Bray et al. 2021; Gaillard et al. 2022; Pretzer 2008).

After birth, canine and feline brains continue to grow and develop with the continuous turnover of neurons as well as the continuous creation and dissipation of neural connections, reaching mature mass and functional states in adulthood (Arant and Gooch 1982; Bray et al. 2021). Compared with herbivores, adult dogs and cats have a relatively large brain, which accounts for 0.85 and 0.80% of the BW, respectively (Latimer 1967). Interestingly, dogs appear to have more neurons in the brain than cats (Jardim-Messede et al. 2017), which may explain a difference in intelligence between the two carnivores. With postnatal development, dogs and cats acquire information from the environment and learn about conceptual skills, as well as memorize, analyze, retrieve, and use the knowledge. This process is defined as cognition. For example, these companion animals can recognize human faces and voices; are sensitive to the emotional, visual, and auditory signals from individuals of the same and other species (including humans); adjust behavior to human emotions; and may experience many signs of mood disorders (e.g., depression, irritation, eating disorders, discomfort, and seclusion) (Andics et al. 2014; Quaranta et al. 2020; Racca et al. 2010). The major objective of this article is to highlight the roles of nutrients in the brain development, cognitive function, and mood of dogs and cats.

8.2 Anatomical and Physiological Bases of Brain Development and Cognitive Functions

Brain development and cognitive function of animals have anatomical and physiological bases (Nieto-Alamilla et al. 2016). This necessitates a brief review of the nervous system in both dogs and cats (Fig. 8.1). The nervous system consists of the central nervous system and the peripheral nervous system. The central nervous system consists of (a) the brain (brain stem, forebrain, and midbrain) that is important for the intelligence and behavior of animals and (b) the spinal cord. The peripheral nervous system includes (a) the involuntary nervous system that controls the internal organs and (b) the voluntary nervous system that controls skeletal muscle. The involuntary nervous system is composed of the norepinephrine-dependent sympathetic nervous system, the acetylcholine-dependent para-sympathetic system, and the non-norepinephrine, non-acetylcholine but nitric oxide (NO)-dependent nervous system (Wu 2022). Norepinephrine is synthesized from tyrosine (the hydroxylation product of phenylalanine), acetylcholine from choline plus acetyl-CoA, NO from arginine, and choline from serine plus palmitate plus methionine (via S-adenosylmethionine) by animals (Wu 2018). Phenylalanine, tyrosine, serine, arginine, and methionine are among the dietary AAs required by both dogs and cats (Li and Wu 2023a).

There are major integrated parts in the brain: forebrain, hippocampus, cerebral cortex, hypophysis, and striatum (Hellier 2014). The forebrain consists of the thalamus (the main information relay station for the brain), hypothalamus (the brain's control coordinating center to keep the body's internal functions in balance), cerebrum (that initiates and coordinates movement and regulates body temperature), and olfactory bulb (that receives neural input about odors from the nasal cavity for sensing smell). The cerebellum is a major structure of the hindbrain located near the brainstem to maintain balance. In addition, the hippocampus has a major role in learning and memory (Sase et al. 2013). The cerebral cortex is the outer surface of the brain and is associated with consciousness, thought, emotion, reasoning, language, and memory. The hypophysis (also known as the pituitary gland) regulates growth, reproduction, metabolism, stress response, and lactation of animals by relaying signals from the

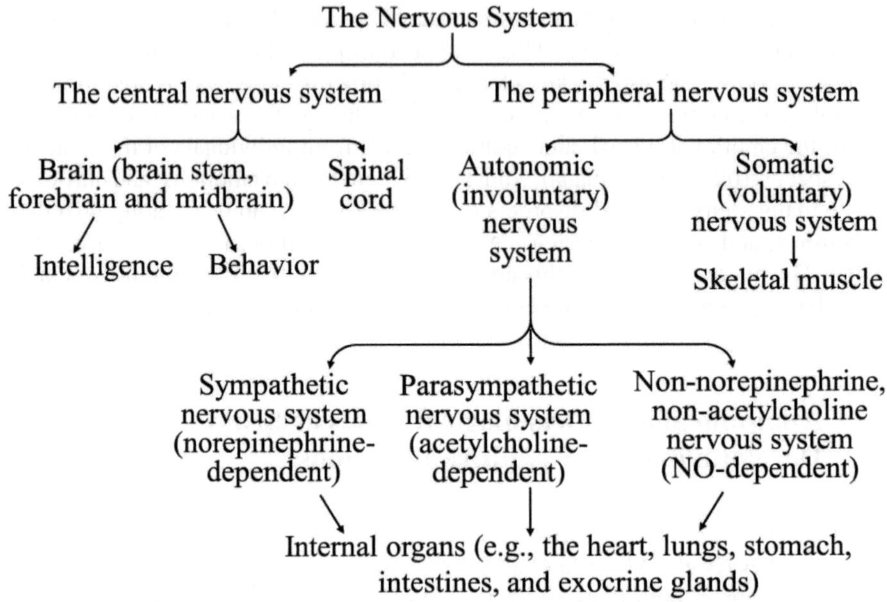

Fig. 8.1 The nervous system in dogs and cats. The nervous system consists of the central nervous system (the brain and spinal cord) and the peripheral nervous system. The brain processes and interprets sensory information transmitted from the spinal cord. Development of the brain and cognitive function affects the intelligence, mood, and behavior of animals

brain to peripheral organs. The striatum coordinates multiple aspects of cognition, which includes motor and action planning, decision-making, motivation, reinforcement, and reward perception. All these components of the brain are integrated for neurological function and supported by proper nutrition (Brodal 2004).

The neural network is the foundation of brain function. It is composed of neurons (the basic unit of the nervous system), neurotransmitters (namely communicators between neurons; Fig. 8.2), and glial cells (namely, non-neuronal cells) that support and nourish neurons, produce myelin, and participate in signal transmission in the nervous system. The distribution of neurotransmitters in the different parts of the brain is summarized in Table 8.1. The neurotransmitters consist of AAs [L-glutamate, L-aspartate, glycine, D-serine, and γ-aminobutyrate (GABA)] and their metabolites (Ajinkya et al. 2016; He and Wu 2020). The latter include gasotransmitters [such as nitric oxide (NO), carbon monoxide, and hydrogen sulfide], also monoamines (such as dopamine, norepinephrine, epinephrine, histamine, agmatine, melatonin, and serotonin), and peptides (such as substance P, opioids, gastrin, cholecystokinin, oxytocin, vasopressin, peptide YY, and neuropeptide Y). In the brain, neurotransmitters are released from presynaptic neurons into the synaptic cleft and then act on receptors in post-synaptic neurons. Note that some non-neurotransmitter metabolites of AAs also play important roles in the brain development and cognitive health of dogs and cats. For example, creatine (synthesized from Arg, Gly, and Met via

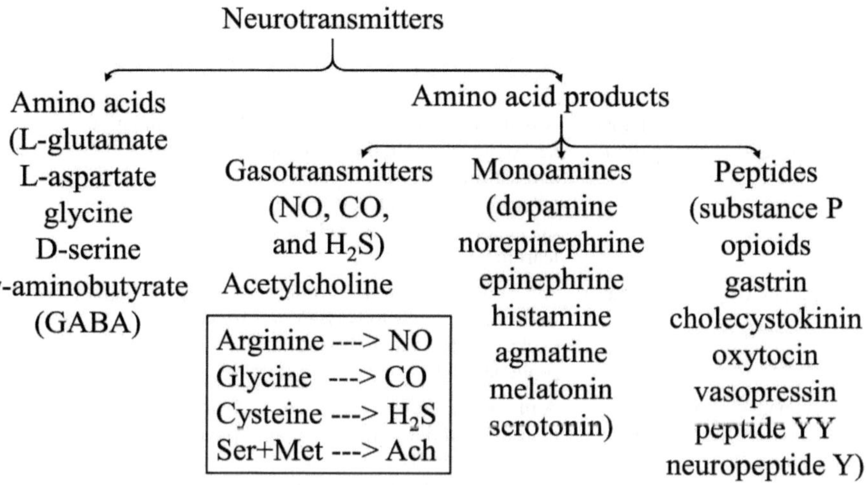

Fig. 8.2 Neurotransmitters in the brain of dogs and cats. These molecules are proteinogenic amino acids (e.g., L-glutamate, L-aspartate, and glycine) and their metabolites (e.g., D-serine and γ-aminobutyrate). The neurotransmitters generated from amino acids include gasotransmitters, monoamines, and peptides. In the brain, neurotransmitters are released from presynaptic neurons into the synaptic cleft and then act on receptors in post-synaptic neurons. CO, carbon monoxide; Ach, acetylcholine; H_2S, hydrogen sulfide; Met, L-methionine; NO, nitric oxide; Ser, L-serine

the interorgan cooperation involving the kidneys and liver) is essential for energy metabolism, as well as the development of the brain and other tissues of the nervous system (Wyss and Kaddurah-Daouk 2000). In addition, carnosine, taurine, and 4-hydroxyproline are abundant in the brain, and these substances also contribute to the health and development of the nervous system (Wu 2020). Adequate synthesis and/or consumption of creatine, anserine, carnosine, and taurine (which are present in animal-sourced foodstuffs but are absent from plant-sourced foodstuffs) is crucial for neurological development and function.

Large neutral AAs (e.g., branched-chain AAs, phenylalanine, tyrosine, and tryptophan), small neutral AAs (e.g., alanine, cysteine, glycine, serine, and taurine), AAs with a side-chain NH_2 group (e.g., glutamine, citrulline, and asparagine), and basic AAs (e.g., arginine, lysine, histidine, and ornithine) in the blood readily cross the blood–brain barrier into the brain (He and Wu 2020). However, the blood–brain barrier restricts the entry of physiological concentrations of imino acids (e.g., proline and hydroxyprolines) and acidic AAs (e.g., glutamate and aspartate) in the blood into the brain (He and Wu 2020). Proper balances and amounts of AAs in diets are crucial for the development and function of the brain as well as the normal mood of animals (Wu 2022).

Table 8.1 Distribution of neurotransmitters in the different parts of brain

Neurotransmitter	Actions	Site of production within the brain
L-Aspartate (from L-glutamate, L-alanine, and L-asparagine)	A major excitatory neurotransmitter in the brain; an agonist for NMDA receptors; enhance synaptic plasticity and the neural network, as well as learning and memory	Neurons and glial cells; spinal cord
D-Aspartate (from L-aspartate)	An excitatory neurotransmitter in the brain; binds to NMDA; evokes L-glutamate release; promotes LTP	Neurons and glial cells; spinal cord
L-Glutamate (from BCAAs, L-glutamine, NH_4^+ + α-KG, and L-alanine)	A major excitatory neurotransmitter in the brain; an agonist for NMDA receptors; enhance synaptic plasticity and the neural network, as well as learning and memory	Neurons and glial cells; spinal cord
Glycine (from serine and 4-hydroxyproline)	A major inhibitory neurotransmitter in the spinal cord and the lower brainstem; a co-agonist with L-glutamate at NMDA receptors	Neurons and glial cells; spinal cord
Aminobutyrate (GABA) (from glutamate)	A major inhibitory transmitter in the mature brain but is primarily excitatory in the developing brain	GABAergic neurons[a]
Carbon monoxide (CO) (from glycine via heme)	Modulates synaptic plasticity and neuronal activity; promotes LTP	Heme oxygenase is localized in many regions of the brain[b]
Acetylcholine (from choline and acetyl-CoA)	Allows impulses to be fired at the synaptic cleft; vital for cognition and perception	Cholinergic neurons in the striatum
Dopamines (from tyrosine)	Critical for movement, mood, addiction, and response to stress	Dopaminergic neurons in midbrain and forebrain[c]
Epinephrine (from tyrosine)	Plays roles in metabolism, attention, focus, panic, and excitement; enhances memory formation process	Medulla oblongata of the brainstem
Norepinephrine (from tyrosine)	Modulates emotion, sleep, attention, focus, learning, and response to ANS	Mainly neurons of locus coeruleus; also the lateral tegmental field

(continued)

Table 8.1 (continued)

Neurotransmitter	Actions	Site of production within the brain
H_2S (from cysteine)	Modulates synaptic plasticity and neuronal activity; facilitates LTP	Astrocytes, microglia, and neurons[d]
Histamine (from histidine)	Plays an essential role in the control of circadian rhythms	Histaminergic neurons[e]
Nitric oxide (NO) (from arginine)	Modulates synaptic plasticity and neuronal activity; inhibits NMDA receptor activity	Various regions[f]
Serotonin (from tryptophan)	Modulates sleep–wakefulness cycles, the processing of afferent inputs, mood, and food intake	Serotonergic neurons mainly in the brainstem raphe complex[g]

[a]These GABAergic neurons are located in the hippocampus, thalamus, basal ganglia, hypothalamus, and brainstem
[b]Including the hypothalamus and olfactory bulb
[c]Dopaminergic neurons extend from the midbrain (e.g., the substantia nigra pars compacta, the ventral tegmental area, the arcuate nucleus, and the retrorubral area) through the forebrain (e.g., the olfactory bulb)
[d]H_2S is produced by cystathionine β-synthase (CBS), cystathionine gamma-lyase (CSE), and 3-mercaptopyruvate sulfurtransferase (3MST); CBS is highly expressed in the astrocytes and microglial cells of the hippocampus and cerebellum; CSE is mainly expressed in the cardiovascular system but is also present in microglial cells, spinal cord, and cerebellar granule neurons; and 3MST occurs in the neurons and astrocytes of the brain
[e]Histaminergic neurons are confined to the tuberomammillary nucleus of the posterior hypothalamus
[f]Neuronal nitric oxide synthase (nNOS) is expressed in the neurons and pineal gland of the brain and in the spinal cord; in the juvenile and mature hippocampus and neocortex, nNOS is primarily expressed in subpopulations of GABAergic interneurons
[g]A midline complex closely associated with the reticular formation within the midbrain, pons, and medulla. ANS, autonomic nervous system; BCAAs, branched-chain amino acids; LTP, long-term potentiation; NMDA, N-methyl-D-aspartate

8.3 Requirements for Arginine to Improve Brain Development and Cognitive Function in Dogs and Cats

Arginine plays an important role in regulating synapse formation/patterning and angiogenesis in the developing brain (Tachikawa et al. 2018; Virarkar et al. 2013), as well as learning and memory retention (Wang et al. 2016). Specifically, arginine is required for the production of NO (serving as both a neurotransmitter and an angiogenic factor) by neurons, as noted previously. The synthesis of NO requires not only arginine but also tetrahydrobiopterin (BH4), NADPH, Ca, FAD, FMN, and calmodulin. Both FAD and FMN are formed from riboflavin (vitamin B_2). BH4 is easily oxidized under stress conditions and must be protected in cells through anti-oxidative reactions such as the synthesis of glutathione from glycine, glutamate and cysteine (Shi et al. 2004). Within cells, BH4 is also stabilized by vitamin C (Meininger and Wu 2002) that can be synthesized from D-glucose by dogs and cats (Wu 2018). Of note, NO synthase is activated by oleic acid, an ω9 monounsaturated fatty acid abundant in beef and pork (Wu and Meininger 2002). In addition, physiological concentrations of arginine and cholesterol increase NO synthesis in the brain by stimulating the expression of endothelial and neuronal NO synthases, as well as GTP cyclohydrolase I for BH4 production (Li et al. 2008; Shi et al. 2004; Wu and Meininger 2002).

Presynaptic neurons synthesize NO from L-arginine by NO synthetase. NO activates cGMP-dependent protein kinase G. The protein kinase G then catalyzes the phosphorylation of a protein called synaptophysin (Fig. 8.3). The phosphorylated synaptophysin promotes the fusion of glutamate-containing granules with the membrane of the presynaptic neuron, resulting in the release of L-glutamate (an excitatory neurotransmitter). The glutamate is taken up by post-synaptic neurons for the generation of long-term potentiation, leading to a long-lasting increase in signal transmission between neurons (Fig. 8.3). Compared with adult humans, adult dogs have a 90% greater rate of whole-body arginine catabolism and consequently a much higher requirement for this AA (Li and Wu 2023a). Despite the endogenous synthesis of arginine, both young and adult dogs must ingest adequate arginine in diets (e.g., 0.4% Arg in the diet) to maintain vital physiological functions (including ammonia detoxification and hemodynamics) beyond nitrogen balance. This means that arginine synthesis in dogs is insufficient to meet the needs and, therefore, can be regarded to be limited. Cats have a more limited ability to synthesize arginine than dogs and, therefore, must also depend on dietary arginine (much greater than dogs) for metabolic utilization, survival, growth, and development (including neurological development) (Li and Wu 2023a; Morris and Rogers 1978). In both dogs and cats, a deficiency of arginine results in hyperammonemia (e.g., occurring within 1–3 h after feeding in cats); neurological signs such as seizures, emesis, ataxia, and muscle spasms; and even death (Czarnecki and Baker 1984; Morris and Rogers 1978).

8.4 Requirements for Taurine to Improve Brain Development and Cognitive Function in Dogs and Cats

Taurine plays major roles in the physiology and nutrition of all mammals, including dogs and cats (Schafer and Kim 2018). This AA serves as: (1) a nutrient to conjugate bile acid, forming bile salt in the liver that facilitate the intestinal absorption of dietary lipids (including lipid-soluble vitamins) and eliminate cholesterol in bile via the fecal route; (2) a major antioxidant, anti-inflammatory, and anti-apoptotic factor in the body; (3) a physiological stabilizer of cell membranes; (4) a regulator of Ca^{2+} signaling, fluid homeostasis in cells, and retinal photoreceptor activity; (5) a contributor to osmoregulation; (6) a key component of nerve and muscle conduction

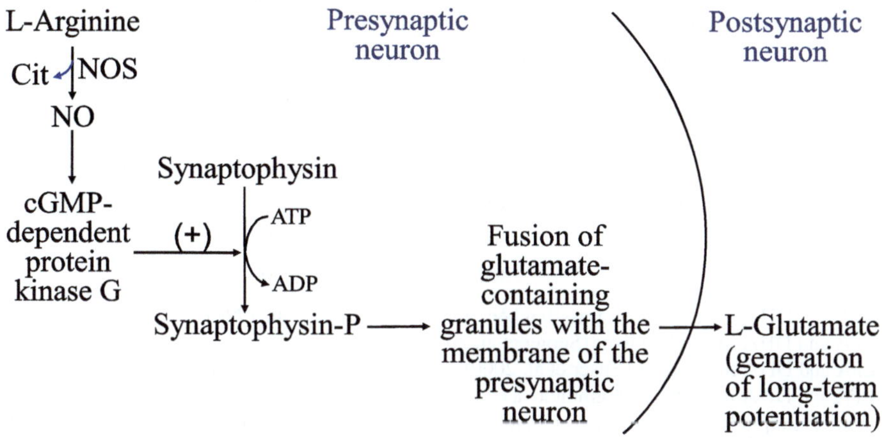

Fig. 8.3 Nitric oxide (NO) as a gasotransmitter in the brain of dogs and cats. Presynaptic neurons synthesize and release NO, which activates cGMP-dependent protein kinase G to catalyze the phosphorylation of a protein called synaptophysin. The phosphorylated synaptophysin promotes the fusion of glutamate-containing granules with the membrane of the presynaptic neuron, resulting in the release of L-glutamate. The latter is taken up by post-synaptic neurons for the generation of long-term potentiation, leading to a long-lasting increase in signal transmission between neurons. Cit, L-citrulline; NOS, nitric oxide synthase

networks; (7) a stimulator of neurological development; (8) an inhibitory neurotransmitter in the central nervous system; and (9) conjugation with bile acid to form bile salt in the liver (Schafer and Kim 2018; Santulli et al. 2023).

Taurine is synthesized sufficiently from cysteine in the liver of most breeds of dogs when fed methionine- or cysteine-adequate diets, but this metabolic pathway is very limited in cats (Knopf et al. 1978). Thus, cats must consume taurine-rich foods (Che et al. 2021). Of note, taurine is particularly abundant in the brain of dogs and cats, ranging from 25 to 30 mM (Lane et al. 1978; Malloy et al. 1981; Sturman et al. 1985). Taurine is essential for the growth, development (including neurological development), structure, and functions (including cognitive function) of the brain, other tissues of the nervous system, and skeletal muscle in dogs and cats (Chen et al. 2019; Wu 2020). Those neurological functions include the generation and regulation of nerve impulses, anti-oxidative effects, and locomotion. For example, the lack of taurine in maternal diets impairs the growth of the brain and the whole body in fetal cats, resulting in an abnormal morphology in the cerebellum and visual cortex, but fortunately these defects can be prevented by dietary supplementation with taurine (e.g., 0.05% taurine) beginning at early pregnancy (Sturman et al. 1985, 1994). In addition, when maternal diets do not provide taurine, cerebellum development in fetal rats is severely impaired, as indicated by a marked decrease in the number of normal granule cells (the most dominant neurons) in the cerebellum (Sturman and Lu 1997). Notably, augmenting the content of taurine in the maternal diet from 0.01 to 0.09% dose-dependently can increase the percentage of normal granule cells in fetal cats on day 45 of gestation from 71 to 99%, while reducing the nuclear area of

Table 8.2 Concentrations of taurine and lipids in the livers of animals

Animal	Concentrations in liver (mg/100 g of fresh tissue)			
	Taurine[e]	Cholesterol[e]	EPA	DHA
Beef[a]	104	402	151	83
Chicken[b]	115	558	16	43
Lamb[a]	101	491	155	171
Pig[c]	136	360	33	90
Rat[d]	118	364	38	612

[a] EPA and DHA in the liver of grass-fed beef and lambs (Enser et al. 1998)
[b] EPA and DHA in chicken liver (Seong et al. 2015)
[c] EPA and DHA in pig liver (Enser et al. 2000)
[d] EPA and DHA in rat liver (Zigmundova et al. 2018)
[e] Wu, G. (unpublished data). Taurine and cholesterol in livers were determined as described by Tekwe et al. (2023)

abnormal cells from 32 to 23 μm^2 (Sturman and Lu 1997). Collectively, these results indicate a crucial role for taurine in the brain development and cognitive function of cats and dogs.

Taurine is abundant in animal-sourced foods such as liver (Table 8.2) and muscle (Wu 2022) but is absent from plant-sourced foods (Li et al. 2021a,b,c; Li and Wu 2020, 2022, 2023b). Among all the animal-sourced foodstuffs that have been analyzed (Li and Wu 2020), taurine is the most abundant in fishmeal. The content of taurine in common foodstuffs (g/kg feed, as-fed basis) is as follows: Southeast Asian fishmeal, 9.86; Peruvian fishmeal, 7.49; menhaden fishmeal, 4.09; poultry by-product meal, 3.88; spray-dried poultry plasma, 2.46; chicken by-product meal, 2.10; porcine mucosal protein, 1.64; chicken visceral digest, 1.32; spray-dried egg product, 0.33; feather meal, 0.14; and black soldier fly larvae meal, 0.09 (Li and Wu 2020). Porcine intestinal-mucosal product contains 1.65 g/kg (as-fed basis; Li and Wu 2022). Note that taurine is absent from plant-sourced foodstuffs (Wu 2022).

8.5 Requirements for PUFAs to Improve Brain Development and Cognitive Function in Dogs and Cats During Gestation and Growing Periods

8.5.1 Insufficient Synthesis of PUFAs in Dogs and Cats

Just like other animals, dogs and cats cannot introduce a double bond beyond the Δ^9 carbon due to the lack of an appropriate desaturase and, therefore, cannot form α-linolenic acid ($C_{18:3}$, $\omega3$) or linoleic acid ($C_{18:2}$, $\omega6$) from palmitate (Wu 2018). In addition, Δ^6-desaturase activity is very low in tissues of young and adult cats

(Bauer 2016). Thus, all age groups of cats have a very limited ability to synthesize (a) 5,8,11,14,17-eicosapentaenoic acid (EPA, $C_{20:5}$, ω3) and 5,8,11,14,17,20-docosahexaenoic acid (DHA, $C_{22:6}$, ω3) from α-linolenic acid and (b) arachidonic acid ($C_{20:4}$, ω6) from linoleic acid (Bauer 2016; Pawlosky et al. 1994, 1997). Although growing and young adult dogs can actively convert α-linolenic acid into EPA and DHA, their synthesis is insufficient to meet their needs (Bauer 2016). In addition, Δ^6-desaturase activity in the tissues of older adult dogs may be substantially reduced when compared with young adults, as reported for rats (Bordoni et al. 1988), thereby decreasing the synthesis of arachidonic acid from linoleic acid, as well as EPA and DHA from α-linolenic acid in aged dogs. All these unsaturated fatty acids are essential components of cell membranes in neurons for nutrient transport and signal transduction and, therefore, are vital for the growth, development, integrity, and function of the central nervous system (Bazinet and Layé 2014; Bryan et al. 2001; Lemaitre et al. 2009). Consequently, canine and feline diets for growth, reproduction, and gestation must contain α-linolenic acid, linoleic acid, arachidonic acid, EPA, and DHA (NRC 2006).

In dogs and cats, dietary long-chain fatty acids (including PUFAs as well as monounsaturated fatty acids and saturated fatty acids) are readily absorbed by the small intestine into the lymphatic vessels, reaching the heart and then the systemic blood circulation (Wu 2018). The unsaturated fatty acids (e.g., α-linolenic acid, eicosapentaenoic acid, docosahexaenoic acid, linoleic acid, γ-linolenic acid, and arachidonic acid) in the blood can be taken up by the brain through either passive diffusion or active transporters, followed by esterification into distinct phospholipids in the cell membrane and subcellular organelles (Lacombe et al. 2018). PUFAs play important roles in the function of the synapse (the basic structural and operational unit for information processing and storage in the brain) by affecting the stabilization and activity of neurotransmitter receptors, the release of neurotransmitter vesicles, and the formation and development of synapses (Miceli et al. 2020). Note that DHA and arachidonic acid are the major ω3 and ω6 PUFAs in the brain, respectively (Pawlosky et al. 1997).

8.5.2 Sources of Fatty Acids for the Diets of Dogs and Cats

Many pet owners prefer to use various ingredients to prepare meals for their dogs and cats. Thus, it is imperative to understand the composition of PUFAs in common foods. Animal-sourced foodstuffs (e.g., beef tallow, pork lard, and poultry fat) and certain plant-sourced foodstuffs (e.g., cottonseed meal and palm oil) contain high content of both palmitate and oleic acid (Rohman et al. 2012; USDA 2022). Palmitate is required for choline synthesis (Wu 2018), and oleic acid stimulates the synthesis of physiological levels of NO from arginine (Wu and Meininger 2002). In addition, animal-sourced foodstuffs (e.g., liver; Table 8.2) are good providers of linoleic acid, α-linolenic acid, EPA, and DHA for the diets of domestic animals including dogs and cats (Rohman et al. 2012), whereas algae and cold-water fish contain high amounts of both EPA and DHA (USDA 2022). Based on the sustainability of petfoods, DHA from microalgae is a viable alternative to fish oil for supplementation to the diets of animals (Remize et al. 2021), including dogs (Dahms et al. 2019; Risolia et al. 2018) and cats (Scheibel et al. 2021; Vuorinen et al. 2020). For comparison, plant-sourced foodstuffs (e.g., corn, soybean meal, and wheat) contain α-linolenic acid and linoleic acid, but little or no EPA and DHA (Wu 2018). This may explain, in part, why plant-sourced ingredients cannot replace most or all animal-sourced ingredients in the diets of cats or aging dogs for brain development or cognitive function.

8.5.3 Requirements for PUFAs to Improve Brain Development and Cognitive Function in Dogs During Gestation and Growing Periods

There is evidence that dietary supplementation with DHA or DHA-rich food ingredients to maternal diets during pregnancy and puppy diets during the neonatal period or after weaning can improve brain development and cognitive function in dogs. For example, dietary supplementation with 0.18% DHA to maternal diets during the entire gestation and 7- or 9-week lactation improves brain development and cognitive function in puppies (Kelley and Lepine 2005; Kelley 2022). Similar results have been reported for the puppies of bitches receiving dietary supplementation with 0.135% DHA (on a DM basis) during the entire gestation and 10- or 15-week lactation (Hoffman et al. 2007; Kelley 2022). In addition, dietary supplementation with DHA-rich fish oils (0.19% DHA, on a DM basis) between 8 and 52 weeks postweaning improves cognitive, memory, psychomotor, immunologic, and retinal functions in growing dogs (Zicker et al. 2012). Similar findings have also been reported for (a) 3-month-old puppies postweaning receiving dietary supplementation with DHA (40 mg/kg BW/day) for 90 days (Rodrigues et al. 2023), (b) 3-week-old puppies receiving dietary supplementation with 1% DHA for 13 weeks (Reynolds et al. 2006), and (c) puppies receiving dietary supplementation with 0.135% DHA (on a DM basis) between weaning and 25 weeks of age (Kelley and Lepine 2014; Kelley 2022). Notably, long-term dietary supplementation with DHA is more efficient than α-linolenic acid in improving electroretinogram responses of puppies exposed to the PUFAs during gestation, lactation, and weaning (Heinemann et al. 2005).

8.5.4 Requirements for PUFAs to Improve Brain Development and Cognitive Function in Cats During Gestation and Growing Periods

Adequate provision of DHA from maternal diets during gestation and lactation periods is crucial for maintaining a desirable status of DHA in the liver, brain, and retina of feline offspring (Pawlosky et al. 1997). At present, no experimental data are available in the literature regarding the effects of dietary DHA or PUFA supplementation alone on feline brain development or cognitive function. This is likely due to practical challenges with such research. However, based on similar nutritional and physiological aspects of mammals (Bazinet and Layé 2014), dietary DHA is expected to play an important role in neurological growth, development, and function in kittens as in puppies. Thus, daily dietary supplementation of with DHA or its rich source is recommended for gestating and lactating cats for the proper brain and retinal development as well as the optimal health of their offspring (Péron 2023). AAFCO (2014) has suggested that the minimum dietary content of α-linolenic acid, EPA, and DHA for the reproduction and growth of cats be 20, 1.2, and 1.2 mg/100 g (on a DM basis), respectively.

8.6 Requirements for Nutrients to Prevent Cognitive Dysfunction in Aged Dogs and Cats

Cognitive function declines with aging in dogs and cats, just like other mammals such as humans (Pan et al. 2013, 2018a, b). Clinical symptoms in affected dogs and cats include disorientation, social distancing from pet owners and other pets, an abnormal sleep–wake cycle, wandering, alterations in activity levels and patterns, as well as increased anxiety, irritability, and aggression. About 28–30% of 11–14-year-old dogs and 48–68% of dogs over 15 years suffer from cognitive dysfunction syndrome (Azkona et al. 2009; Milgram et al. 1994). Similarly, 28% of 11- to 14-year-old cats and 50% of cats over 15 years have cognitive dysfunction syndrome (Gunn-Moore et al. 2007; Landsberg et al. 2010). The cognitive dysfunction syndrome appears to be irreversible after it is established and is associated with severe brain atrophy (including the loss of neurons). One of the pathogenic factors is the accumulation of a protein called β-amyloid, which creates toxic conditions for neurons. The development of this brain disorder may be alleviated or prevented by improving nutrition (e.g., adequate intakes of anti-oxidative AAs, minerals, and vitamins, as well as PUFAs) in dogs and cats (May and Laflamme 2019; Pan 2021).

Pan et al. (2018a) published a study to determine the effects of dietary supplementation with a nutrient blend [arginine plus vitamins plus fish oil; named as the brain protection blend (BPB)] for 6 months on cognitive function in old (9.1- to 11.5-year-old) dogs. The fish oils were EPA and DHA, and their dietary content was increased from < 0.1 to 0.24% and 0.21%, respectively. The content of arginine in the diet was increased from 1.91 to 2.52%, namely a 32% increase. The cognitive function in dogs was performed by using a landmark discrimination learning protocol to assess the spatial learning skill of the animals. These authors reported that dietary supplementation with the BPB for 6 months improved cognitive function in the aging dogs, as fewer behavioral errors were made in the treatment (BPB) group, when compared with the control group. In addition, based on the measurement of egocentric discrimination, which assessed spatial learning based on internal body-centered cues, dietary supplementation with the BPB for 6 months improved cognitive function in aged dogs, as these animals made fewer behavioral errors in egocentric reversal tests in phase 1 and phase 2, when compared with the control group. Similarly, Pan et al. (2018b) reported that dietary supplementation with the BPB plus 6.5% medium chain triglycerides (substrates for the hepatic production of ketone bodies as metabolic fuels for the brain) for 90 days improved cognitive function in aged dogs as indicated by decreases in scores of all the six categories measured: disorientation, altered social interaction, anxiety, sleep–wake cycle disturbance, house training plus learning plus memory, and altered physical activity. Likewise, dietary supplementation with DHA-rich alga *Schizochytrium* sp. for 25 weeks enhances shape discrimination learning associated with visual processing in aged dogs (Hadley et al. 2017). This is consistent with the report that dietary supplementation with γ-linolenic acid ameliorated glycation-induced memory impairment in aged rats (Khan et al. 2017). Furthermore, dietary supplementation with antioxidant nutrients (vitamins E and C, a mixture of fruits and vegetables, α-lipoic acid, and L-carnitine) alleviates cognitive dysfunction in aged dogs (Cotman et al. 2002).

In another study, Pan et al. (2013) determined the effects of dietary supplementation with the BPB for 6 months on cognitive function in old (5.5- to 8.7-year-old) cats. The fish oils were EPA and DHA and were increased from 0.04 to 0.28% and 0.27%. The content of arginine in the diet was increased from 1.48 to 2.3%, namely a 55% increase. The cognitive function in cats was performed by determining egocentric discrimination on days 31–74 of the feeding experiment to assess spatial learning based on internal body-centered cues; and also by using a landmark discrimination learning protocol on days 82–177 of the feeding experiment to assess a spatial learning skill of the animals, and also by determining size discrimination and reversal learning ability of the animals on days 201–281 of the feeding experiment. On the last 41 days of the feeding experiment, the authors re-tested the animal's delayed non-matching-to-position (DNMP), which is a test of spatial memory. Dietary supplementation with the BPB for 6 months improved cognitive function in the aging cats, as these animals had a greater learning ability score and a greater DNMP score, when compared with the control group. Thus, in aged dogs and cats, increasing the dietary intakes of arginine, vitamins, and fish oil can improve their neurological function, thereby preventing abnormal behaviors associated with cognitive dysfunction syndrome.

8.7 Requirements for Tryptophan to Improve Behavior in Dogs and Cats

The catabolism of tryptophan in dogs and cats produce kynurenine, serotonin, and melatonin (Wu 2022). Kynurenine, which is formed after tryptophan is oxidized by indoleamine 2,3-dioxygenase or tryptophan 2,3-dioxygenase, has been reported to positively improve neurological function in animals (Xue et al. 2023). As neurotransmitters, serotonin and melatonin can also affect the behavior of dogs and cats (Bosch et al. 2007). Thus, dietary tryptophan confers mood-stabilizing and calming effects in the animals. Note that (a) the catabolism of tryptophan to serotonin and melatonin requires BH4 as an essential cofactor for the initial reaction catalyzed by tryptophan hydroxylase; (b) L-arginine stimulates the synthesis of BH4 from GTP in animal cells; and (c) BH4 is stabilized by vitamin C (Shi et al. 2004). This indicates a crucial role for dietary AAs and vitamins in the behavior of the companion animals.

In support of the foregoing view, a low dietary intake of tryptophan (e.g., primarily due to the consumption of foods consisting of a large amount of plant-sourced ingredients; e.g., 37% corn for dogs) results in low concentrations of serotonin in the brain and is associated with aggressive behaviors in dogs and cats (DeNapoli et al. 2000; Landsberg et al. 2017; Tynes and Landsberg 2021). Thus, dietary tryptophan supplementation can reduce the aggression of dogs and cats, while improving their ability to cope with stress DeNapoli et al. 2000; Landsberg et al. 2017; Pereira 2010; Pereira et al. 2010; Tynes and Landsberg 2021). For example, in adult dogs fed a diet containing 18.8% protein with a basal content of 0.18% tryptophan, dietary supplementation with 0.12% tryptophan reduced the territorial aggression score by 14% (DeNapoli et al. 2000). In addition, dietary supplementation with tryptophan (12.5 mg/kg BW/day) reduced barking and staring behaviors in multi-housed dogs (Pereira 2010), as well as vocalization, agonistic behavior, exploring, scratching, and agonistic interactions in multi-housed cats (Pereira et al. 2010). Interestingly, in a study to identify the behavioral patterns of healthy female adult dogs fed a control diet (with 0.18% tryptophan, as-fed basis) or one of several experimental diets with graded concentrations of tryptophan (0.05, 0.1, and 0.15%, as-fed basis), Templeman et al. (2018) reported that dietary supplementation with tryptophan did not affect behavioral scores for activity, distance, confidence, and ear position. It remains to be determined whether the content and balance of nutrients (e.g., large neutral AAs, PUFAs, and vitamin B_6) other than tryptophan in the basal diet may affect the response of the dogs to dietary supplementation with tryptophan.

8.8 Requirements for Tyrosine to Prevent Stress-Induced Cardiac, Behavioral, and Cognitive Abnormality in Dogs and Cats

Tyrosine is used for the synthesis of catecholamines in dogs and cats (Wu 2022). This metabolic pathway occurs in the brain and adrenal gland. BH4 is required by tyrosine hydroxylase to oxidize tyrosine into dopa. Dopa is then decarboxylated by the vitamin B_6-dependent dopa decarboxylase to form dopamine. In the presence of vitamin C and copper, dopamine is oxidized to norepinephrine. The latter is methylated by *S*-adenosylmethionine to form epinephrine. There is evidence that tyrosine can ameliorate stress (epinephrine infusion)-induced increases in ventricular vulnerability to arrhythmias in conscious dogs. For example, Pinto et al. (1990) assessed the effect of tyrosine on the heart rate in dogs by measuring the repetitive extra-systole threshold (an index of ventricular fibrillation, a type of abnormal heart rhythm or irregular heart beat). Epinephrine infusion reduced the repetitive extra-systole threshold, but this adverse effect of the stress was prevented by tyrosine treatment with the outcome of improving the output of blood from the heart (Pinto et al. 1990). Likewise, tyrosine supplementation to dogs ameliorates the vulnerability of their heart to ventricular fibrillation by protecting against severe mental fatigue, therefore preventing sudden stress-induced cardiac arrest (Scott et al. 1981). Similarly, oral administration of tyrosine can improve cognitive performance in humans with environmental stress such as high noise levels (Banderet and Lieberman 1989; Deijen and Orlebeke 1994). Furthermore, intraperitoneal administration of tyrosine (200 mg/kg BW) protected rats from the acute behavioral depression brought about by tailshock-induced stress (Lehnert et al. 1984). To date, little information is available regarding the effects of tyrosine supplementation on stressed cats, but this nutritional treatment is expected to alleviate stress-induced behavioral abnormalities and improving cognitive function in cats as in dogs and humans. Of note, among common animal- and plant-sourced foodstuffs studies, black soldier flyer larvae meal provides the highest amount of tyrosine (Li and Wu 2020).

8.9 Requirements for Nutrients to Improve Mood in Dogs and Cats

8.9.1 Overview

Mood is an emotional state of mind or feeling in an individual. About 150 years ago, Charles Darwin (1872) published a book entitled "*The Expression of Emotions in Man and Animals*", recognizing that both humans and animals (including dogs and cats) have mood. Mood can be classified as positive or negative (Fig. 8.4). The positive mood can be mildly pleasant or strongly pleasant. Relaxation and social

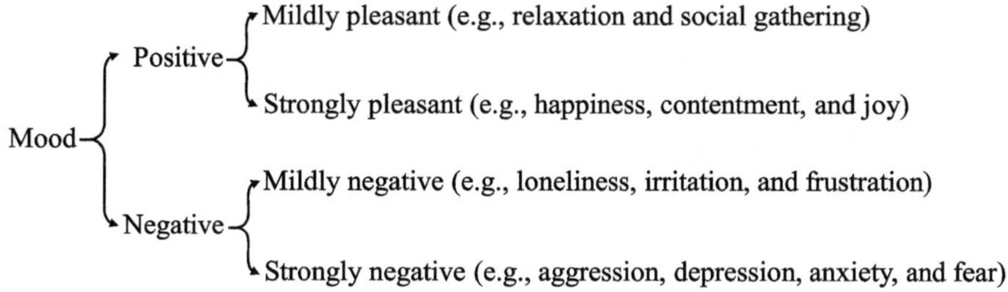

Fig. 8.4 Classification of mood in mammals. Mood is an emotional state of mind or feeling in an individual. Both dogs and cats have mood, which can be classified as positive or negative. The positive mood can be mildly or strongly pleasant, whereas the negative mood can be mildly or strongly negative

gathering are examples of being mildly pleasant. Happiness, contentment, and joy are examples of being strongly pleasant. The negative mood can be mildly negative or strongly negative. Loneliness, irritation, and frustration are examples of being mildly negative. Aggression, depression, anxiety, and fear are examples of being strongly negative.

Much research over the past three decades has shown that both dogs and cats can distinguish human faces and voices (Andics et al. 2014; González-Ramírez and Landero-Hernández 2021; Quaranta et al. 2020; Racca et al. 2010). In addition, dogs and cats are sensitive to emotional, visual, and auditory signals from individuals of the same species and other species (including humans). Of particular note, dogs and cats can adjust their behaviors to human emotions. This is why dogs and cats are the excellent companions of humans for comfort, mental health, and protection. As in humans (Chao 2024), dogs and cats can experience many of the same signs of mood disorders. For example, dogs and cats with depression may either stop eating or eat excessively as a form of comfort and may avoid people or hide in secluded places (e.g., a closet or basement).

8.9.2 Factors Affecting Mood in Dogs and Cats

Evidence from both experimental and clinical research has shown that many factors can affect mood in dogs and cats (Dawson et al. 2019; Gee et al. 2017; Miklósi et al. 2005; Notari et al. 2020; Pickersgill et al. 2023). These diverse factors include the lack of sleep, environment, social interactions, human behavior and voice, levels of physical activity, hormones, and nutrition (e.g., meat vs grain; protein versus carbohydrates) (Fig. 8.5). Either complete or even partial sleep deprivation can result in mood disorders, such as anger and depression, depending on individuals and situations. Dogs and cats will be happy in a clean, spacious, and a thermoneutral environment but will be uncomfortable in a dirty, crowded, hot, or noisy environment. Dogs and cats will enjoy being together with friendly companions and caring owners. In response to angry and loud voices, dogs will be frightening and become hostile to strangers. Both dogs and cats can happily do moderate exercise but will become obese and sad if they have a sedentary life style. Good nutrition, such as consuming adequate amounts of either meat or protein plus lipids along with vitamins and minerals, will make dogs and cats happy. Happy animals release small neuropeptides called endorphins, and these peptides make dogs and cats feel relaxed and joyful. By contrast, stressed animals have high levels of glucocorticoids, and these catabolic hormones make dogs and cats feel irritable, anxious, and depressed. The effects of all these factors on mood in animals are mediated through neurotransmitters produced by the brain. The hippocampus of the brain is mainly responsible for mood in animals. Both dogs and cats naturally avoid whole grains and high-carbohydrate diets, although grains can be a useful source of nutrients in canine diets (Kronfeld 1976; Volhard and Brown 1995). In these carnivores, particularly cats, high intakes of protein-inadequate and carbohydrate-rich diets can have adverse effects on their metabolic and mental health.

Creatine is another important molecule required for normal mood in animals. Creatine is synthesized from arginine, glycine, and methionine through interorgan metabolism involving the kidneys, liver, and pancreas (Wu 2022). All of these three AAs are abundant in animal-sourced foods but are generally low in vegetables and

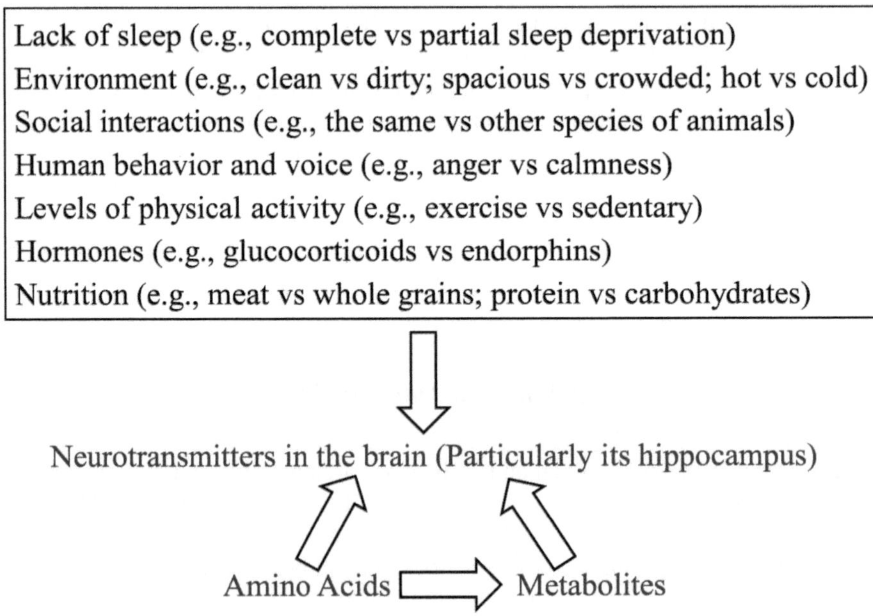

Fig. 8.5 Factors affecting mood in dogs and cats. Nutritional, physiological, and environmental factors affect mood in animals through changes in the synthesis of neurotransmitters from amino acids and the release of the neurotransmitters [including amino acids (e.g., L-glutamate and L-aspartate) and their metabolites (e.g., γ-aminobutyrate and nitric oxide)] by neurons of the brain. The hippocampus of the brain is mainly responsible for mood in animals

grains (Li et al. 2021a). It is noteworthy that creatine is present in animal-sourced products but is absent from plant-sourced foodstuffs (Li and Wu 2020). In the brain, creatine is a potent and abundant antioxidant by scavenging free radical species, and creatine reacts with ATP to form phosphocreatine for storing energy (Wyss and Kaddurah-Daouk 2000). In response to physiological needs, such as NO synthesis and nutrient transport, phosphocreatine releases energy in the form of ATP. Much evidence shows that creatine is essential for the energy metabolism, development, and function of the brain and other tissues of the nervous system (Wyss and Kaddurah-Daouk 2000). Thus, a deficiency of creatine results in mood disorders in animals including dogs and cats, and creatine can be used to alleviate depression in animals by providing energy and removing oxidants. In support of this view, dietary supplementation with creatine ameliorates depression in mammals (e.g., rats, mice, and humans) by providing energy and removing oxidants (Kim et al. 2020; Kious et al. 2019).

Pet owners are generally very concerned about cholesterol in the diets of their dogs and cats. However, dogs and cats require cholesterol for the development and function of the brain and other tissues of the nervous system (Barenholz 2002). Specifically, cholesterol is an essential component of all cell membranes to maintain their integrity, fluidity, and function (e.g., nutrient transport) (Wu 2018). Cholesterol is also the substrate for the synthesis of bile acid and steroid hormones. Bile acid in conjugation with taurine is required for the digestion and absorption of lipids, including EPA and DHA, as well as lipid-soluble vitamins (Wu 2018). Steroid hormones participate in metabolic regulation (e.g., the synthesis of neurotransmitters), thereby affecting the mood and behavior of animals. There is evidence that low concentrations of cholesterol impair serotonin signaling in neurons, leading to depression and other types of abnormal mood in animals (Papakostas et al. 2004). We can surmise that a major reason why dogs and cats prefer to eat visceral organs of their prey may be that such foods contain not only high concentrations of taurine and glutamate but also lipids including saturated fats, PUFAs and cholesterol. It should be borne in mind that high concentrations of cholesterol in the blood increase risks for cardiovascular disorders in animals (including dogs and cats) and, therefore, must be avoided in nutritional practice.

8.10 Conclusion and Perspectives

The nervous system consists of the central nervous system (brain stem, forebrain, midbrain, and spinal cord) and the peripheral nervous system (including the involuntary and voluntary nervous systems) and is the foundation for the cognitive function and behavior of animals including dogs and cats. Development of the brain (which is very rapid during gestation) and cognitive health critically depend on its neural network, which is composed of neurons, glial cells (non-neuronal cells), and neurotransmitters. The latter consist of proteinogenic AAs (e.g., L-glutamate, L-aspartate, and glycine) and their metabolites [e.g., D-aspartate, D-serine, GABA,

NO, carbon monoxide, hydrogen sulfide, and monoamines (e.g., dopamine, norepinephrine, epinephrine, and serotonin)]. The neurotransmitters are released from presynaptic neurons to act on post-synaptic receptors, thereby allowing the neurons to communicate with each other throughout the brain and the rest of the body. In addition, some non-neurotransmitter metabolites of AAs, such as taurine, creatine, and carnosine, also play important roles in the brain development, cognitive health, behavior, and mood of dogs and cats through protecting against oxidative stress, modulating energy metabolism, and affecting ion channels of cell membranes. As essential components of cell membranes in the nervous systems, EPA, DHA, and cholesterol also play crucial roles in the growth, development, and function of the brain and spinal cord.

Both dogs and cats have a limited ability to synthesize de novo arginine from glutamine/glutamate and proline likely due to low activities of pyrroline-5-carboxylate synthase, proline oxidase, and ornithine aminotransferase, and carbamoylphosphate synthetase I in enterocytes of the small intestine (Li and Wu 2023a). Although most breeds of dogs can generate sufficient taurine from cysteine in their livers when fed methionine- or cysteine-adequate diets, this metabolic pathway is limited in all cats (Sturman et al. 1985) and a few breeds of dogs. Just like other mammals, dogs and cats are not able to introduce a double bond beyond a Δ^9 carbon and, therefore, cannot form α-linolenic acid (ω3 PUFA) or linoleic acid (ω6 PUFA) from palmitate. Growing and young adult dogs can actively synthesize EPA and DHA from α-linolenic acid, as well as arachidonic acid from linoleic acid. In contrast, all age groups of cats have a very limited ability to convert α-linolenic acid into EPA and DHA or linoleic acid into arachidonic acid. In addition, the rates of syntheses of EPA, DHA, and arachidonic acid from their respective substrates are substantially reduced in aged dogs. Thus, cats and dogs require dietary ω3 (particularly EPA and DHA) and ω6 PUFAs (particularly linoleic acid and arachidonic acid) for neurological development and cognition. Therefore, inadequate protein or lipid nutrition will lead to mood disorders and abnormal behavior in both dogs and cats. Good nutrition (including sufficient and balanced amounts of AAs, carbohydrates, PUFAs, vitamins, minerals, and water) during pregnancy and lactation is critical for the optimal development of the brain that enables the lifelong learning and cognitive health of offspring and their future generations. Importantly, some AAs (e.g., arginine, glycine, methionine, serine, taurine, tryptophan, and tyrosine) along with EPA, DHA, antioxidant minerals, and antioxidant vitamins can prevent mood disorders (e.g., instability, depression, anxiety, and aggression) and maintain normal behavior in the companion animals. As abundant providers of all these functional AAs, lipids, vitamins, and minerals, animal-sourced foods can play important roles in brain development, cognitive function, and mood of both dogs and cats. Dietary supplementation with the synthetic forms of these nutrients is also expected to have the same benefits.

Acknowledgements The author thanks Dr. Peng Li (North American Renderers Association, Alexandria, VA, USA) and Prof. Nancy Ing (Texas A&M University) for helpful discussions. This work was supported by Texas A&M AgriLife Research (H-8200).

Conflict of Interest The author declares no conflict of interest.

References

AAFCO (2014) AAFCO dog and cat food nutrient profiles. Association of American Feed Control Officials (AAFCO), Champaign, IL
Ajinkya S, Nawaratna G, Hu SD, Wu G, Lubec G (2016) Decreased hippocampal homoarginine and increased nitric oxide and nitric oxide synthase levels in rats parallel training in a radial arm maze. Amino Acids 48:2197–2204
Andics A, Gácsi M, Faragó T, Kis A, Miklósi A (2014) Voice-sensitive regions in the dog and human brain are revealed by comparative fMRI. Curr Biol 24:574–578
Arant BS Jr, Gooch WM III (1982) Developmental changes in the mongrel canine brain during postnatal life. Early Hum Dev 7:179–194
Azkona G, García-Belenguer S, Chacón G, Rosado B, León M, Palacio J (2009) Prevalence and risk factors of behavioural changes associated with age-related cognitive impairment in geriatric dogs. J Small Anim Pract 50:87–91
Banderet LE, Lieberman HR (1989) Treatment with tyrosine, a neurotransmitter precursor, reduces environmental stress in humans. Brain Res Bull 22:759–762
Barenholz Y (2002) Cholesterol and other membrane active sterols: from membrane evolution to "rafts." Prog Lipid Res 41:1–5
Bauer JE (2016) The essential nature of dietary omega-3 fatty acids in dogs. J Am Vet Med 249:1267–1272
Bazinet RP, Layé S (2014) Polyunsaturated fatty acids and their metabolites in brain function and disease. Nat Rev Neurosci 15:771–785
Bordoni A, Biagi PL, Turchetto E, Hrelia S (1988) Aging influence on delta-6-desaturase activity and fatty acid composition of rat liver microsomes. Biochem Int 17:1001–1009
Bosch G, Beerda B, Hendriks H, van der Poel AFB, Verstegen MWA (2007) Impact of nutrition on canine behaviour: current status and possible mechanisms. Nutr Res Rev 20:180–194
Bray EE, Gruen ME, Gnanadesikan GE, Horschler DJ, Levy KM, Kennedy BS, Hare BA, MacLean EL (2021) Dog cognitive development: a longitudinal study across the first two years of life. Anim Cogn 24:311–328
Brodal P (2004) The central nervous system: structure and function. Oxford University Press, Cambridge UK
Bryan DL, Hart P, Forsyth K, Gibson R (2001) Incorporation of alpha-linolenic acid and linoleic acid into human respiratory epithelial cell lines. Lipids 36:713–717
Chao S (2024) Overview of depression. Emerg Med Clin North Am 42:105–113
Che DS, Nyingwa PS, Ralinala KM, Maswanganye GMT, Wu G (2021) Amino acids in the nutrition, metabolism, and health of domestic cats. Adv Exp Med Biol 1285:217–231
Chen C, Xia S, He J, Lu G, Xie Z, Han H (2019) Roles of taurine in cognitive function of physiology, pathologies and toxication. Life Sci 231:116584
Cotman CW, Head E, Muggenburg BA, Zicker S, Milgram NW (2002) Brain aging in the canine: a diet enriched in antioxidants reduces cognitive dysfunction. Neurobiol Aging 23:809–818
Czarnecki GL, Baker DH (1984) Urea cycle function in the dog with emphasis on the role of arginine. J Nutr 114:581–590
Dahms I, Bailey-Hall E, Sylvester E, Parenteau A, Yu S, Karagiannis A, Roos F, Wilson J (2019) Safety of a novel feed ingredient, algal oil containing EPA and DHA, in a gestation-lactation-growth feeding study in Beagle dogs. PLoS ONE 14:e0217794
Darwin C (1872) The expression of the emotions in man and animals. Cambridge University Press, Cambridge, UK

Dawson LC, Cheal J, Niel L, Mason G (2019) Humans can identify cats' affective states from subtle facial expressions. Anim Welf 28:519–531
de Morais-Pinto L, da Veiga ML, da Anunciação ARA (2021) Central nervous system development of cats (*Felis catus* L. 1758). Res Vet Sci 141:81–94
Deijen JB, Orlebeke JF (1994) Effect of tyrosine on cognitive function and blood pressure under stress. Brain Res Bull 33:319–323
DeNapoli JS, Dodman NH, Shuster L, Rand WM, Gross KL (2000) Effect of dietary protein content and tryptophan supplementation on dominance aggression, territorial aggression, and hyperactivity in dogs. J Am Vet Med Assoc 4:504–508
Enser M, Hallett KG, Hewett B, Fursey GA, Wood JD, Harrington G (1998) The polyunsaturated fatty acid composition of beef and lamb liver. Meat Sci 49:321–327
Enser M, Richardson RI, Wood JD, Gill BP, Sheard PR (2000) Feeding linseed to increase the n-3 PUFA of pork: fatty acid composition of muscle, adipose tissue, liver and sausages. Meat Sci 55:201–212
Fontaine E (2012) Food intake and nutrition during pregnancy, lactation and weaning in the dam and offspring. Reprod Domest Anim 47(Suppl 6):326–330
Fox MW (1971) Integrative development of brain and behaviour in the dog. The University of Chicago Press, Chicago, IL
Gaillard V, Chastant S, England G, Forman O, German AJ, Suchodolski JS, Villaverde C, Chavatte-Palmer P, Péron F (2022) Environmental risk factors in puppies and kittens for developing chronic disorders in adulthood: a call for research on developmental programming. Front Vet Sci 9:944821
Gee NR, Mueller MK, Curl AL (2017) Human-animal interaction and older adults: an overview. Front Psychol 8:1416
González-Ramírez MT, Landero-Hernández R (2021) Pet–human relationships: dogs versus cats. Animals 11:2745
Gunn-Moore DA, Moffat K, Christie L-A, Head E (2007) Cognitive dysfunction and the neurobiology of ageing in cats. J Small Anim Pract 48:546–553
Hadley KB, Bauer J, Milgram NW (2017) The oil-rich alga Schizochytrium sp. as a dietary source of docosahexaenoic acid improves shape discrimination learning associated with visual processing in a canine model of senescence. Prostaglandins Leukot Essent Fat Acids 118:10–18
He WL, Wu G (2020) Metabolism of amino acids in the brain and their roles in regulating food intake. Adv Exp Med Biol 1265:167–185
Heinemann KM, Waldron MK, Bigley KE, Lees GE, Bauer JE (2005) Long-chain (n-3) polyunsaturated fatty acids are more efficient than alpha-linolenic acid in improving electroretinogram responses of puppies exposed during gestation, lactation, and weaning. J Nutr 135:1960–1966
Hellier J (2014) The brain, the nervous system, and their diseases. ABC-Clio, Santa Barbara, CA
Hoffman L, Kelley R, Waltz D (2007) For smarter, more trainable puppies: effect of docosa hexaenoic acid on puppy trainability. IAMS Company
Jardim-Messeder D, Lambert K, Noctor S, Pestana FM, de Castro Leal ME, Bertelsen MF, Alagaili AN, Mohammad OB, Manger PR, Herculano-Houzel S (2017) Dogs have the most neurons, though not the largest brain: trade-off between body mass and number of neurons in the cerebral cortex of large carnivoran species. Front Neuroanat 11:118
Kelley R (2022) DHA for puppies. Vet Focus 32:1
Kelley R, Lepine A (2005) Improving puppy trainability through nutrition. Advances in puppy and kitten health care. The IAMS Company, Geneva, pp 28–33
Kelley R, Lepine J (2014) Supplementation with docosahexaenoic acid (DHA) improves cognitive performance in growing puppies. The Veterinarian 12:10–12
Khan SA, Haider A, Mahmood W, Roome T, Abbas G (2017) Gamma-linolenic acid ameliorated glycation-induced memory impairment in rats. Pharm Biol 55:1817–1823
Kim S, Hong K-B, Kim S, Suh HJ, Jo K (2020) Creatine and taurine mixtures alleviate depressive-like behaviour in Drosophila melanogaster and mice via regulating AKT and ERK/BDNF pathways. Sci Rep 10:11370

Kious BM, Kondo DG, Renshaw PF (2019) Creatine for the treatment of depression. Biomolecules 9:406
Knopf K, Sturman JA, Armstrong M, Hayes KC (1978) Taurine: an essential nutrient for the cat. J Nutr 108:773–778
Kronfeld DS (1976) Canine and feline nutrition. Part 2: Protein, carbohydrate, and fat. Mod Vet Pract 57:95–98
Lacombe RJS, Chouinard-Watkins R, Bazinet RP (2018) Brain docosahexaenoic acid uptake and metabolism. Mol Aspects Med 64:109–134
Lane JD, Smith JE, Hall PV, Campbell RL (1978) Distribution of taurine and putative amino acid neurotransmitters in eight areas of the canine lumbar spinal cord. Brain Res 152:386–390
Latimer HB (1967) Variability in body and organ weights in the newborn dog and cat compared with that in the adult. Anat Red 157:449–456
Landsberg GM, Denenberg S, Araujo JA (2010) Cognitive dysfunction in cats: a syndrome we used to dismiss as 'old age.' J Feline Med Surg 11:837–848
Landsberg G, Milgram B, Mougeot I, Kelly S, de Rivera C (2017) Therapeutic effects of an alpha-casozepine and L-tryptophan supplemented diet on fear and anxiety in the cat. J Feline Med Surg 19:594–602
Lehnert H, Reinstein DK, Wurtman RJ (1984) Tyrosine reverses the depletion of brain norepinephrine and the behavioral deficits caused by tail-shock stress in rats. In: Usdin E, Kvetnansky R, Axelrod J (eds) Stress: the role of the catecholamines and other neurotransmitters. Gordon and Beach, New York, pp 81–91
Lemaitre RN, King IB, Sotoodehnia N, Rea TD, Raghunathan TE, Rice KM, Lumley TS, Knopp RH, Cobb LA, Copass MK, Siscovick DS (2009) Red blood cell membrane alpha-linolenic acid and the risk of sudden cardiac arrest. Metabolism 58:534–540
Li P, Kim SW, Li XL, Datta S, Pond WG, Wu G (2008) Dietary supplementation with cholesterol and docosahexaenoic acid increases the activity of the arginine-nitric oxide pathway in tissues of young pigs. Nitric Oxide 19:259–265
Li P, Wu G (2020) Composition of amino acids and related nitrogenous nutrients in feedstuffs for animal diets. Amino Acids 52:523–542
Li P, Wu G (2022) Functional molecules of intestinal mucosal products in animal nutrition and health. Adv Exp Med Biol 1354:263–277
Li P, Wu G (2023a) Amino acid nutrition and metabolism in domestic cats and dogs. J Anim Sci Biotechnol 14:19
Li P, Wu G (2023b) Composition of nutrients in rendered animal-sourced feedstuffs. In: Phillips C (ed) Encyclopedia of animal nutrition, 2nd edn. CABI, Wallingford, Oxon, UK, pp 576–579
Li P, He WL, Wu G (2021a) Composition of amino acids in foodstuffs for humans and animals. Adv Exp Med Biol 1332:189–210
Li XY, Zheng SX, Wu G (2021b) Nutrition and functions of amino acids in fish. Adv Exp Med Biol 1285:133–168
Li XY, Han T, Zheng SX, Wu G (2021c) Nutrition and functions of amino acids in aquatic crustaceans. Adv Exp Med Biol 1285:169–197
Malloy MH, Rassin DK, Gaull GE, Heird WC (1981) Development of taurine metabolism in beagle pups: effects of taurine-free total parenteral nutrition. Biol Neonate 40:1–8
May KA, Laflamme DP (2019) Nutrition and the aging brain of dogs and cats. J Am Vet Med Assoc 255:1245–1254
Meininger CJ, Wu G (2002) Regulation of endothelial cell proliferation by nitric oxide. Methods Enzymol 352:280–295
Miceli MD, Bosch-Bouju C, Layé S (2020) PUFA and their derivatives in neurotransmission and synapses: a new hallmark of synaptopathies. Proc Nutr Soc 79:388–403
Milgram NW, Head E, Weiner E, Thomas E (1994) Cognitive functions and aging in the dog: acquisition of nonspatial visual tasks. Behav Neurosci 108:57–68

Miklósi Á, Pongrácz P, Lakatos G, Topál J, Csányi V (2005) A comparative study of the use of visual communicative signals in interactions between dogs (*Canis familiaris*) and humans and cats (*Felis catus*) and humans. J Comp Psychol 119:179–186
Morris JG, Rogers QR (1978) Arginine: an essential amino acid for the cat. J Nutr 108:1944–1953
National Research Council (NRC) (2006) Nutrient requirements of dogs and cats. National Academies Press, Washington DC
Nelson NS, Cooper J (1975) The growing conceptus of the domestic cat. Growth 39:435–451
Nieto-Alamilla G, Márquez-Gómez R, García-Gálvez AM, Morales-Figueroa G-E, Arias-Montaño JW (2016) The histamine H3 receptor: structure, pharmacology, and function. Mol Pharmacol 90:649–673
Notari L, Cannas S, Di Sotto YA, Palestrini C (2020) A retrospective analysis of dog–dog and dog–human cases of aggression in northern Italy. Animals 10:1662
Packard GC (2021) Allometric growth in mass by the brain of mammals. Anat Red 304:1551–1561
Pan Y (2021) Nutrients, cognitive function, and brain aging: what we have learned from dogs. Med Sci 9:72
Pan Y, Araujo JA, Burrows J, de Rivera C, Gore A, Bhatnagar S, Milgram NW (2013) Cognitive enhancement in middle-aged and old cats with dietary supplementation with a nutrient blend containing fish oil, B vitamins, antioxidants and arginine. Br J Nutr 110:40–49
Pan Y, Kennedy AD, Jönsson TJ, Milgram NW (2018a) Cognitive enhancement in old dogs from dietary supplementation with a nutrient blend containing arginine, antioxidants, B vitamins and fish oil. Br J Nutr 119:349–358
Pan Y, Landsberg G, Mougeot I, Kelly S, Xu H, Bhatnagar S, Gardner CL, Milgram NW (2018b) Efficacy of a therapeutic diet on dogs with signs of cognitive dysfunction syndrome (CDS): a prospective double blinded placebo controlled clinical study. Front Nutr 5:127
Papakostas GI, Ongür D, Iosifescu DV, Mischoulon D, Fava M (2004) Cholesterol in mood and anxiety disorders: review of the literature and new hypotheses. Eur Neuropsychopharmacol 14:135–142
Pawlosky R, Barnes A, Salem N Jr (1994) Essential fatty acid metabolism in the feline: relationship between liver and brain production of long-chain polyunsaturated fatty acids. J Lipid Res 35:2032–2040
Pawlosky RJ, Denkins Y, Ward G, Salem N Jr (1997) Retinal and brain accretion of long-chain polyunsaturated fatty acids in developing felines: the effects of corn oil-based maternal diets. Am J Clin Nutr 65:465–472
Pereira GG (2010) L-Tryptophan supplementation and its effect on multi-housed cats and working dogs. Improv Vet Pract. https://www.veterinary-practice.com/article/l-tryptophan-supplementation-and-its-effect-on-multi-housed-cats-and-working-dogs. Accessed 24 Nov 2023
Pereira GG, Fragoso S, Pires E (2010) Efect of dietary intake of L-tryptophan supplementation on multi-housed cats presenting stress related behaviours. In: Proceedings of the 53rd British small animal veterinary association congress. Birmingham, p 446
Péron F (2023) How can DHA dietary supplementation support brain development in pets? Vet Pract. https://www.veterinary-practice.com/article/dha-supplementation
Pickersgill O, Mills DS, Guo K (2023) Owners' beliefs regarding the emotional capabilities of their dogs and cats. Animals 13:820
Pinto JM, Kirby DA, Lown B (1990) Epinephrine-induced decrease in repetitive extrasystole threshold is reversed by tyrosine in conscious dogs. Life Sci 47:917–923
Pretzer D (2008) Canine embryonic and fetal development: a review. Theriogenology 70:300–303
Quaranta A, d'Ingeo S, Amoruso R, Siniscalchi M (2020) Emotion recognition in cats. Animals 10:1107
Racca A, Amadei E, Ligout S, Guo K, Meints K, Mills D (2010) Discrimination of human and dog faces and inversion responses in domestic dogs (Canis familiaris). Anim Cogn 13:525–533
Remize M, Brunel Y, Silva JL, Berthon JY, Filaire E (2021) Microalgae n-3 PUFAs production and use in food and feed industries. Mar Drugs 19:113

Reynolds AJ, Waldron M, Wilsson E, Leavitt Y, Dunlap A, Bailey K (2006) Effect of long-chain polyunsaturated fatty acid supplementation on mental stability, problem-solving ability, and learned pattern retention in young, growing dogs. In: Proceedings Nestlé Purina nutrition forum, p 74

Risolia DW, Menezes CMM, Lima DC, Santos AAM, Montiani-Ferreira F, Félix AP (2018) Supplementation in the diet of adult dogs with docosahexaenoic acid (DHA) from the microalgae *Schizochytrium* sp. and evaluation of their effects on the retina. Arch Vet Sci 23:19–20

Rohman A, Triyana K, Sismindari S, Erwanto Y (2012) Differentiation of lard and other animal fats based on triacylglycerols composition and principal component analysis. Int Food Res J 19:475–479

Rodrigues RBA, Zafalon RVA, Rentas MF, Risolia LW, Macedo HT, Perini MP, da Silva AMG, Marchi PH, Balieiro JCdC, Mendes WS, Vendramini THA, Brunetto MA (2023) The supplementation of docosahexaenoic acid-concentrated fish oil enhances cognitive function in puppies. Animals 13:2938

Santulli G, Kansakar U, Varzideh F, Mone P, Jankauskas SS, Lombardi A (2023) Functional role of taurine in aging and cardiovascular health: an updated overview. Nutrients 15:4236

Sase A, Dahanayaka S, Höger H, Wu G, Lubec G (2013) Changes of hippocampal β-alanine and citrulline levels parallel early and late phase of retrieval in the Morris Water maze. Behav Brain Res 249:104–108

Schafer S, Kim HW (2018) Efects and mechanisms of taurine as a therapeutic agent. Biomol Ther (seoul) 26:225–241

Scheibel S, de Oliveira CAL, de Alvarenga BM, Pereira LC, Rodrigues DM, Barion MRL, Carbonera F, Visentainer JV, Huppes RR, Bueno Ribeiro L, Vasconcellos RS (2021) DHA from microalgae *Schizochytrium* spp. (Thraustochytriaceae) modifies the inflammatory response and gonadal lipid profile in domestic cats. Br J Nutr 126:172–182

Scott NA, DeSilva RA, Lown B, Wurtman RJ (1981) Tyrosine administration decreases vulnerability to ventricular fibrillation in the normal canine heart. Science 211:727–729

Seong PN, Cho SH, Park KM, Kang GH, Park BY, Moon SS, Ba HV (2015) Characterization of chicken by-products by mean of proximate and nutritional compositions. Korean J Food Sci Anim Resource 35:179–188

Shi W, Meininger CJ, Haynes TE, Hatakeyama K, Wu G (2004) Regulation of tetrahydrobiopterin synthesis and bioavailability in endothelial cells. Cell Biochem Biophys 41:415–433

Stiles J, Jernigan TL (2010) The basics of brain development. Neuropsychol Rev 20:327–348

Sturman JA, Lu P (1997) Role of feline maternal taurine nutrition in fetal cerebellar development: an immunohistochemical study. Amino Acids 13:369–377

Sturman JA, Moretz RC, French JH, Wisniewski HM (1985) Taurine deficiency in the developing cat: persistence of the cerebellar external granule cell layer. J Neurosci Res 13:405–416

Sturman JA, Lu P, Xu Y, Imaki H (1994) Feline maternal taurine deficiency: effects on visual cortex of the offspring. A morphometric and immunohistochemical study. Adv Exp Med Biol 359:369–384

Tachikawa M, Hirose S, Akanuma SI, Matsuyama R, Hosoya KI (2018) Developmental changes of l-arginine transport at the blood-brain barrier in rats. Microvasc Res 117:16–21

Tekwe CD, Luan YY, Meininger CJ, Bazer FW, Wu G (2023) Dietary supplementation with L-leucine reduces nitric-oxide synthesis by endothelial cells of rats. Exp Biol Med 248:1537–1549. https://doi.org/10.1177/15353702231199078

Templeman JR, Davenport GM, Cant JP, Osborne VR, Shoveller A-K (2018) The effect of graded concentrations of dietary tryptophan on canine behavior in response to the approach of a familiar or unfamiliar individual. Can J Vet Res 82:294–305

Tynes VV, Landsberg GM (2021) Nutritional management of behavior and brain disorders in dogs and cats. Vet Clin Small Anim 51:711–727

USDA (2022) Nutrient database. https://fdc.nal.usda.gov/fdc-app.html#/food-details/171400/nutrients

Virarkar M, Alappat L, Bradford PG, Awad AB (2013) L-arginine and nitric oxide in CNS function and neurodegenerative diseases. Crit Rev Food Sci Nutr 53:1157–1167
Volhard W, Brown K (1995) The holistic guide for a healthy dog. Howell Book House, NY
Vuorinen A, Bailey-Hall E, Karagiannis A, Yu S, Roos F, Sylvester E, Wilson J, Dahms I (2020) Safety of algal oil containing EPA and DHA in cats during gestation, lactation and growth. J Anim Physiol Anim Nutr 104:1509–1523
Wang S, Qi Y, Yu L, Zhang L, Chao F, Huang W, Huang R, Li H, Luo Y, Xiu Y, Tang Y (2016) Endogenous nitric oxide regulates blood vessel growth factors, capillaries in the cortex, and memory retention in Sprague-Dawley rats. Am J Transl Res 8:5271–5285
Wu G (2018) Principles of animal nutrition. CRC Press, Boca Raton, Florida
Wu G (2020) Important roles of dietary taurine, creatine, carnosine, anserine and 4-hydroxyproline in human nutrition and health. Amino Acids 52:329–360
Wu G (2022) Amino acids: biochemistry and nutrition. CRC Press, Boca Raton, Florida
Wu G, Meininger CJ (2002) Regulation of nitric oxide synthesis by dietary factors. Annu Rev Nutr 22:61–86
Wyss M, Kaddurah-Daouk R (2000) Creatine and creatinine metabolism. Physiol Rev 80:1107–1213
Xue C, Li G, Zheng Q, Gu X, Shi Q, Su Y, Chu Q, Yuan X, Bao Z, Lu J, Li L (2023) Tryptophan metabolism in health and disease. Cell Metab 35:1304–1326
Zicker SC, Jewell DE, Yamka RM, Milgram NW (2012) Evaluation of cognitive learning, memory, psychomotor, immunologic, and retinal functions in healthy puppies fed foods fortified with docosahexaenoic acid–rich fish oil from 8 to 52 weeks of age. J Am Vet Med 241:583–594
Zigmundova V, Komprda T, Rozikova V (2018) Changes of fatty acids content in rat liver after different diet. Mendel Net 25:314–318

Chapter 9
Nutrition and Aging in Dogs and Cats

Jonathan Stockman

Abstract Aging is often associated with chronic inflammation and declining health. Both veterinarians and owners of aging dogs and cats are interested in nutritional solutions and strategies to prevent signs of age-related disease, increase longevity, and improve quality of life. Physiological decreases in muscle mass, decreased immunity, and a decrease in sense acuity are some of the changes often seen in otherwise healthy senior pets; however, there may also be an increase in risk for pathologies such as renal, cardiovascular, musculoskeletal, and neoplastic diseases. Aging may also lead to cognitive decline and even cognitive dysfunction. Some nutritional strategies that may be helpful with the prevention and treatment of age-related diseases include supplementation with ω3 polyunsaturated fatty acids and antioxidant nutrients that can help modulate inflammation and benefit osteoarthritis, renal disease, cancer, and more. Supplementation with medium-chain triglycerides shows promise in the treatment of canine cognitive dysfunction as these may be metabolized to ketone bodies that are utilized as an alternative energy source for the central nervous system. Additionally, a high intake of dietary phosphorus in soluble and bioavailable forms can lead to renal disease, which is of greater concern in senior pets. There are no published guidelines for nutritional requirements specific to senior pets and as a result, products marketed for senior dogs and cats are highly variable.

Keywords Dog · Cat · Aging · Senescence · Senior · ω3-fatty acids · Medium-chain triglycerides · Phosphorous

J. Stockman (✉)
Department of Veterinary Clinical Sciences, College of Veterinary Medicine, Long Island University, Old Brookville, NY 11548, US
e-mail: Jonathan.stockman@liu.edu

Department of Clinical Sciences, College of Veterinary Medicine and Biomedical Sciences at Colorado State University, Fort Collins, CO 80523, US

G. Wu (ed.), *Nutrition and Metabolism of Dogs and Cats*, Advances in Experimental Medicine and Biology 1446, https://doi.org/10.1007/978-3-031-54192-6_9

Abbreviations

EPA	Eicosatetraenoic acid
DHA	Docosahexaenoic acid
MCT	Medium-chain triglycerides
SAMe	S-adenosyl–L-methionine

9.1 Introduction

Nutrition is vital for the normal development and function of every animal. The nutritional needs of dogs and cats change throughout their lives (as in people). The nutritional needs of a growing animal and requirements during reproduction in females are different to those of an adult animal during maintenance. Growing animals have higher requirements of protein, specific essential amino acids, and fatty acids as well as altered mineral requirements and tolerance for minerals required for skeletal growth and maturation (NRC 2006). During reproduction, nutritional requirements in females reflect their metabolic investment in fetal development and lactation (NRC 2006). Similarly, as animals age, there are many notable changes that occur in physiology and metabolism that are likely reflected in altered nutritional needs. Additionally, with aging, there is an increase in the prevalence of certain age-related diseases that may impact the animal's dietary requirements and tolerance. Very little research is available regarding the specific nutritional requirements of senior pets although some information may be applicable from research regarding aging in humans. This article aims to review and explore what is currently known about the nutritional needs of senior dogs and cats and look at current dietary recommendations and products marketed for older companion dogs and cats.

9.2 What is Aging?

Aging affects the organism, its organs, tissues, and cells. It can be defined as a progressive time-dependent decline in function, and it is also characterized by an increase in senescent cells. Senescence is defined on the cellular level as cellular cell cycle arrest (Di Micco et al. 2021), and includes a chronic loss of regenerative and bioprotective mechanisms that progresses over time (Sándor and Kubinyi 2019). Senescence occurs throughout life (and even before birth). Senescent cell proliferation would generally slow down until it gradually ceases. There may also be changes in protein production and function of the ubiquitin system which removes defective proteins. Some cells may have an increase in apoptotic resistance which will ultimately increase the number of cells that are not functioning at an optimal level and may even increase the risk for pathologies such as neoplastic disease (Sándor and

Kubinyi 2019). The increase in senescent cells contributes to the physical changes that are associated with aging including increased inflammation and decreased normal immune function, neuronal flexibility, and tissue regeneration.

Aging ultimately affects all body systems and reduces the organism's ability to cope with environmental changes. The nervous system's compensatory mechanisms are decreased, and aging organisms often have decreases in sensory acuity for vision, smell, and taste. The changes in the canine or feline nervous system can also lead to behavioral and cognitive declines that range in severity (Cummings et al. 1996).

The aged gastrointestinal tract may show altered motility and structural mucosal changes that can result in changes in the digestive function that may lead to dietary intolerance, and some may argue that these changes in physiology indicate a need to alter the diet to increase digestibility in senior dogs and cats (Baum et al. 2007). The renal system may have a decrease in glomerular filtration due to a decrease in the number of functional nephrons or an increase in renal lesions (Cianciolo et al. 2016). As a result, the kidney's glomerular filtration may be reduced. Aging-related changes in the cardiovascular system can reduce vascular compliance and degenerative changes that can impact cardiac function and the animal's ability to perform physical work. The respiratory system may also have a decrease in elasticity and fibrosis contributing to exercise intolerance. The skin may become thinner and may lose some of its protective function. Both muscle and skeleton change with aging and degenerative joint changes and sarcopenia can contribute to old-age frailty. Physiological changes may eventually lead to disease which negatively affects the animal's quality of life and many illnesses in dogs and cats can be considered as age-related as they are more common in aging pets. These include renal disease, cardiac disease, neoplasia, many endocrine diseases, osteoarthritis, and more.

9.2.1 When Do Dogs and Cats Become Senior?

There is no accepted age threshold to be considered 'old,' 'senior,' or 'geriatric'; nor is there a clear distinction between these terms in the context of pet aging and health. Some define dogs as being senior at the age of 7 years, which is an arbitrary age considering the important differences in maturation and longevity between different dog breeds and dog sizes. Although small dog breeds may complete skeletal maturation and growth much sooner than large and giant breed dogs (Hawthorne et al. 2004), the longevity of small breed dogs is generally longer than large breed dogs and develops age-related physiological and pathological changes later (Creevy et al. 2016).

There may be several purposes for defining an age cutoff for an age of being 'senior.' One possible goal is to define an age when animals may become more susceptible to disease. Another goal may be to identify an age in which some interventions may be applied to improve quality of life and increase longevity. It may also be that establishing the senior life stage would help pet owners recognize when animals are approaching the end of their life expectancy. It may be argued that two sub-stages should be recognized as part of the senior life stage. The first stage is the 'healthy senior' age, in which there are physiological declines in function, but not much increase in risk for age-related diseases compared with the advanced senior stage in which the risk for disease is increased and often some age-related diseases are manifested. Epidemiological data may be helpful to establish these ages as they pertain to different species and sizes and perhaps also to sex and neuter status. The rate of aging has environmental influences as well. High caloric intake and body condition score influence longevity and onset of age-related disease in dogs. This was demonstrated in paired Labrador retriever puppies that were followed throughout their lives with one being fed caloric intake than the other, and dogs that were calorically restricted had a two-year average increase in longevity as well as delayed signs of aging (Kealy et al. 2002). Similarly, overweight and obese animals generally have shorter life spans than those in an ideal bodyweight; this was demonstrated in domestic pet cats (Teng et al. 2018b) and laboratory rodents (Johnson et al. 1997). Obesity is associated with multiple diseases in dogs and cats (as it is in humans) which may partly explain this association. These include increased risk of cardiovascular disease, joint disease, cancer, dermatological disease, dyslipidemia, insulin resistance, etc. (Tropf et al. 2017; Teng et al. 2018a).

Canine aging is often referred to as being an equivalence between seven years in a human life to a single canine year. This subjective observation is not supported by scientific research but does describe a similarity in the aging process between people and animals. DNA methylation patterns are used as an 'epigenetic clock' to create formulas that describes the relationship between human and dog aging or human and cat aging (Wang et al. 2020; Horvath et al. 2022; Raj et al. 2021). There are limitations to these models as some formulas are based on data from a small number of breeds and many health and lifestyle variables are not considered; but these provide initial tools to provide a more accurate conversion between human age and dog or cat age. Cat life stages are often determined subjectively, for example, mature adult cats may be defined as between the ages of 7 and 10 years and cats above the age of 10 years may be considered senior (Quimby et al. 2021). This approach divides older cats into two age groups (sometimes referred to as senior and super-senior, or senior and geriatric) (Sordo et al. 2021; Knies et al. 2023). It is generally expected that cats in the younger group would show physiological changes related to age but would be at lesser risk for age-related morbidity.

9.3 Nutritional Considerations in Senior Dogs and Cats

9.3.1 Age-Related Changes in Gut Morphology and Digestive Function

It was previously thought that aging leads to a universal decrease in digestive function, although it is now known that some functions are left relatively intact (Russell 1992). Structural changes in the morphology of the gut have been described in aging animals. It was described that the number of villi and crypts increases over time in older rats and that the tight junctions in the epithelium were widened and discontinuous (Ecknauer et al. 1982; Ren et al. 2014). In dogs, it was also shown that colonic crypt depth was greater in 12-year-old beagles compared with 1-year-old dogs, whereas the jejunal and ileal villus heights were smaller (Kuzmuk et al. 2005). Colonic microbial population shifts and changes in fecal fermentation products have also been documented in senior beagles compared with young counterparts (Benno et al. 1992; Gomes et al. 2011).

While these changes could correspond with a decline in digestive function in older dogs, data for this is limited and often conflicting. It appears that such a decline is not universal. A recent publication evaluated diet digestibility in aging dogs and cats fed different diets with varying fiber content (Schauf et al. 2021). Findings from this study indicated that dogs under the age of 14 years and cats under the age of 13 do not show a consistent age-related decline in nutrient digestibility. Ether extract analysis (an approximation of fat digestibility) was not impacted by dog age. Digestibility of crude protein, calcium and phosphorus improved in older dogs on the lowest fiber diet, and no effect was seen with other diets. In cats, there was an increase in total dietary fiber and calcium digestibility with aging for all diets. This study had important limitations as it was conducted in beagles and Brittany spaniels but other breeds (large or small breeds) were not included. Similarly, cats older than 13 years were not included. Despite the limitations, it can be concluded that the relationship between digestibility and age is more complex than previously stated and older dogs and cats may not always show reduced diet digestibility.

9.3.2 Frailty

Aging changes body composition, as there is typically an age-related decline in muscle mass known as sarcopenia. The term sarcopenia should be differentiated from cachexia which is the loss of muscle tissue attributed to disease including age-related diseases such as kidney disease or cancer. Sarcopenia may become a pathology if it exceeds normal levels and severely impacts quality of life, ability to perform physical activity and if it increases risk for concurrent morbidities. Sarcopenia is thought to be the result of several age-related factors, including physical inactivity, cytokine production, changes in hormonal activity, and a decreased protein synthesis

(Freeman 2012). Muscle mass loss is often associated with a reduction in strength and decreased immune function or a condition termed 'frailty' which increases the risk of comorbidities and mortality. While sarcopenia is influenced by several factors, malnutrition is often a contributing factor as decreased energy and protein intake may accelerate and exacerbate sarcopenia. Changes in appetite in older animals may result from a variety of causes such as a decrease in taste and smell sense acuity, cognitive impairment, and changes in metabolic rate as well as alterations in appetite-regulating peptides (increased sensitivity to cholecystokinin for example) (Balaskó et al. 2013). The resulting decrease in energy intake can lead to a catabolic state where substrates such as amino acids that originate from muscle catabolism are oxidized for energy, thereby leading to a decrease in muscle mass. This may be exacerbated by a diet that is low in protein and may even lead to protein malnutrition when the animal's intake is reduced. Therefore, the daily protein intake in senior dogs and cats should be assessed to ensure that protein requirements are met (Table 9.1) (NRC 2006).

Table 9.1 Daily protein intake recommended allowance for healthy adult dogs and cats in maintenance

Dog	3.28 g × $\text{Bodyweight}_{(kg)}{}^{0.75}$
Cat	4.96 g × $\text{Bodyweight}_{(kg)}{}^{0.67}$

Example: What is the protein requirement for a healthy adult dog in maintenance that weighs 30 kg?

- $30(\text{kg})^{0.75} = 12.81$
- $12.81 \times 3.28 = 42$ g protein/day

The guaranteed analysis provided on the pet food label should indicate the minimum percent of protein in the food. This can be used to calculate the amount of daily protein intake for the patient if the amount of food consumed each day is known. Protein intake is often not be a concern in a dog that consumes enough calories to meet their energy requirements if those are provided by a complete and balanced diet. The concern is greater for dogs that either consume a reduced number of calories or those that do not eat a complete and balanced diet. Protein malnutrition may be detrimental as it can result in decreased wound healing, decreased immune function, and even hypoalbuminemia and ascites in addition to a decrease in muscle mass (Che et al. 2021; Oberbauer and Larsen 2021). These signs should be monitored even in animals that are meeting the calculated protein intake as individual requirements and diet digestibility may differ. The true protein requirements in senior dogs and cats have not been established although it is likely that these are higher than those of younger animals due to increased muscle turn-over, amino acid catabolism, and factors that may be mitigated with a higher protein intake such as reduced delivery of amino acids to peripheral tissues and impaired uptake of amino acids into muscle cells (Murton 2015). Reduced protein digestibility in senior cats is hypothesized to increase the protein requirement, although more data are needed to substantiate this (Fahey et al. 2008). It can be generally recommended that senior cats should not be fed a low-protein diet unless medically indicated. Sufficient protein diets is needed to support muscle mass and immune functions, both critically important in senior dogs and cats (Li and Wu 2023).

There are various interventions that could help increase food intake in aging dogs and cats to help meet energy and protein requirements. Nutritional intervention should always be individualized for the patient and consider any relevant medical history such as food adverse response, renal or liver disease, and chronic enteropathies. These interventions include adding aroma and flavor to address decreases in sensory acuity, increasing caloric and protein density in the diet, and changing the food temperature as a recent study identified that senior cats prefer heated food (Eyre et al. 2022).

9.3.3 Supplementation with Long-Chain Omega-3 Fatty Acids

Marine-sourced long-chain omega-3 fatty acids [eicosapentaenoic acid (EPA) and docosahexaenoic acid (DHA)] benefit senior dogs and cats in several ways as they as their metabolism produces eicosanoids with decreased inflammatory potency compared with omega-6 fatty acids and decreases cytokine production. In addition to their anti-inflammatory function, EPA and DHA also have an anabolic effect on muscle through activation of the mTOR signaling and reduction of insulin resistance (Jeromson et al. 2015). For these reasons, it could be argued that supplementation with a source of long-chain omega-3 fatty acids (such as a fish oil supplement) would be beneficial for any senior dog and cat that can safely receive them, with the exceptions being dogs with pancreatitis or fat intolerance or if the animal is overweight and additional calories in the form of oil are unwanted. Suggested dosages of EPA and DHA for dogs have been published (Table 9.2) (Bauer 2011); however, less data is available for cats. EPA and DHA amounts of 200–300 mg per cat per day have been suggested in the literature (Lascelles et al. 2010; Parker 2021). Since polyunsaturated fatty acids are prone to oxidation which negates their beneficial effect and can even be harmful (Wu 2018), it is advised to supplement diets with antioxidants such as vitamin E.

Table 9.2 Suggested doses of EPA and DHA for dogs

Orthopedic disease: up to 310 mg × $BW_{(kg)}{}^{0.75}$
Renal disease: 140 mg × $BW_{(kg)}{}^{0.75}$
Cardiovascular disease: 115 mg × $BW_{(kg)}{}^{0.75}$

9.3.4 Age-Related Cognitive Decline

Cognitive decline often occurs with aging. While mild cognitive impairment usually has low impact on the animal's function or quality of life, 20–68% of senior dogs are estimated to suffer from canine cognitive dysfunction syndrome which manifests in attention deficit, decreased ability to solve problems, changes in sleep patterns and unwanted behaviors such as pacing and vocalization. Long-chain omega-3 fatty acids have been shown to have a beneficial effect on cognitive function in senior dogs (Hadley et al. 2017). Middle-aged beagles (between ages 8 ½ and 11 years) fed a diet with 0.1% DHA (as fed) from algal source had improved performance in cognitive visuospatial memory tests. A possible explanation for the benefit of DHA on cognition is that DHA is important for phosphatidylserine production in the central nervous system which is found in the neuron membrane and plays a critical role in activating signaling pathways.

Medium-chain triglycerides (MCT) have been studied as a food supplement in senior dogs with cognitive dysfunction, and it was found that supplementation 5.5% MCT as fed to a complete and balanced diet resulted in improved results in cognitive tests in senior beagles (Pan et al. 2018b). MCTs supplementation leads to an increase in circulating ketone bodies which in turn can be used as alternative energy sources for neuronal cells (Studzinski et al. 2008). Other supplements that may be of benefit for cognitive function are listed in Table 9.3 (Pan et al. 2018a).

Table 9.3 Nutritional supplements that have been proposed to benefit dogs with age-related cognitive dysfunction

- Arginine- precursor of nitric oxide, regulation of glucose metabolism
- Alpha lipoic acid- cofactor for mitochondrial enzymes, antioxidant
- B vitamins- improve energy metabolism, cofactors for production of neurotransmitters
- Long-chain omega-3 fatty acids (EPA and DHA)- may reduce inflammation, help with neural function
- Medium chain triglycerides- increase blood ketones which are alternative energy source to glucose for neuron cells
- Vitamin E- antioxidant
- L-carnitine- transports fatty acids through the mitochondrial membrane
- S-adenosyl–L-methionine (SAMe)- maintaining membrane integrity, synthesis and inactivation of neurotransmitters

9.3.5 Nutritional Considerations Related to Risk of Renal Disease in Senior Pets

Aging dogs and cats are at higher risk of metabolic diseases such as kidney disease, which was estimated in some studies as 80.9% prevalence in cats above the age of 15 years (Marino et al. 2014; Brown et al. 2016; Willems et al. 2017). Recent data suggest that there is a deleterious effect for the inclusion of dietary phosphorus exceeding its nutritional requirement on kidney health in adult cats. While the risk is not exclusive to senior cats, kidney disease is highly common in senior cats and older animals often had a prolonged high intake of high phosphorus diets in addition to having some natural reduction in normal organ function and resilience to compensate for organ damage. High dietary phosphorus in a bioavailable form (typically inorganic water-soluble salts) or a low calcium to phosphorus ratio can cause dose-dependent kidney injury in previously healthy adult cats (Coltherd et al. 2019, 2021, 2022; Alexander et al. 2019). Furthermore, the total phosphorus in commercial cat foods that may be fed to adult and senior cats is highly variable but may exceed the intakes that have been shown to lead to kidney disease in adult cats (Summers et al. 2020a, 2020b).

9.4 Nutritional Guidelines and Products Marketed for Senior Dogs and Cats

While guidelines regarding the nutritional requirements of dogs and cats in various life stages exist, these pertain to the requirements during growth, reproduction, or during maintenance but no special guidelines exist for senior dogs and cats. As a result, diets that are marketed for senior pets are typically formulated to meet the requirements for maintenance and do not differ from maintenance diets in parameters such as calcium, phosphorus or sodium content or the percentage of protein or fat (Hutchinson et al. 2011; Summers et al. 2020a). Certainly, there are diets that are formulated with modified specifications aimed to address aging and therefore veterinarians should not make generalized recommendations regarding 'senior diets' but evaluate the adequacy of specific products to the patient and consider their unique medical history and exam findings.

9.5 The Gut Microbiome and Aging

There is an increasing understanding that the gut microbiome has paramount importance in nutritional physiology as it may affect both micronutrient and macronutrient metabolism in a manner that has a huge impact on the hosts health and may increase or decrease susceptibility to clinical or sub-clinical disease. In older people, changes in the gut microbiome appear to be good indicator of health and age-related disease (Ghosh et al. 2022).

While data paucity exists regarding age-related microbiome changes in dogs, one study looking at a colony of Shiba-Inu dogs found that there is a decrease in microbial diversity which is often considered a sign of declining health, as well as some reduction in lactic acid producing bacterial populations (Mizukami et al. 2019). The minor populations of microbes decreased in aging dogs, whereas the major populations were less often changed. While limited in terms of age distribution, number of dogs and the use of a single dog breed, this study provides some initial insight on some changes that occur with dog aging. The effect of aging on the gut microbiome in dogs and cats continues to be investigated as less is known about age-related microbiome changes in these species compared with people.

9.6 Summary

Aging dogs and cats are an important demographic to consider in veterinary medicine as the proportion of senior pets is high and could increase with advances in veterinary medicine and increased longevity. While senior pets should not be fed diets that are nutrient deficient, low in protein or excessive in minerals such as phosphorus, there is insufficient data to inform aging-specific nutritional guidelines. Moreover, further research is needed to investigate the role of the microbiome in healthy aging and to evaluate the benefits of supplements in improving quality of life, longevity, or addressing age-related disease.

References

Alexander J, Stockman J, Atwal J et al (2019) Effects of the long-term feeding of diets enriched with inorganic phosphorus on the adult feline kidney and phosphorus metabolism. Br J Nutr 121:249–269
Balaskó M, Rostás I, Füredi N et al (2013) Age and nutritional state influence the effects of cholecystokinin on energy balance. Exp Gerontol 48:1180–1188
Bauer JE (2011) Therapeutic use of fish oils in companion animals. J Am Vet Med Assoc 239:1441–1451
Baum B, Meneses F, Kleinschmidt S et al (2007) Age-related histomorphologic changes in the canine gastrointestinal tract: a histologic and immunohistologic study. World J Gastroenterol WJG 13:152
Benno Y, Nakao H, Uchida K (1992) Impact of the advances in age on the gastrointestinal microflora of Beagle dogs. J Vet Med Sci 54:703–706
Brown C, Elliott J, Schmiedt C et al (2016) Chronic kidney disease in aged cats: clinical features, morphology, and proposed pathogeneses. Vet Pathol 53:309–326
Che DS, Nyingwa PS, Ralinala KM, Maswanganye GMT, Wu G (2021) Amino acids in the nutrition, metabolism, and health of domestic cats. Adv Exp Med Biol 1285:217–231
Cianciolo R, Benali S, Aresu L (2016) Aging in the canine kidney. Vet Pathol 53:299–308
Coltherd JC, Alexander JE, Pink C et al (2021) Towards establishing no observed adverse effect levels (NOAEL) for different sources 1 of dietary phosphorus in feline adult diets: results from a 7 month feeding study. Br J Nutr 126:1626–1641
Coltherd JC, Staunton R, Colyer A et al (2019) Not all forms of dietary phosphorus are equal: an evaluation of postprandial phosphorus concentrations in the plasma of the cat. Br J Nutr 121:270–284
Coltherd JC, Staunton R, Colyer A et al (2022) Dietary calcium to phosphorus ratio affects postprandial phosphorus concentrations in feline plasma. Br J Nutr 128:1689–1699
Creevy KE, Austad SN, Hoffman JM et al (2016) The companion dog as a model for the longevity dividend. Cold Spring Harb Perspect Med 6:a026633
Cummings BJ, Head E, Ruehl W et al (1996) The canine as an animal model of human aging and dementia. Neurobiol Aging 17:259–268
Di Micco R, Krizhanovsky V, Baker D et al (2021) Cellular senescence in ageing: from mechanisms to therapeutic opportunities. Nature Rev Mol Cell Biol 22:75–95
Ecknauer R, Vadakel T, Wepler R (1982) Intestinal morphology and cell production rate in aging rats. J Gerontol 37:151–155
Eyre R, Trehiou M, Marshall E et al (2022) Aging cats prefer warm food. J Vet Behav 47:86–92
Fahey G, Barry KA, Swanson KS (2008) Age-related changes in nutrient utilization by companion animals. Annu Rev Nutr 28:425–445
Freeman LM (2012) Cachexia and sarcopenia: emerging syndromes of importance in dogs and cats. J Vet Int Med/Am Coll Vet Intern Med 26:3–17
Ghosh TS, Shanahan F, O'Toole PW (2022) Toward an improved definition of a healthy microbiome for healthy aging. Nat Aging 2:1054–1069
Gomes M, Beraldo M, Putarov T (2011) Old Beagle dogs have lower faecal concentrations of some fermentation products and lower peripheral lymphocyte counts than young adult beagles. Br J Nutr 106:S187–S190

Hadley K, Bauer J, Milgram N (2017) The oil-rich alga Schizochytrium sp. as a dietary source of docosahexaenoic acid improves shape discrimination learning associated with visual processing in a canine model of senescence. Prostaglandins Leukot Essent Fatty Acids 118:10–18
Hawthorne AJ, Booles D, Nugent PA et al (2004) Body-weight changes during growth in puppies of different breeds. J Nutr 134:2027S-2030S
Horvath S, Lu AT, Haghani A et al (2022) DNA methylation clocks for dogs and humans. Proc Nalt Acad Sci USA 119:e2120887119
Hutchinson D, Freeman LM, Schreiner KE et al (2011) Survey of opinions about nutritional requirements of senior dogs and analysis of nutrient profiles of commercially available diets for senior dogs. Int J Appl Res Vet Med 9:68–79
Jeromson S, Gallagher IJ, Galloway SD et al (2015) Omega-3 fatty acids and skeletal muscle health. Mar Drugs 13:6977–7004
Johnson PR, Stern JS, Horwitz BA et al (1997) Longevity in obese and lean male and female rats of the Zucker strain: prevention of hyperphagia. Am J Clin Nutr 66:890–903
Kealy RD, Lawler DF, Ballam JM et al (2002) Effects of diet restriction on life span and age-related changes in dogs. J Am Vet Med Assoc 220:1315–1320
Knies M, Kooistra HS, Teske E (2023) Prevalence of persistent hypertension and situational hypertension in a population of elderly cats in The Netherlands. J Feline Med Surg 25:1098612X231172629
Kuzmuk KN, Swanson KS, Tappenden KA et al (2005) Diet and age affect intestinal morphology and large bowel fermentative end-product concentrations in senior and young adult dogs. J Nutr 135:1940–1945
Lascelles B, DePuy V, Thomson A et al (2010) Evaluation of a therapeutic diet for feline degenerative joint disease. J Vet Intern Med 24:487–495
Li P, Wu G (2023) Amino acid nutrition and metabolism in domestic cats and dogs. J Anim Sci Biotechnol 14:19
Marino CL, Lascelles BDX, Vaden SL et al (2014) Prevalence and classification of chronic kidney disease in cats randomly selected from four age groups and in cats recruited for degenerative joint disease studies. J Feline Med Surg 16:465–472
Mizukami K, Uchiyama J, Igarashi H et al (2019) Age-related analysis of the gut microbiome in a purebred dog colony. FEMS Microbiol Lett 366:fnz095
Murton AJ (2015) Muscle protein turnover in the elderly and its potential contribution to the development of sarcopenia. Proc Nutr Soc 74:387–396
National Research Council (NRC (2006) Nutrient requirements of dogs and cats. National Academies Press, Washington DC
Oberbauer AM, Larsen JA (2021) Amino acids in dog nutrition and health. Adv Exp Med Biol 1285:199–216
Pan Y, Kennedy AD, Jönsson TJ et al (2018a) Cognitive enhancement in old dogs from dietary supplementation with a nutrient blend containing arginine, antioxidants, B vitamins and fish oil. Br J Nutr 119:349–358
Pan Y, Landsberg G, Mougeot I et al (2018b) Efficacy of a therapeutic diet on dogs with signs of cognitive dysfunction syndrome (CDS): a prospective double blinded placebo controlled clinical study. Front Nutr 5:127
Parker VJ (2021) Nutritional management for dogs and cats with chronic kidney disease. Vet Clin: Small Anim Pract 51:685–710
Quimby J, Gowland S, Carney HC et al (2021) 2021 AAHA/AAFP feline life stage guidelines. J Am Anim Hosp Assoc 57:51–72
Raj K, Szladovits B, Haghani A et al (2021) Epigenetic clock and methylation studies in cats. GeroScience 43:2363–2378
Ren W-y, Wu K-f, Li X et al (2014) Age-related changes in small intestinal mucosa epithelium architecture and epithelial tight junction in rat models. Aging Clin Exp Res 26:183–191
Russell RM (1992) Changes in gastrointestinal function attributed to aging. Am J Clin Nutr 55:S1203–S1207

Sándor S, Kubinyi E (2019) Genetic pathways of aging and their relevance in the dog as a natural model of human aging. Front Genet 10:948

Schauf S, Stockman J, Haydock R et al (2021) Healthy ageing is associated with preserved or enhanced nutrient and mineral apparent digestibility in dogs and cats fed commercially relevant extruded diets. Animals 11:2127

Sordo L, Martini AC, Houston EF et al (2021) Neuropathology of aging in cats and its similarities to human Alzheimer's disease. Front Aging 2:684607

Studzinski CM, MacKay WA, Beckett TL et al (2008) Induction of ketosis may improve mitochondrial function and decrease steady-state amyloid-β precursor protein (APP) levels in the aged dog. Brain Res 1226:209–217

Summers SC, Stockman J, Larsen JA et al (2020a) Evaluation of nutrient content and caloric density in commercially available foods formulated for senior cats. J Vet Intern Med 34:2029–2035

Summers SC, Stockman J, Larsen JA et al (2020b) Evaluation of phosphorus, calcium, and magnesium content in commercially available foods formulated for healthy cats. J Vet Intern Med 34:266–273

Teng K, McGreevy P, Toribio J et al (2018a) Associations of body condition score with health conditions related to overweight and obesity in cats. J Small Anim Pract 59:603–615

Teng KT, McGreevy PD, Toribio J-AL et al (2018b) Strong associations of nine-point body condition scoring with survival and lifespan in cats. J Feline Med Surg 20:1110–1118

Tropf M, Nelson O, Lee P et al (2017) Cardiac and metabolic variables in obese dogs. J Vet Intern Med 31:1000–1007

Wang T, Ma J, Hogan AN et al (2020) Quantitative translation of dog-to-human aging by conserved remodeling of the DNA methylome. Cell Syst 11:176–185

Willems A, Paepe D, Marynissen S et al (2017) Results of screening of apparently healthy senior and geriatric dogs. J Vet Intern Med 31:81–92

Wu G (2018) Principles of Animal Nutrition. CRC Press, Boca Raton, Florida

Chapter 10
Association Between Household Location (Urban versus Rural) and Fundamental Care Provided to Domestic Dogs (*Canis Familiaris*) in Northern Ireland

Violetta Naughton, Teresa Grzelak, and Patrick J. Naughton

Abstract This study aimed to generate information regarding living conditions (e.g. indoors and outdoors, living space available), fundamental care (e.g. feeding and exercise) and owners' perceptions of dog's preferences (e.g. living conditions) in the urban and rural areas of Northern Ireland via a cross-sectional survey questionnaire. The responses were collected from May to August 2022, at 24 locations across Northern Ireland, including 15 agricultural shows and livestock markets and nine large supermarkets (single supermarket chain) located in an area of the show or market grounds. In all, 548 questionnaires were collected and after exclusion of questionnaires with missing or incomprehensible responses, 507 questionnaires were included in the final database. Out of 507 questionnaires, 264 respondents resided in a city while 243 respondents resided on farms. The majority of respondents from the city locations perceived their dogs as pets, while those living on farms regarded their dogs as working animals. The populations of dogs in the city locations and on the farms in this study were similar regarding the age range and numbers of the animals, but more female dogs were spayed in the city locations than on the farms. Most respondents in this study, regardless of their household location, declared that they did not monitor their animal's body weight or body condition. When feeding their dogs, the respondents from the city locations were predominantly following veterinary advice or instructions on food labels. On the other hand, the respondents from farm locations mostly reported that they fed their dogs based on 'a visual inspection of dog condition'; this type of feeding was associated with a certain type of household occupancy (more frequent in single and adults only households) and respondents' employment status (more frequent by retired and those managing the home). The living conditions of dogs in city and farm locations in this study were different, namely dogs in the city were kept predominantly indoors with access to

V. Naughton (✉) · P. J. Naughton
School of Biomedical Sciences, University of Ulster, Cromore Road, Coleraine, Northern Ireland
e-mail: v.naughton@ulster.ac.uk

T. Grzelak
Department of Physiology, Poznan University of Medical Sciences, Poznan, Poland

G. Wu (ed.), *Nutrition and Metabolism of Dogs and Cats*, Advances in Experimental Medicine and Biology 1446, https://doi.org/10.1007/978-3-031-54192-6_10

outdoors while dogs from farm locations were kept predominantly outdoors. The dogs were reported to be walked daily for a shorter time (up to 1 h/day) in the city locations and longer on the farm locations (1–2 h/day). Regardless of household location (city versus farm) respondents believed that exercise needs depend on animal age, body condition and medical condition, that dogs need to be kept active by owners to keep them fit, that dogs cannot self-regulate the amount of food they eat daily, and finally that walking with a dog a few times a day is difficult due to other commitments. On the other hand, the respondents from farm locations more often believe that dogs can get all the exercise they need by themselves if kept outdoors and they are happier with living outdoors, while the respondents from city locations believed that dogs are happier with living indoors. In conclusion, the results of this study have shown a number of differences in basic care and perception of dogs kept in city locations and on farm locations. Further studies are required to understand the provision of health care and fulfilments of all welfare needs of the dogs kept on farms.

Keywords Pet dog · Working dog · Farm dog · Dog welfare · Feeding · Exercise · Body condition score

10.1 Introduction

Domestic dogs (*Canis familiaris*) play an important function in human society, both as pets providing predominantly emotional support to owners (Holland et al. 2022) and in various 'working' roles as service dogs, guide dogs, etc. (Udell et al. 2010, 2014). Literature suggests that people's perception of pet dogs and 'working' dogs as a function of dogs' relationship with humans differs (Lazarowski et al. 2020). Ample literature is available regarding the selection of dogs to suit the owners needs as pets or working animals (e.g. Hall 2017; Pongrácz et al. 2001; Rooney et al. 2004; Scandurra et al. 2015, 2017; Wobber et al. 2009) as well as domestic dog populations in western countries (e.g. Bedford 2022 in Europe; PDSA 2022 and Anderson et al. 2023 for UK 2022; AVMA 2017–18 for USA). However, the western reports of domestic dog populations normally refer to pet dogs, and do not differentiate if the owners live in urban or rural areas. Studies from Mexico and Chile that assessed urban and rural domestic dog populations showed that in Mexico it was more common to find a dog-owning household in a city than in the rural areas (Ortega-Pacheco et al. 2007), while in Chile a higher proportion of households owning dogs was located in rural areas than in towns and cities (Acosta-Jamett et al. 2010). Ortega-Pacheco et al. (2007) reported that ~ 25% of rural dogs in Mexico were used for company (25%), 20% both for company and as guardians while the majority (~ 55%) were used as working animals, e,g.: guardians, for hunting or herding. The same report by Ortega-Pacheco et al. (2007) showed that in a city the majority of dogs were used for company (50%), ~ 29% both for company and guarding purposes, with a lesser proportion as working (~ 23%) or breeding animals (1.3%). Similar results were reported by Acosta-Jamett et al. (2010) for Chile, specifically the majority of dogs in rural areas performed specific working tasks such as guarding, herding or hunting (~ 91%), while the dogs in a city and town area were kept predominantly (63–73%) as pets. Collectively these two reports would indicate that the function of dogs depends on the location of the households owning the dogs, i.e. urban versus rural areas.

A report from the U.S.A. by Shore et al. (2006) investigated care provided by owners to either 'yard' or house dogs, and it showed that both house and yard dogs received basic care; however, fewer yard dogs received as high a level of attention to some of their physical and safety needs as house dogs. The Shore et al. (2006) study recruited owners from Wichita State University, which is located in the largest city in the state indicating that all dogs in that study were from the urban area and that the living environment of dogs (indoors or outdoors) affects the care and attention to animal needs that the owners provide.

To our knowledge, there is no data on domestic dogs living on farms (versus cities) in western countries. Understanding living conditions, care and attention provided by the owners is crucial to assess animal health and welfare in any location. Furthermore, adequate background information is required for the understanding of the epidemiology of animal diseases and the planning and implementation of control programs in both urban and rural areas. Therefore, this study aimed to generate information regarding living conditions (e.g. indoors and outdoors, living space available), fundamental care (e.g. feeding and exercise practices) and the perception of dogs' preferences by dog owners from urban and rural areas of Northern Ireland.

10.2 Materials and Methods

An anonymous cross-sectional survey questionnaire was developed to investigate fundamental care practices of dog owners from the city and farm locations. The draft questionnaire was developed and piloted in-house ($n = 16$) using one-to-one interviews with volunteers (DeMaio et al. 1998). Their feedback in relation to each part of the questionnaire was collected, and the questionnaire was revised accordingly. The final questionnaire consisted of questions which were grouped into the following sections: (i) owners demographics (age group, gender, occupation, education, employment) and type of household (single versus adults only versus adults with children), (ii) information about dogs (number, age, sex and physiological status, e.g. intact versus spayed) and the function of the dog as perceived the owner, i.e. pet and companion versus working animal, (iii) monitoring body weight and body condition, (iv) rationale behind the amount of food fed daily, (v) dogs living conditions (indoor versus outdoors; space available), (vi) information about the frequency and duration of walks with dogs, and finally (vii) owners perception of dogs preferences with regards to living indoors versus outdoors and best care related to exercising.

The sample size of 354 completed questionnaires was estimated to be required to represent the population. The sample size was based on data from the latest CENSUS (2021) regarding the total number of 768,810 households in Northern Ireland (NI) and the proportion of households in urban (64%) versus rural (36%) locations, and including 95% confidence interval and 5% margin of error. To account for missing and incomprehensible responses while assuring required representation, the target sample size was increased by 5% to 372 questionnaires to be collected. In order to minimize the influence of cultural differences on respondent data, the survey was made available to responders residing across all of NI, i.e. the responses were collected from May to August 2022, at 24 locations across all of NI, including 15 agricultural shows and livestock markets and nine large supermarkets (single supermarket chain) located in an area of the show or market grounds. At the time of the distribution of the questionnaires, the prospective participants were asked to complete the questionnaires on site. However, the questionnaires were also supplied with pre-paid envelopes to those who were unable to complete the questionnaire at a given time. Each questionnaire was accompanied with a consent letter that informed the respondents that only one questionnaire per household was required. The completion of questionnaires was treated as a consent to participate in the study and all responses were anonymous. Survey respondents received no monetary or any other compensation for completing the questionnaire. Ethical approval for the study was granted by the School of Biomedical Sciences Research Ethics Filter Committee, Ulster University.

In total 548 questionnaires were collected. The data from all collected questionnaires were transferred into Excel (Microsoft), and 10% of transferred data were blind-checked for transfer accuracy. Questionnaires which included > 20% of missing or incomprehensible responses were excluded from the database, and thus 507 questionnaires were included in the final database. Out of 507 questionnaires, 264 respondents resided in a city; thus, their animals were categorized as the dogs living in a city (the city-location category), while 243 respondents resided on farms and thus their animals were categorized as the dogs living on a farm (the farm-location category).

Data in the final database were subsequently analysed with IBM SPSS Statistics v.25 (SPSS, Inc., Armonk, NY, USA) and Statistica 13 software with Medical Set (StatSoft, Tulsa, OK, USA) as follows: (i) frequency tables and chi-square tests were used to evaluate possible multicollinearity among risk factors significantly associated with the outcomes; if the expected cell frequencies were between 5 and 10, then chi-square test with Yates' continuity correction was used; in the case of the smaller expected cell frequencies (< 5), we employed the exact Fisher test of independence and 2×2 contingency table, and the exact Fisher–Freeman–Halton test and Cramer's V coefficient for bigger tables; (ii) the logistic regression analyses (progressive step method) were conducted to determine the predictors of monitoring body weight and the rationale for the amount of food fed daily to dogs in the city-dog and farm-dog categories; the significance of variables in the logistic regression models was tested using Wald Chi-squared statistics and corresponding p-value; the goodness-of-fit of the logistic regression models were assessed using the Hosmer–Lemeshow statistic; finally, the 95% confidence interval (95% CI) was used to estimate the precision of the odds ratio (OR) and a two-tailed p-value < 0.05 was considered statistically significant.

10.3 Results

10.3.1 Characteristics of the Respondents

Age of the Respondents. The majority of all respondents (70%, $n = 354$) were over 26 years of age, while 18–25 years of age accounted for 30% ($n = 153$). Within each location, the age distribution was as follows, 18–25 years of age: 102 versus 51, in the city and farm location, respectively, and over 26 years of age: 162 versus 192, in the city and farm location, respectively. There was a significant relationship between the household location and the age of the owners (X^2 (df1), $n = 507$) = 18.705, exact $p < 0.001$); with the owners categorized as 18–25 years old were more frequent (Odds Ratio = 1.84; 95% Cl: 1.38–2.45) in the households located in a city as compared to the farm locations.

Gender of the Respondents. The owner gender was represented analogously in both locations, with 168 females versus 96 males from the city locations and 170 females versus 73 males from the farm locations. A chi-square test of independence was performed to evaluate the relationship between the household location (city versus farm) and the owner's gender. The relationship between these variables was not significant (X^2 (df = 1, $n = 507$) = 2.276, exact $p = 0.157$).

Occupation of the Respondents. The majority of all respondents (97%) were classified as those having or not having a job related to animal health, husbandry or nutrition. The percentage of animal-related jobs to non-animal-related jobs was similar among city (1.5% versus 99%—animal versus non-animal-related, respectively) and farm locations (4% versus 96%—animal versus non-animal-related, respectively). There was no relationship between the household location (city versus farm) and owner occupation (exact Fisher test ($n = 507$), exact $p = 0.161$).

Employment Status of the Respondents. Out of 507 respondents, 405 (80%) declared to be in full or part-time employment, while 102 (20%) were either homemakers (managing a home), not employed, retired or studying. There was a significant relationship between the household location (city versus farm) and the owners' employment status (X^2 (df = 1, $n = 507$) = 17.543, exact $p < 0.001$); there was a higher probability of the dogs being owned by working owners on the farm locations (Odds Ratio = 2.21; 95% Cl: 1.498—3.258) versus the city locations.

Education of the Respondents. Approximately half of the respondents (53%) reported having educational qualifications at International Standard Classification of Education (ISCED 2011) levels 1–4, while 37% and 10% reported having educational qualifications at ISCED (2011) levels 5–6 and 7–8, respectively. Table 10.1 shows the counts (and %) of the respondents' educational qualifications in both the city and the farm locations. There was a significant relationship between the household location (city versus farm) and the respondent educational qualifications (exact Fisher–Freeman–Halton test: 11.134; exact $p = 0.006$, Cramer's V coefficient =

Table 10.1 Educational qualification of the respondents owning dogs in city and farm locations surveyed (May to August 2022) in Northern Ireland

Location		Educational qualifications—ISCED (2011) level*				
		1–4	5–6	7–8	Not sure	Total
City	Count (%)	154 (58)	91 (35)	17 (6)	2 (0.8)	264 (100)
Farm	Count (%)	115 (47)	96 (40)	32 (13)	0 (0)	243 (100)
	Total (%)	269 (53)	187 (37)	49 (10)	2 (0.4)	507 (100)

* ISCED—International Standard Classification of Education; level 1–4: from primary to post-secondary; level 5–6: short-cycle tertiary education and Bachelor's or equivalent; level 7–8: Masters or equivalent and Doctoral or equivalent

0.15). The educational qualifications ISCED (2011) at level 1- 4 was more frequent (Odds Ratio = 1.37; 95% Cl: 1.112–1.675), while level 7–8 were less frequent (Odds Ratio = 0.5; 95% Cl: 0.279–0.858) in the city location household than in the farm location.

10.4 Characteristics of the Households

Single versus Multiple Occupancies. Out of 507 respondents, 51 (10%) reported living in a single occupancy household, and 456 (90%) residing in households with multiple occupancies. Single occupancy households were reported by 7.2% (n = 19) and 13% (n = 32) of respondents from the city and farm locations, respectively. Further analysis showed a significant relationship between the household location (city versus farm) and the household occupancy (X^2 (df = 1, n = 507) = 4.988, exact p = 0.027) with lower chances of a dog being kept in a single occupancy household on a city location (Odds Ratio = 0.55; 95% Cl: 0.318–0.938) as compared to a farm location.

Type of Household Occupancy: Adults only versus Adults with Children. The households were occupied by either adult(s)-only (56%, n = 282) or adult(s) with children (44%, n = 225). There was a significant relationship between the household location (city versus farm) and the type of household occupancy (X^2 (df = 1, n = 507) = 3.987, exact p = 0.049), with a dog having a higher chance of being owned by adult owner(s) in a city location (Odds Ratio = 1.17; 95% Cl: 1.002–1.373) as compared to a farm location.

Number of Dogs Own in One Household. Among the households located in a city (n = 264), 144 owned one dog, 74 owned two dogs and 46 owned at least three or more dogs. Similarly, the farm households (n = 243) predominantly owned one dog, i.e. 135 households, while 70 and 38 households owned two or at least three or more dogs, respectively. A chi-square test of independence was performed to evaluate the relationship between household location (city versus farm) and the number of dogs

Table 10.2 Function of dogs as declared by respondents owning dogs in city and farm locations surveyed (May to August 2022) in Northern Ireland

Location		Declared 'function' of dogs			
		Pet/companion	Working	Both*	Total
City	Count (%)	241 (91)	11 (4)	12 (5)	264 (100)
Farm	Count (%)	2 (0.8)	199 (82)	42 (17)	243 (100)
	Total (%)	243 (50)	210 (41)	54 (11)	507 (100)

* Considered as a pet and working animal

owned in a household (one, two, three or more). The relationship between these variables was not significant, (X^2 (df $= 2$, $n = 507$) $= 0.294$, exact $p = 0.865$).

10.5 Function of Dogs Perceived by the Respondents

Out of all 507 respondents, 243 (50%) categorized their dogs as pets and companions, 210 (41%) regarded their dogs as working animals, while 54 (11%) reported having some dogs as pets and others as working dogs. Table 10.2 shows the counts (and %) of the reported function of dogs in both the city and farm locations. The relationship between the city versus farm locations and the perceived dogs' 'function' was significant (exact Fisher–Freeman–Halton test: 530.003, exact $p = 0.001$, Cramer's V coefficient = 0.91). The results showed that the number of dogs kept as working animals is significantly lower in the city location (Odds Ratio = 0.05; 95% Cl: 0.028–0.091) than on the farm location.

10.6 Animal Age and Sex

Total number of dogs owned by the study respondents was 885.

Age of Animals. The majority of all respondents (61%, $n = 310$) reported to own adult dogs (1–7 years old). Young (0–12 months old) and senior (older than 7 years old) dogs were reported by 18% ($n = 91$) and 19% ($n = 97$) of households, respectively. A few respondents were not sure of their animal age (2%, $n = 9$). A chi-square test of independence showed no relationship between household location (city versus farm) and the age of dogs own (exact Fisher–Freeman–Halton test: 2.738, exact $p = 0.433$, ($n = 507$), Cramer's V coefficient = 0.07).

Sex of Animals and Their Physiological Status. Out of 885 dogs in this study, 27% ($n = 236$) were intact female (not spayed), 26% ($n = 227$) were spayed female, 22% ($n = 190$) were intact male (not neutered), 26% (n = 228) were neutered males, while 0.4% ($n = 4$)—sex and physiological status was not declared. The relationship between the city versus farm locations and the sex and physiological status of dogs was significant (exact Fisher–Freeman–Halton test: 13.389; $n = 885$, exact $p = 0.007$, Cramer's V coefficient = 0.12) with higher chances of the female being spayed in the city location (Odds Ratio = 1.77; 95% Cl: 1.206–1.610, X^2 (df = 1, $n = 463$) = 9.324, exact $p = 0.002$) as compared to ones on farms.

10.7 Fundamental Care Related to Animal Health and Welfare: Body Weight and Body Condition Monitoring

Monitoring Body Weight of Dog(s). The highest number of respondents reported that they do not monitor the body weight of their dogs (44%, $n = 224$); 197 respondents (39%) reported taking the body weight of their animals only occasionally, e.g. once a year, while 86 (17%) of respondents reported monitoring the body weight of their dogs on a regular basis (at least every 3–6 months). A chi-square test of independence showed no relationship between household location (city versus farm) and the frequency of animal body weight monitoring (X^2 (df = 2, $n = 507$) = 2.968, exact $p = 0.231$).

Owner Familiarity with Body Condition Scoring (BCS) and its Practical Use. The majority of respondents ($n = 413$, 82%) reported not being familiar with and not using BCS to assess the body condition of their dogs. About 70 respondents (14%) reported to be familiar with BCS but not sure how to use it. Finally, only 24 (5%) of respondents reported being familiar with and using BCS to assess their dog's body condition. There was a significant difference between city and farm locations with regard to familiarity and use of BCS (exact Fisher–Freeman–Halton test: 14.088; exact $p < 0.001$, Cramer's V coefficient = 0.16), i.e. those who monitored BCS (combined answers 'Familiar with and using BCS' and 'Familiar with BCS but not sure how to use it') were mainly from a city location (see Table 10.3 for details).

Table 10.3 Monitoring animal body weight and knowledge, and use of body condition score (BCS) reported by the dog owners surveyed (May to August 2022) in city and farm locations in Northern Ireland

Location		Body condition scoring (BCS): familiarity and use			
		Familiar with and using BCS	Familiar with BCS but not sure how to use it	Not familiar with BCS and not using it	Total
City	Count (%)	21 (8)	38 (14)	205 (78)	264 (100)
Farm	Count (%)	3 (1)	32 (13)	208 (86)	243 (100)
	Total (%)	24 (5)	70 (14)	413 (82)	507 (100)

Table 10.4 Factors associated with body condition scoring (BCS), familiarity and use by owners from the city locations in Northern Ireland

Factors	b	SE (b)	OR	95% CI	*p*
Constant (y-intercept)	−2.233	0.395			
Owner's age group [years] 18–25	0		1	Reference	
Owner's age group [years] 26 +	0.686	0.328	1.99	1.05–3.78	0.036
Education ISCED1 2011 level 5–6	0		1	Reference	
Education Combined data: 'ISCED1 2011 level 1–4, 7–8 and 'not sure'	0.779	0.345	2.18	1.11–4.29	0.024

The Hosmer–Lemeshow statistic was 0.122, $p = 0.941$, and therefore, the model was judged to fit the observed data well (Hosmer et al. 2013); ISCED—International Standard Classification of Education; b—regression coefficient in Wald-statistic; SE (b)—standard error in Wald-statistic; OR—Odds Ratio; 95% CI—95% confidence interval; *p*—level of statistical significance

Factors Associated with BCS Monitoring. The logistic regression analysis was performed to identify which of the sociodemographic factors and household characteristics, i.e. (i) age, (ii) employment status, (iii) educational qualifications, (iv) single versus multiple occupancies households and (v) type of occupancy: adults only versus adults with children were determinants associated with BCS familiarity and use in dogs. Two factors, i.e. the age and the educational qualifications of respondents, were shown to be the predictive factors associated with familiarity and use of BCS by the owners from city locations, i.e. owners with age 26 + years were two times more likely than owners 18 to 25 to be familiar with BCS; also owners with education level 5–6 were two times more likely to be familiar with BCS than all other levels (see Table 10.4 for details). No associations between BCS familiarity and use in dogs on farms were determined.

10.8 Rationale Behind Daily Amount of Food Fed to Dog(S)

The rationale for the amount of food the respondents fed daily to their animals varied, but the highest number of respondents belonged to two categories, i.e. 145 (29%) reported following the feeding instructions on the dog food labels, while 131 (26%) followed veterinary advice. Others, i.e. 73 (14%) followed the advice of a breeder or other dog owner, while 97 (19%) fed their dogs based on the animal's appetite. Other respondents [61 (12%)] declared to feed their dogs based on a visual inspection of the animal. There was a significant relationship between household location (city versus farm) and the rationale behind the daily amount of food fed to a dog (X^2 (df

Table 10.5 Declared rationale behind the daily amount of food to fed to dog(s) by owners from the city and farm locations surveyed (May to August 2022) in Northern Ireland

Location	Rationale behind the daily amount of food to fed to dog(s)—count (%)					
	Dog's appetite	Following breeder/ other dog owner advice	Veterinary advice	Following feeding instructions on the dog food	Visual inspection of dog condition	Total
City	58 (22)	35 (13)	78 (30)	71(27)	22 (8)	264 (100)
Farm	39 (16)	38 (16)	53 (22)	74 (31)	39 (16)	243 (100)
Total	97 (19)	73 (14)	131 (26)	145 (29)	61 (12)	507 (100)

$= 4, n = 507) = 12.567$, exact $p = 0.013$). The respondents from the city locations were predominantly following veterinary advice or instructions on food labels. On the other hand, the highest number of respondents from farm locations reported that they fed their dogs based on a guide on feed labels. Notably however, feeding based on 'a visual inspection of dog condition' by respondents from farm locations was associated with a certain type of household occupancy (more frequent in single and adults only households) and respondents' employment status (more frequent by retired and those managing the home). Results obtained from the city and farm locations are shown in Table 10.5.

Factors Associated with Rationale Behind Amount of Food Fed to Dog(s). The logistic regression analysis was performed to identify which of the sociodemographic actors and household characteristics, i.e. (i) age, (ii) employment status, (iii) educational qualifications, (iv) single versus multiple occupancies households and (v) type of occupancy: adults only versus adults with children were determinants associated with a rationale for the daily amount of food fed to dogs. The employment status and household occupancy (adults only and single occupancy) were shown as significant predictive factors associated with declaring feeding dogs based on a visual inspection of dog condition in the case of farm location (see Table 10.6 for details). No associations between the rationale behind amount of food fed to dogs and tested sociodemographic factors and household characteristics in city locations were determined.

10.9 Where Dogs Are Kept—Living Conditions and Space

Out of all 507 respondents, 54 (11%) reported that they keep their dogs indoors and take the dog(s) for walks, 212 (42%) keep their dog(s) indoors with free access to outdoors, 100 (20%) keep their dog(s) in an outdoor pen (with a shelter), while 141 (28%) keep their dog(s) in a large yard/garden (with a shelter). There was a significant difference between the city and farm locations regarding where the dogs

Table 10.6 Factors associated with declared rationale, i.e. 'a visual inspection of dog condition' behind the amount of food fed to dog/s by respondents from the farm locations surveyed (May to August 2022) in Northern Ireland

Factors	b	SE (b)	OR	95% CI	*p*
Constant (y-intercept)	−2.937	0.419			
Household adults with children	0		1	Reference	
Household combined data: 'Single person' and 'Adults only'	1.424	0.507	4.15	1.54–11.22	0.005
Household multiple occupancies households	0		1	Reference	
Household single occupancies households	1.001	0.463	2.72	0.1–6.75	0.031
Employment Combined data: 'Employed or self-employed full-time (40 or more hours per week)' and 'Employed part-time (up to 39 h per week)' and 'Unemployed' and 'Student'	0		1	Reference	
Employment Combined data: 'Retired' and 'Homemaker (managing a home)'	2.205	0.762	9.07	2.03–40.41	0.004

The Hosmer–Lemeshow statistic was 0.001, $p = 0.999$, and therefore, the model was judged to fit the observed data well (Hosmer et al. 2013); b—regression coefficient in Wald-statistic; SE (b)—standard error in Wald-statistic; OR—Odds Ratio; 95% CI—95% confidence interval; *p*—level of statistical significance

are kept (exact Fisher–Freeman–Halton test: 599.940; exact $p < 0.001$, Cramer's V coefficient = 0.961), with dogs in the city locations being predominantly kept indoors with/without free access to the outdoors (Odds Ratio = 39.89; 95% Cl: 18.096–87.917), while dogs from the farm locations being predominantly kept outdoors (in a pen or a yard) (Odds Ratio = 64.37; 95% Cl: 24.33–170.27), see Table 10.7 for details.

10.10 Walks, the Frequency and Duration

Frequency of Walks. The majority of respondents, i.e. 205 (40%) reported that they walk their dog(s) once a day. About 142 (28%) reported to walk their dogs occasionally (e.g. once or twice a week). Respondents also reported walking their dogs more frequently, i.e. 120 (24%) owners declared to take a dog for a walk twice

Table 10.7 Animals living environment and space as declared by respondents from city and farm locations surveyed (May to August 2022) in Northern Ireland

	Living conditions—count (%)				
Location	Indoors with no access to outdoors (just taken out for walks)	Indoors with free access to outdoors (a patio or a garden)	Outdoors, in a pen with a shelter	Outdoors, in a large area (yard/ large garden) with a shelter	Total
City	54 (21)	206 (78)	4 (2)	0 (0)	264 (100)
Farm	0 (0)	6 (3)	96 (40)	141 (58)	243 (100)
Total	54 (11)	212 (42)	100 (20)	141 (28)	507 (100)

a day, 24 (5%) three times a day and 16 (3%) on animal demand. A chi-square test of independence showed no relationship between household location (city versus farm) and the frequency of daily walks provided to the dogs (X^2 (df = 4, n = 507) = 1.797, exact $p = 0.782$).

Duration of Walks. Among all respondents (n = 507), the two most frequently reported duration of walking a dog(s) were 15–30 min (34%, n = 174) and 30–60 min (37%, n = 186). A chi-square test of independence showed a significant relationship between household location (city versus farm) and the durations of walking the dogs (X^2 (df = 4, n = 507) = 23.574, exact $p < 0.001$). The shorter walks (combined data for 10–15 min and 15–30 min) were more frequent (X^2 (df = 1, n = 507) = 18.043, exact $p < 0.001$) in the city location (Odds Ratio 1.42; 95% Cl: 1.203–1.684) than the farm location. Further analysis showed that the longer walks (combined data for 30–60 min and 60–120 min) were more frequent (X^2 (df = 1, n = 507) = 18.043, exact $p < 0.001$) on the farm location (Odds Ratio 2.07; 95% Cl: 1.427–3.015) than the city location (see Table 10.8 for details). Among the respondent who selected the option 'normally short walks but occasionally for 1 h or longer', the answer 'once a month' was the most frequently reported (32%, n = 17 out 53) and no differences between the city and farm locations were detected (X^2 test with Yates' continuity correction: 0.973 (df = 1, n = 56); $p = 0.323$).

Table 10.8 Duration of walking a dog/s as declared by respondents from city and farm locations surveyed (May to August 2022) in Northern Ireland

Location	Duration of walks with a dog/s [min/one walk]—count (%)					Total
	10–15	15–30	30–60	60–120	Varied*	
City	28 (11)	110 (42)	84 (32)	13 (5)	29 (11)	264 (100)
Farm	20 (8)	64 (26)	102 (42)	33 (14)	24 (10)	243 (100)
Total	48 (10)	174 (34)	186 (37)	46 (9)	53 (11)	507 (100)

* i.e. normally short walks but occasionally for 1 h or longer

10.11 Perceptions of Dogs' Preferences and Best Care Provision

(a) We asked if pet dogs can get all the exercise they need themselves if they are kept outdoors. The majority of respondents, i.e. 61% ($n = 309$) disagreed with the statement, while 23% ($n = 116$) were neutral or not sure, and 16% ($n = 82$) agreed with the statement. There were significant relationships between household location (city versus farm) and the answer to this question (X^2 (df = 2, $n = 507$) = 8.098, exact $p = 0.015$). Owners from the city were significantly more likely to disagree than agree with the statement as compared to the owners from the farm locations (Odds Ratio 2.04; 95% Cl: 1.204–3.483; X^2 (df = 1, $n = 391$) = 7.979, exact $p = 0.005$).

(b) We asked if the pet dogs need to be kept active by their owner to keep them fit. The majority of the respondents (95%, $n = 479$) agreed with the statement and no effects of city versus farm locations were detected.

(c) We asked if pet dog(s) are 'happier' being kept outdoors. The majority of respondents, i.e. 49% ($n = 248$) were neutral and not sure about this, while others either agreed (27%, $n = 137$) or disagreed (24%, $n = 122$). Further analysis has shown a significant relationship between household location (city versus farm) and the frequency of answers to this statement (X^2 (df = 2, $n = 507$) = 59.247, exact $p < 0.001$). The respondents from city locations, in comparison to the farm location, were more frequently 'neutral' than 'agreeing' (Odds Ratio 2.85; 95% Cl: 1.790–4.601; X^2 (df = 1, $n = 385$) = 21.912, exact $p < 0.001$) or 'disagreeing' (Odds Ratio 2.81; 95% Cl: 1.695–4.756; X^2 (df = 1, $n = 370$) = 18.158, exact $p < 0.001$) with this statement.

(d) We asked if pet dog(s) are happier being kept indoors. As it was in the case of the previous statement, the majority of respondents, i.e. 59% ($n = 297$) were neutral and not sure about this, while others either agreed 22% ($n = 112$) or disagreed 19% ($n = 98$). Significant relationship has been detected between household location (city versus farm) and the frequency of answers to this statement (X^2 (df = 2, $n = 507$) = 49.175, exact $p < 0.001$). Respondents from city locations were more likely to agree than to disagree with this statement than those from farm locations (Odds Ratio 7.83; 95% Cl: 4.082–15.485; X^2 (df = 2, $n = 210$) = 46.896, exact $p < 0.001$).

(e) We asked if walking a dog once a day is sufficient to keep it fit and healthy. Although 44% ($n = 222$) agreed with this statement, 30% ($n = 152$) were not sure about it while 26% ($n = 133$) disagreed with this statement. Significant relationship was detected between household location (city versus farm) and the frequency of answers to this statement (X^2 (df = 2, $n = 507$) = 12.325, exact $p = 0.002$) with more respondents from farm location ($n = 126$, 52%) agreeing than disagreeing with the statement as compared to city locations ($n = 96$, 36%) (Odds Ratio 1.92; 95% Cl: 1.213–3.044, X^2 (df = 1, $n = 355$) = 8.685, exact $p = 0.003$).

(f) We asked if pet dogs require a minimum 2-h walk a day to keep healthy. About 42% ($n = 213$) of respondents were not sure about this, while nearly the same number of respondents, i.e. 41% ($n = 210$) disagreed with this statement. Only 17% ($n = 84$) agreed with this statement. Significant relationship was detected between household location (city versus farm) and the type of answers to this statement (X^2 (df = 2, $n = 507$) = 18.873, exact $p < 0.001$), with respondents from city locations being more likely to be neutral than 'agreeing' with this statement (Odds Ratio 2.3; 95% Cl: 1.336–3.992; X^2 (df = 1, n = 297) = 10.388, exact p = 0.002), while respondents from farm location were more likely to disagree than declare to be neutral regarding this statement (Odds Ratio 2.17; 95% Cl: 1.448–3.277; X^2 (df = 1, $n = 423$) = 15.516, exact $p <$ 0.001).
(g) We asked if it is difficult to get pet dogs to exercise a few times a day due to other (work and/or family) commitments. The majority of respondents (66%, $n = 336$) agreed with this statement, while 20% ($n = 101$) were neutral and not sure and 14% ($n = 70$) of respondents disagreed with this statement. There was no relationship between the household location (city versus farm) and the frequency of answers to this statement.
(h) We also asked if the amount of exercise a pet dog needs depends on its age, body condition and medical condition. The majority of the respondents 94% ($n = 477$) agreed with this statement and no relationship between the household location (city versus farm) and the frequency of answers to this statement was detected.
(i) Finally, we asked if pet dogs can self-regulate amount of food they eat daily. The majority of respondents 56% ($n = 283$) disagreed with this statement, while 26% ($n = 134$) of respondents were not sure about it and 19% ($n = 90$) declared that dogs can self-regulate amount of food consumed per day. No differences between the city and farm locations were detected in responses to this statement.

10.12 Discussion

The most prominent difference between respondents from the city and the farms was their perception of their dogs. The majority of respondents from the city locations perceived their dogs as pets, while those living on farms regarded their dogs as working animals. The results of a study conducted by Holland et al. (2022) showed that majority of people in their study keep dogs for companionship, however some acquired dogs to perform specific tasks (e.g. security or assistance). As pets, dogs lack a practical function, and although our study did not investigate the roles that dogs fulfil on farms (e.g. sheepherding or yard guarding), it can be postulated that dogs on farms are kept mainly for practical purposes and thus are perceived as having specific working functions.

The population of dogs in the city locations and on the farms in this study were similar regarding animals age range and animal numbers; however, noticeably more female dogs were spayed in the city locations as compared to females kept on the farms. Downes et al. (2009) reported that 47% of pet dogs on the island of Ireland were neutered, findings similar to our current results and past investigation (Naughton et al. 2021). The literature indicates that a number of factors affect owners' choice of not-neutering their pets, such as perceived lack of benefits of neutering to the pet owner (PSDA 2017, 2019), young age of the pet (PDSA 2019) and the cost of the procedure (Frank et al. 2007).

The current results showed that more dogs were owned by single households on farms (versus city locations) but more dogs were owned by adults only households in the city as compared to the farms. Downes et al. (2009) have found that dog ownership on the island of Ireland is associated with households of multiple occupancies. Westgarth et al. (2007) have shown that multiple occupancy households, especially those with older children (6–19 years), are more likely to own a dog. The discrepancies between the current study and previously published results are probably due to the methods used to gather the data. In this study, volunteers were recruited through a variety of face-to-face advertising and data collection methods, and in places (e.g. agricultural shows or supermarkets) visited by dog owners, while the authors of the cited reports have used computer-assisted telephone technology to recruit respondents and to collect data. However, it is worth noting that single ownership of a dog on farm locations observed in this study may be related to the psychological needs of the owners. Holland et al. (2022) have shown that a majority of respondents in their study indicated that their decision to get a dog was influenced by a desire for companionship. The Northern Ireland Statistics and Research Agency (NISRA) Report (2020) for Northern Ireland showed that ~ 60% of those who live alone feel lonely more often than people living with others. However, while single person households are equally common in urban and rural locations in Northern Ireland, the separation from other humans, due to distances between rural households and lone working conditions are more common on farms as compared to any city location. Thus, despite the fact that dogs were being classified as working animals on farms in this study, it can be speculated that their role on farm locations is not only practical, but also for companionship.

The majority of respondents in this study, regardless of their household location, declared that they do not monitor their animals body weight and body condition. Recent results by German et al. (2018) have shown that 715 out of 1100 assessed adult dogs (65%) were overweight (body condition score of 6/9 to 9/9) and 99 (9%) were obese (body condition score of 8/9 or 9/9). A lack of monitoring of body weight and body condition of dogs has direct negative implications for the health and welfare of the animals (Chandler et al. 2017). Further analysis of the results of this study showed those reported to be familiar with body condition scoring (BCS) and using BCS were mainly from the city locations. We have tested sociodemographic factors as predictors of familiarity and use of BCS with resulting age (younger owners, 18–25 years of age) and the educational qualifications (higher level of education) showing to be two significant determinants of the respondents from city locations in monitoring their dog's body condition. Previously published studies indicated different associations between the education of the owners and care provided to dogs. For example, a study by Marinelli et al. (2007) and older publications (Johnson et al. 1992; Lago et al. 1987) reported the negative influence of having a college education on care provided to dogs. If we consider that monitoring BCS belongs to health care provision, our results are further in contrast with previously published results by Marinelli et al. (2007) indicating that the level of medical care increases with the level of education until High School, while it decreases for dogs belonging to University graduates. It is rather difficult to rationalize the differences between current and previously published studies due to methodological differences; however, since it is acknowledged that proactive monitoring of body weight and body condition throughout an animal's life is fundamental to any obesity preventive plan (German, 2016) further research into BCS monitoring is required.

Body weight and conditions are two primary factors influencing daily feeding amounts and thus daily energy intake (Wu 2018). Results from the current study show that respondents from the city locations were predominantly following veterinary advice or instructions on food labels when feeding their dogs. On the other hand, responses by the respondents from farm locations varied but noticeably those who reported that they fed their dogs based on 'a visual inspection of dog condition' were associated with the type of household occupancy (more frequent in single and adults only households) and the respondents' employment status (more frequent by retired and those managing home). These results could indicate that single, adult owners living on a farm, may interact with their dogs more and thus feed their animals based on a subjective impression of animal appearance and not objective assessment of the animal's needs. This feeding type combined with a lack of a body weight and body condition assessment may lead directly to animal overfeeding and consequent obesity. Indeed, misjudged body condition score (BCS) of the dog by its owner has been shown as one of the factors contributing to overfeeding and obesity and overweight status in dogs (Colliard et al. 2006; Courcier et al. 2011; Kienzle et al. 1998).

The living conditions of dogs in city and farm locations in this study were different, namely dogs in the city were kept predominantly indoors with access to outdoors while dogs from farm locations were kept predominantly outdoors. We have not investigated if the respondents of this study from city locations live in apartments or reside in semi-detached or detached dwellings; however, the majority of responses collected, i.e. keeping dogs indoors with free access to outdoors, would indicate that the respondents live in in semi-detached or detached dwellings with access to a garden, which may facilitate keeping a dog at home with access to outdoors. Notably, only 6 out of 243 respondents in this study from farm locations declared keeping a dog indoors (with free access to outdoors). Similar results were previously reported by Slater et al. (2008) regarding dogs kept in central Italy, namely 62% (112/180) of dogs were kept entirely outdoors despite the fact that 82% (147/180) were considered companions rather than working dogs. Furthermore, Carvelli et al. (2020) in a larger study also reported that in Italy the majority of dogs in rural areas lived outdoors. Critical interpretations of the current results regarding keeping dogs solely outdoors on farm locations is unattainable due to lack of published literature on any associated factors but it could be speculated that keeping a dog outdoors is related to general perception of animals by farmers, i.e. farmers' perspectives appear to be more closely aligned with the traditional welfare science views of minimizing harm, e.g. a yard may be considered as a self-exercise area by the owners as opposed to maximizing positive experiences, e.g. actively walking a dog (Vigors et al. 2021). However, when comparing responses by respondents from city location versus farm location in this study, it was revealed that the dogs were walked daily for a shorter time (up to 1 h) in the city and longer on the farms (1–2 h). To our knowledge, there are no published results that have directly assessed associations between household locations (urban versus rural) and frequency and duration of walks with dogs. Our previous study (Naughton et al. 2021) for NI and the PAW report (PSDA, 2019) for the UK both have shown that walking a dog once a day for less than 1 h was the most prevalent type of physical activity provided to dogs in an urban location. It can be speculated that the differences between place and type of work performed by respondents in a city (away from home) versus on a farm (at home and able to take a dog along when working) directly affect the differences in walk frequency and duration.

Aside from factual information on similarities and differences of care provided to dogs in the city or on the farm locations, we have also collected the views of the respondents on perceived dogs' preferences for living indoors or outdoors, and the animals' ability to self-regulate feed intake or exercise. The results of the current study showed that regardless of household location (city versus farm) respondents believed that dogs exercise needs depend on animal age, body condition and medical condition, that dogs need to be kept active by owners to keep them fit, that dogs cannot self-regulate the amount of food they eat daily, and finally that walking with a dog a few times a day is difficult due to other commitments. On the other hand, the respondents of the current study from the farms locations more often believed that dogs can get all the exercise they need themselves if are kept outdoors and they are happier living outdoors, while the respondents from the city locations believed that dogs are happier to live indoors. Published literature suggests that dogs kept indoors are at reduced risk of relinquishment (Patronek et al. 1996). Furthermore, Kotrschal el al. (2009) showed that the more perceived social support the dog was providing to the owner, the less time the owner spent together with the dog on a daily basis. This would partially explain the differences between opinions of the respondents from farm locations versus city locations in this study. On the other hand, Shore et al. (2006) showed that living in close proximity to the owners at most times at home was associated with dogs having opportunities to socialize more, and proposed educating dog owners on the social needs of dogs. Namely, the benefits of keeping dogs indoors could result in richer lives for the animals.

To our knowledge, this is the first study to investigate fundamental care and perception of dogs kept in cities versus on the farms in a developed country. Our methods generated qualitative data that built on the statistical analyses of the quantitative survey data. However, this study has specific limitations, i.e. the self-selection process of the survey sample. Although the study population showed similar spread of gender and type of household occupation by location (city versus farm), we have noted differences in age (younger owners in the city while older on the farms), employment status (higher probability of the dogs being owned by working owners on the farm locations versus city locations) and education (different profile of education of respondents in city versus farm locations). Thus, caution should be taken if attempting to generalize our results to other populations.

In conclusion, the current results have shown a number of differences in basic care and perception of dogs kept in city locations and on farm locations. Further studies are required to understand the provision of health care and fulfilments of all welfare needs of working farm dogs. Also, it would be advisable to include specific household locations in any domestic dog population studies carried out in developed countries because of the differences identified in this report.

Acknowledgements Financial support for this study has been provided by the School of Biomedical Sciences, Ulster University. The authors declare that they have no competing interests.

References

Acosta-Jamett G, Cleaveland S, Cunningham AA, Bronsvoort BD (2010) Demography of domestic dogs in rural and urban areas of the Coquimbo region of Chile and implications for disease transmission. Pre Vet Med 94:272–281

American Veterinary Medical Associations (AVMA). U.S. pet owners statistics 2017–2018. Available at: https://www.avma.org/resources-tools/reports-statistics/us-pet-ownership-statistics

Anderson KL, Casey RA, Cooper B, Upjohn MM, Christley RM (2023) National dog survey: describing UK dog and ownership demographics. Animals 13:1072

Bedford E (2022) Number of pet dogs in Europe from 2010 to 2021. Available at: https://www.statista.com/statistics/515579/dog-population-europe/#:~:text=In%202021%2C%20the%20dog%20population,million%20in%20the%20previous%20year

Carvelli A, Scaramozzino P, Iacoponi F, Condoleo R, Della Marta U (2020) Size, demography, ownership profiles, and identification rate of the owned dog population in central Italy. PLoS ONE 15:e0240551

Census (2021) Population and household estimates for Northern Ireland (May 2022) https://www.nisra.gov.uk/statistics/2021-census/results

Chandler M, Cunningham S, Lund EM, Khanna C, Naramore R, Patel A, Day MJ (2017) Obesity and associated comorbidities in people and companion animals: a one health perspective. J Comp Pathol 156:296–309

Colliard L, Ancel J, Benet JJ, Paragon BM, Blanchard G (2006) Risk factors for obesity in dogs in France. J Nutr 136:1951S-1954S

Courcier EA, Mellor DJ, Thomson RM, Yam PS (2011) A cross sectional study of the prevalence and risk factors for owner misperception of canine body shape in first opinion practice in Glasgow. Prev Vet Med 102:66–74

DeMaio TJ, Rothgeb J, Hess J (1998) Improving survey quality through pretesting. US Bureau of the Census, Washington, DC, pp 50–58

Downes M, Canty MJ, More SJ (2009) Demography of the pet dog and cat population on the island of Ireland and human factors influencing pet ownership. Prev Vet Med 92(1–2):140–149

Frank JM, Carlisle-Frank PL (2007) Analysis of programs to reduce overpopulation of companion animals: Do adoption and low-cost spay/neuter programs merely cause substitution of sources? Ecol Econ 62:740–746

German AJ (2016) Obesity prevention and weight management after loss. Vet Clin N Amer 46:913–929

German AJ, Woods GR, Holden SL, Brennan L, Burke C (2018) Dangerous trends in pet obesity. Vet Rec 182:25

Hall NJ (2017) Persistence and resistance to extinction in the domestic dog: basic research and applications to canine training. Behav Processes 141:67–74

Holland KE, Mead R, Casey RA, Upjohn MM, Christley RM (2022) Why do people want dogs? A mixed-methods study of motivations for dog acquisition in the United Kingdom. Front Vet Sci 9

Hosmer DW, Lemeshow Jr SA, Sturdivant RX (2013) Applied logistic regression, 3rd edn. Wiley: Hoboken, NJ, USA, pp 153–226

Johnson TP, Garrity TF, Stallones L (1992) Psychometric evaluation of the Lexington attachment to pets scale (LAPS). Anthrozoös 5:160–175

Kienzle E, Bergler R, Mandernach A (1998) A comparison of the feeding behavior and the human–animal relationship in owners of normal and obese dogs. J Nutr 128:2779S-2782S

Kotrschal K, Schöberl I, Bauer B, Thibeaut AM, Wedl M (2009) Dyadic relationships and operational performance of male and female owners and their male dogs. Behav Processes 81:383–391

Lago D, Kafer R, Delaney M, Connell C (1987) Assessment of favorable attitudes toward pets: development and preliminary validation of self-report pet relationships scales. Anthrozoös 1:240–254

Lazarowski L, Thompkins A, Krichbaum S, Waggoner LP, Deshpande G, Katz JS (2020) Comparing pet and detection dogs (*Canis familiaris*) on two aspects of social cognition. Learn Behav 48:432–443

Loneliness in Northern Ireland 2019/20 (2020) Northern Ireland statistics and research agency Colby house, Stranmillis Court, Belfast. www.nisra.gov.uk

Marinelli L, Adamelli S, Normando S, Bono G (2007) Quality of life of the pet dog: influence of owner and dog's characteristics. Appl Anim Behav Sci 108:143–156

Naughton V, Grzelak T, Mulhern MS, Moffett RC, Naughton PJ (2021) Caring practices of pet cat and dog owners in Northern Ireland versus potential implications for animals health and welfare. Anim Welf 30:131–144

Ortega-Pacheco A, Rodriguez-Buenfil JC, Bolio-Gonzalez ME, Sauri-Arceo CH, Jiménez-Coello M, Forsberg CL (2007) A survey of dog populations in urban and rural areas of Yucatan, Mexico. Anthrozoös 20:261–274

Patronek GJ, Glickman LT, Beck AM, McCabe GP, Ecker C (1996) Risk factors for relinquishment of dogs to an animal shelter. J Amer Vet Med Ass 209:572–581

PDSA (2017) Animal wellbeing (PAW) report 2017. pdsa.org.uk

PDSA (2019) Animal wellbeing (PAW) report 2019. pdsa.org.uk

PDSA (2022) Animal wellbeing (PAW) report 2022. pdsa.org.uk

Pongrácz P, Miklósi Á, Kubinyi E, Gurobi K, Topál J, Csányi V (2001) Social learning in dogs: the effect of a human demonstrator on the performance of dogs in a detour task. Anim Behav 62:1109–1117

Rooney NJ, Bradshaw JWS, Almey H (2004) Attributes of specialist search dogs—a questionnaire survey of UK dog handlers and trainers. J Forensic Sci 49:300–306

Scandurra A, Prato-Previde E, Valsecchi P, Aria M, D'Aniello B (2015) Guide dogs as a model for investigating the effect of life experience and training on gazing behaviour. Anim Cogn 18:937–944

Scandurra A, Alterisio A, Marinelli L, Mongillo P, Semin GR, D'Aniello B (2017) Effectiveness of verbal and gestural signals and familiarity with signal-senders on the performance of working dogs. Appl Anim Behav Sci 191:78–83

Shore ER, Riley ML, Douglas DK (2006) Pet owner behaviors and attachment to yard versus house dogs. Anthrozoös 19:325–334

Slater MR, Di Nardo A, Pediconi O, Dalla Villa P, Candeloro L, Alessandrini B, Del Papa S (2008) Cat and dog ownership and management patterns in central Italy. Prev Vet Med 85:267–294

Udell MAR, Dorey NR, Wynne CDL (2010) What did domestication do to dogs? A new account of dogs sensitivity to human actions. Biol Rev Biol Proc 85:327–345

Udell MAR, Ewald M, Dorey NR, Wynne CDL (2014) Exploring breed differences in dogs (*Canis familiaris*): does exaggeration or inhibition of predatory response predict performance on human-guided tasks? Anim Behav 89:99–105

Vigors B, Ewing DA, Lawrence AB (2021) The importance of farm animal health and natural behaviors to livestock farmers: findings from a factorial survey using vignettes. Anim Front 2:638782

Westgarth C, Pinchbeck GL, Bradshaw JW, Dawson S, Gaskell RM, Christley RM (2007) Factors associated with dog ownership and contact with dogs in a UK community. BMC Vet Res 3:5

Wobber V, Hare B, Koler-Matznick J, Wrangham R, Tomasello M (2009) Breed differences in domestic dogs (*Canis familiaris*) comprehension of human communicative signals. Interact Stud 10:206–224

Wu G (2018) Principles of animal nutrition. CRC Press, Boca Raton, Florida

Chapter 11
Analysis of Gizzerosine in Foodstuffs by HPLC Involving Pre-column Derivatization with *o*-Phthaldialdehyde

Peng Li and Guoyao Wu

Abstract Gizzerosine [2-amino-9-(4-imidazolyl)-7-azanonanoic acid] is a toxic amino acid formed from histamine and lysine at high temperatures, and may be present in foodstuffs (e.g., fishmeal and meat-bone meal) for animals including cats and dogs. Here we developed a simple, rapid, sensitive, specific, and automated method for the analysis of gizzerosine in foodstuffs by high-performance liquid chromatography (HPLC) involving pre-column derivatization with *o*-phthaldialdehyde (OPA) in the presence of *N*-acetylcysteine (instead of the usual 2-mercaptoethanol or ethanethiol reagent). OPA reacted immediately (within 1 min) with gizzerosine in an autosampler at room temperatures (e.g., 20–25 °C), and their derivative was directly injected into the HPLC column. The highly fluorescent gizzerosine-OPA derivative was well separated from the OPA derivatives of all natural amino acids known to be present in physiological fluids (e.g., plasma), proteins and foodstuffs, and was detected at an excitation wavelength of 340 nm and an emission wavelength of 450 nm. The total time for chromatographic separation (including column regeneration) was 20 min per sample rather than 40 min and longer in previous HPLC methods. The detection limit for gizzerosine was at least 6 pmol/ml in an assay solution (HPLC vial) or at least 0.09 pmol per injection into the HPLC column. The analysis of gizzerosine was linear between 1 and 100 pmol per injection. When gizzerosine was extracted from foodstuffs, its detection limit was at least 875 pmol/g foodstuff or at least 0.21 mg/kg foodstuff. Our routine HPLC technique does not require any cleanup of samples or the OPA derivatization products (including the OPA-gizzerosine adduct), and is applicable for the analysis of gizzerosine in both foodstuffs and animal tissues.

Keywords Amino acids · Animals · Gizzerosine · HPLC · Foodstuffs · *o*-Phthadilaldehyde

P. Li
North American Renderers Association, Alexandria, VA 22314, USA

G. Wu (✉)
Department of Animal Science, Texas A&M University, College Station, TX 77845, USA
e-mail: g-wu@tamu.edu

G. Wu (ed.), *Nutrition and Metabolism of Dogs and Cats*, Advances in Experimental Medicine and Biology 1446, https://doi.org/10.1007/978-3-031-54192-6_11

Abbreviations

AA	Amino acid
HPLC	High-performance liquid chromatography
LC	Liquid chromatography
OPA	*o*-Phthaldialdehyde

11.1 Introduction

Gizzerosine [2-amino-9-(4-imidazolyl)-7-azanonanoic acid; an amino acid (AA)] is not a natural substance but is formed from histamine and lysine at high temperatures (Fig. 11.1) (Tao et al. 2019). Both histamine (a product of histidine decarboxylation) and lysine may be present at relatively high concentrations in animal-sourced foodstuffs (Li and Wu 2022) and can react spontaneously to generate gizzerosine during ingredient processing and transportation as well as product transportation and storage at elevated temperatures (e.g., 80 °C for 24 h or 100 °C for 2 h) (Sugahara et al. 1992). Gizzerosine was first discovered in fishmeal as a substance to cause gizzard erosion (black vomiting) in chicks by stimulating the secretion of gastric acid (a histamine H_2-receptor agonist) (Masumura et al. 1985; Gjevre et al. 2013). The presence of even 2.2 ppm gizzerosine in the diet can result in severe gizzard erosion in chicks within a week (Okazaki et al. 1983). Fishmeal (Aldrich 2018; Funaba et al. 2001) and other animal-sourced foodstuffs (Li et al. 2021; Li and Wu 2022, 2023b) are excellent sources of AAs, minerals and polyunsaturated fatty acids for cats and dogs. Thus, it is imperative to determine gizzerosine in foodstuffs for animals including cats and dogs.

Several liquid chromatography (LC)-based methods have been developed for the analysis of gizzerosine (Table 11.1), including LC–mass spectrometry (MS)–MS;

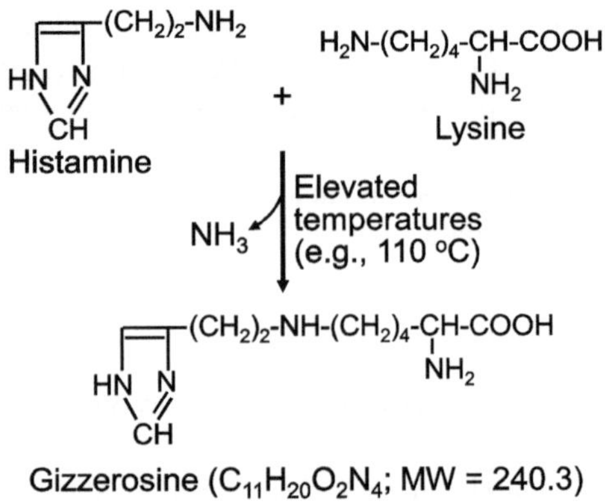

Fig. 11.1 Reaction of histamine with lysine at high temperatures to form gizzerosine. The latter may be present in overheated foodstuffs such as fishmeal and meat-bone meal. Gizzerosine is stable in water, ethanol or acid solutions

pre-column derivatization with *o*-phthaldialdehyde (OPA) in the presence of *N*-acetylcysteine (instead of 2-mercaptoethanol or ethanethiol), sulfanilic acid (Pauly's reagent), or dansyl chloride; or post-column derivatization with OPA. Major disadvantages of all these previous techniques except for the OPA method are the need for a tedious step of cleanup of food samples, derivatization products, or both and a long running time for the analysis of one sample (35 to > 65 min). Other approaches to the determination of gizzerosine, which are less commonly used in laboratories, include radioimmunoassays with monoclonal or polyclonal antibodies (Rosselot et al. 1996; Torres et al. 1999), enzyme-linked immunoassays (Becker et al. 1998; Manosalva et al. 2004), and paper (Tao et al. 2012) or microchip capillary (Xiao et al. 2017) electrophoresis.

The objective of this study was to develop a simple, rapid, sensitive, specific, reproducible, accurate, and automated method for the analysis of gizzerosine in

Table 11.1 Liquid chromatography-based analysis of gizzerosine in foodstuffs

Separation	Sample/DV cleanup	LC Column	Derivatization	Detection	Running time (min)	Authors
LC-MS-MS	Yes	250 × 2.0 mm TSKgel Amide-80 (5 μm)	N/A	MS	35	Uhlig et al. (2011)
HPLC	Yes	ODS column (250 × 4.6 mm, C_{18})	Sulfanilic acid (pre-column)	UV	35	Tao et al. (2019)
HPLC	Yes	Waters Pico-Tag	Dansyl chloride (pre-column)	Fluorescence	> 65	Wagener et al. (1991)
HPLC	Yes	ODS (5 μm, 200 × 6 mm)	OPA (ET) (pre-column)	Fluorescence	40	Ito et al. (1985)
HPLC	Yes	Amberlite IR CG-50 (200 × 15 mm and 70 × 10 mm)	OPA (ME) (post-column)	Fluorescence	50	Ohta et al. (1988)
HPLC	Yes	Shim-pack ISC-07/S1504 (150 mm × 4 mm)	OPA (post-column)	Fluorescence	50	Murakita and Gotoh (1990)

DV derivatives; *ET* ethanethiol; *HPLC* high-performance liquid chromatography; *LC* liquid chromatography; *MS* mass spectrometry; *ODS* octadecylsilica; *OPA* *o*-phthaldialdehyde (also known as *o*-phthalaldehyde); *UV* ultra-violet foodstuffs by HPLC. The technique was based on our routine work on the determination of physiological AAs via pre-column derivatization with OPA (Hu et al. 2023; Li and Wu 2020; Wu and Meininger 2008) with some modifications. Namely, *N*-acetylcysteine, rather than 2-mercaptoethanol, was used for the reaction between AAs and OPA to facilitate the separation of the gizzerosine-OPA from the derivatives of OPA with other AAs on the HPLC column and to reduce the total running time, as we reported previously for the analysis of homoarginine (Hou et al. 2015) and polyamines (Dai et al. 2014b).

11.2 Materials and Methods

11.2.1 Chemicals

All chemicals used in the protocol described herein except for HPLC-grade methanol and H_2O were obtained from Sigma Chemicals (St. Luis, MO) and were of the highest purity. All of the prepared solutions can be stored at room temperatures (20–25 °C) except for the OPA-*N*-acetylcysteine reagent solution as well as gizzerosine and physiological AA standards, unless specified otherwise.

(a) HPLC-grade methanol and H_2O: These were obtained from Fisher Scientific (Houston, TX, USA) and used for preparing all reagents and mobile phase solutions, as well as making dilution for AA standards and food samples. Methanol and H_2O are stable at 25 °C for 6 months.
(b) 6 M HCl: Add slowly 49.1 mL of concentrated HCl (37–38%) to 45 mL H_2O in a glass flask. Make up to a final volume of 100 mL with H_2O. This solution is stable at 25 °C for 2 years.
(c) 6 M NaOH: Dissolve 12 g NaOH in 40 mL H_2O. Make up to a final volume of 50 mL with H_2O. This solution is used freshly.
(d) 1.5 M $HClO_4$: Add slowly 32.2 mL of 70% $HClO_4$ to 150 mL H_2O. Make up to a final volume of 250 mL with H_2O. This solution is stable at 25 °C for 2 years.
(e) 2 M K_2CO_3: Dissolve 69.11 g K_2CO_3 in 150 mL H_2O. Make up to a final volume of 250 mL with H_2O. This solution is stable at 25 °C for 2 years.
(f) Benzoic acid (1.2%): Dissolve 8.4 g benzoic acid (an antibiotic) in 525 mL H_2O, followed by the addition of 175 mL of saturated $K_2B_4O_7$ (prepared in H_2O). This solution is stable at 25 °C for 2 years. [Do not add directly $K_2B_4O_7$ powder to H_2O that contains benzoic acid powder.]
(g) Sodium borate (40 mM, pH 9.5): Dissolve 30.51 g $Na_2B_4O_7.10H_2O$ in 2 L H_2O. This solution is stable at 25 °C for 2 years.
(h) OPA-*N*-acetylcysteine reagent solution (30 mM OPA, 23.7 mM *N*-acetylcysteine, 40 mM sodium borate, and 3.1% Brij-35; pH 9.5): Add 50 mg OPA (stored at 4 °C), 50 mg N-acetyl-cysteine, and 1.25 mL methanol to a brown bottle, followed by addition of 11.2 mL of 40 mM sodium borate buffer (pH 9.5), and 0.4 mL Brij-35. The solution is mixed thoroughly, stored at 4 °C, and

used within 36 h after preparation. The use of *N*-acetyl-cysteine in the derivatization solution is based on the work of Hou et al. (2018) for the analysis of glutathione.

(i) Gizzerosine (1 mM): Dissolve 1.3 mg gizzerosine-2Cl [MW = 313.22; a kind gift from Prof. Hiromasa Kiyota, Tohoku University, Sendai, Japan (Shimasaki et al. 2005)] in 4.15 mL HPLC-grade water. Store aliquots of the standard solution at – 80 °C.

11.2.2 Foodstuffs

The sources of seven U.S. poultry meal samples, three fishmeal samples, three U.S. soybean meal samples, and three U.S. corn grain samples were the same as we described previously (Li and Wu 2020). They were stored at – 20 °C.

11.2.3 Extraction of Gizzerosine from Foodstuffs

(a) ***Water extraction***

Gizzerosine is a highly water-soluble substance (Wu 2022). Thus, water is an excellent solvent to extract free gizzerosine from foodstuff samples. Each food sample (1.0 g) was added to 10 ml of HPLC-grade water, followed by vortexing for 1 min. An aliquot (0.2 mL) of the supernatant fluid was mixed with 0.1 mL of 1.5 M $HClO_4$, followed by neutralization with 0.05 mL of 2 M K_2CO_3. The extract was then used for the analysis of gizzerosine.

(b) ***Ethanol extraction***

Gizzerosine may be associated with lipid-soluble materials. Thus, we also used ethanol to extract gizzerosine from foodstuff samples. Each food sample (1.0 g) was added to 10 ml of HPLC-grade ethanol, followed by vortexing for 1 min. An aliquot (0.2 mL) of the supernatant was evaporated under a stream of nitrogen gas. The residue was suspended with 0.2 mL of HPLC-grade water, mixed with 0.1 mL of 1.5 M $HClO_4$, and neutralized with 0.05 mL of 2 M K_2CO_3. The extract was then used for the analysis of gizzerosine.

(c) ***Protein hydrolysates***

The ε-amino group of lysine residues in protein may react with histamine at high temperatures to yield protein- or peptide-bound gizzerosine (Wu 2022). Thus, gizzerosine may be covalently linked with AAs in foodstuff proteins and peptides. For this reason, we also hydrolyzed proteins and peptides in foodstuff samples under standard conditions to extract gizzerosine. Each food sample (500 mg) was hydrolyzed, under nitrogen gas, with 6 mL of 6 M HCl at 110 °C for 24 h (Dai et al. 2014a).

Thereafter, the solution was neutralized with 6 M NaOH, adjusted to a final volume of 15 ml with HPLC-grade water, mixed thoroughly, and then used for the analysis of gizzerosine.

11.2.4 HPLC Apparatus

The Waters HPLC apparatus consisted of the following: (1) a Model 600E Powerline multi-solvent delivery system with 100-μl heads, (2) a Model 600E system controller, (3) a Model 717plus autosampler (or a Model 712 WISP autosampler), (4) an analytical column (Supelco 3-μm C18 column, 150 × 4.6 mm I.D.) protected by a guard column (Supelco 40-μm C_{18} column, 50 × 4.6 mm I.D.), (5) a Model 2475 Multi λ fluorescence detector (or a Model 420-AC fluorescence detector), and (6) a Millenium-32 Workstation or a Model 810 Baseline Workstation (Waters Inc., Milford, MA) (Wu and Meininger 2008). Fluorescence was monitored at excitation and emission wavelengths of 340 and 455 nm, respectively. All mobile phase solutions were degassed with helium for at least 30 min before use. To prevent accumulation of salts in the HPLC system, both columns and pumps were thoroughly washed with water for at least 20 min and then with methanol for at least 20 min before and after running a sample set. Each sample set should not contain more than 34 vials (including water blanks, gizzerosine/AA standards, and samples) to ensure the chemical stability of the OPA-N-acetylcysteine reagent solution. Additionally, the autosampler was purged with water twice before starting the run of each sample set.

11.2.5 Mobile Phase Solutions

Mobile phase A (0.1 M sodium acetate, pH 7.2; Solvent A): Add 27.3 g sodium acetate trihydrate to 1.6 L H_2O in a glass bottle, followed by the sequential addition of 96 μL of 6 M HCl, 180 mL methanol, and 10 mL tetrahydrofuran (Wu and Meininger 2008). Mixing of the solution was performed after each addition. Adjust the final volume of the solution to 2 L with H_2O. Mobile phase B (Solvent B) was 100% methanol in a brown glass bottle.

11.2.6 HPLC Analysis

(a) Preparation of standard or sample vials: To a micro-insert tube within a 4-mL glass vial, add the following:

10 μL of 1.2% benzoic acid

Table 11.2 Solvent gradient for the analysis of gizzerosine

Time (min)	Solvent A (%)	Solvent B (%)
0	86	14
8	86	14
10	0	100
12	0	100
12.1	86	14
20	86	14

Combined flow rate = 1.1 ml/min
Solvent A = 0.1 M sodium acetate buffer (pH. 7.2) containing 9% methanol and 0.5% tetrahydrofuran; Solvent B = 100% methanol

10 μL of saturated $K_2B_4O_7$
10 μL of gizzerosine standard (0.1 to 100 μM) or 10 μL of sample
140 μL of HPLC H_2O

(b) Cap the tubes and vortex them before placing them onto an autosampler (one vial containing a reference standard at both the start and the end, with samples in the middle).
(c) Conditioning of HPLC columns. Before running a sample set, the guard column and the connecting analytical column are washed sequentially with water, methanol, and Solvent A for 15, 20, and 20 min, respectively, at a flow rate of 1.1 mL/min.
(d) The autosampler was programmed to mix 15 μL of a standard or sample solution with 15 μL of the OPA reagent solution for 1 min in the in-line derivatization loop.
(e) After the AA-OPA mixtures were injected into the HPLC column, a solvent gradient (a total running time of 20 min, including the time for column regeneration between samples) as given in Table 11.2 is started, with the combined flow rate for mobile phases A and B being 1.1 mL/min.
(f) Integration of AA peaks. Chromatographic peaks for gizzerosine and natural AAs were identified and integrated by the workstation software (Li and Wu 2022; Wu and Meininger 2008).

11.3 Results and Discussion

11.3.1 Analysis of Gizzerosine Standard

As reported for natural AAs (Wu and Meininger 2008), gizzerosine reacted readily with OPA at room temperatures (20–25 °C) in the presence of *N*-acetylcysteine (instead of the usual 2-mercaptoethanol or ethanethiol reagent) to yield a fluorescent derivative for separation by HPLC and detection by a fluorometer (340 nm excitation wavelength and 455 nm emission wavelength). We found that when gizzerosine was derivatized with OPA in the presence of 2-mercaptoethanol under the usual conditions (Li and Wu 2020; Wu and Meininger 2008), a total running time of 49 min was required to separate gizzerosine from all natural AAs present in plasma, proteins, and foodstuffs. Notably, the use of *N*-acetyl-cysteine in the OPA derivatization solution greatly facilitated the separation of gizzerosine from all natural AAs, thereby reducing the time for the analysis. A representative HPLC chromatogram for gizzerosine is illustrated in Fig. 11.2. The total running time (including column regeneration) for a sample was 20 min. The detection limit (indicated by a signal to noise ratio of 3) was 6 pmol/ml in an assay solution (the HPLC vial). Because only 15 μL of the solution was injected onto the HPLC column, the amount of gizzerosine detected was 0.09 pmol per injection into the HPLC column. The analysis of gizzerosine was linear between 1 and 100 pmol per injection ($R^2 = 0.998$, $n = 6$). When the volume of gizzerosine standard was 150 μl rather than 10 μl of gizzerosine standard and 140 μl water that were added to the assay solution in the HPLC vial, the detection limit for gizzerosine standard was improved to 0.4 pmol/ml in an assay solution (the HPLC vial) or 0.006 pmol per injection into the HPLC column. Thus, we succeeded at developing a simple, rapid, sensitive, and automated HPLC method for analysis of gizzerosine.

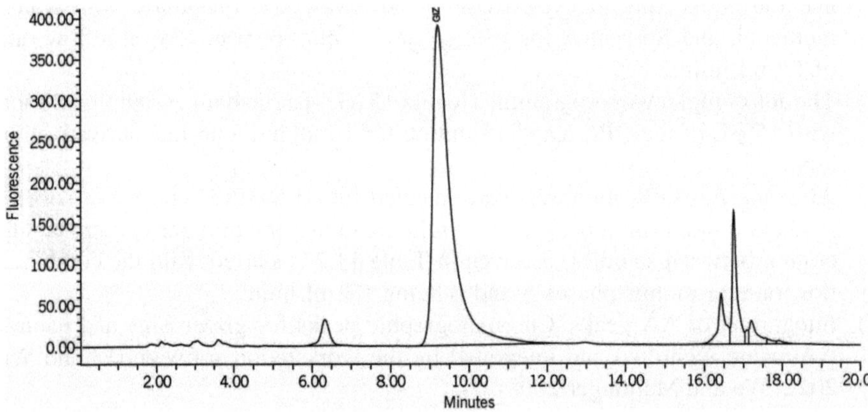

Fig. 11.2 An HPLC chromatogram for the analysis of gizzerosine. Gizzerosine standard (100 nmol/ml) reacted with *o*-phthaldialdehyde (OPA; also known as *o*-phthalaldehyde) in the presence of *N*-acetylcysteine to a highly yield fluorescent derivative. The gizzerosine-OPA derivative was separated on a reversed-phase C_{18} column and detected at 340 nm excitation wavelength and 455 nm emission wavelength. Giz = Gizzerosine

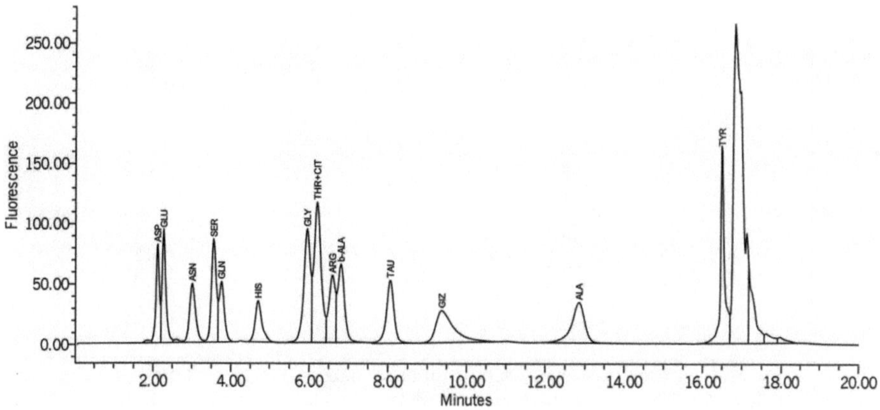

Fig. 11.3 An HPLC chromatogram for the analysis of gizzerosine and a mixture of standard amino acids (Ala, β-alanine, Arg, Asn, Asp, citrulline, Cys, Gln, Glu, Gly, His, Ile, Leu, Lys, Met, ornithine, Phe, Pro, 4-OH-Pro, Ser, Thr, Trp, Tyr, Val, and taurine). Gizzerosine standard (10 nmol/ml for each) and a mixture of standard amino acids (10 nmol/ml for each amino acid) reacted with *o*-phthaldialdehyde (OPA) in the presence of *N*-acetylcysteine to highly yield fluorescent derivatives. The OPA derivatives were separated on a reversed-phase C_{18} column and detected at 340 nm excitation wavelength and 455 nm emission wavelength. Giz = Gizzerosine

11.3.2 Separation of Gizzerosine from Physiological AAs

When gizzerosine was mixed with all natural AAs found in plasma and proteins, gizzerosine was well separated from the physiological AAs present in foodstuffs such as fishmeal (Fig. 11.3). When gizzerosine was absent from the mixture of the natural AA standards, gizzerosine did not appear on the HPLC chromatogram (Fig. 11.4). The lack of interference with gizzerosine analysis by natural AAs present in foods (Li et al. 2021; Li and Wu 2022, 2023a, b) indicates the specificity and the good performance of our HPLC method.

11.3.3 Separation of Gizzerosine from Natural AAs in Foodstuffs

When gizzerosine was added to foodstuff samples such as fishmeal and chicken meat-bone meal, gizzerosine was well separated from all natural AAs in the foodstuffs. HPLC chromatograms for the separation of added gizzerosine from natural AAs in acid extracts from fishmeal and chicken meat-bone meal are shown in Figs. 11.5 and 11.6, respectively. Gizzerosine was stable under strong acid (e.g., 1.5 M $HClO_4$ and 6 M HCl) and high temperatures (e.g., 110 °C) conditions for 24 h. The detection limit for gizzerosine in foodstuffs was at least 875 pmol/g foodstuff or at least 0.21 mg/kg foodstuff (i.e., 0.21 ppm). The detection limit for gizzerosine in foodstuffs

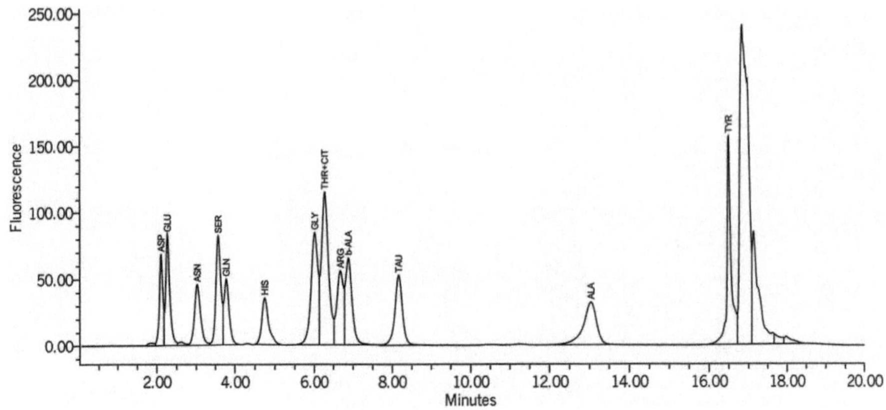

Fig. 11.4 An HPLC chromatogram for the analysis of a mixture of amino acids (Ala, β-alanine, Arg, Asn, Asp, citrulline, Cys, Gln, Glu, Gly, His, Ile, Leu, Lys, Met, ornithine, Phe, Pro, 4-OH-Pro, Ser, Thr, Trp, Tyr, Val, and taurine) without gizzerosine. A mixture of standard amino acids (10 nmol/ml for each amino acid) without gizzerosine reacted with *o*-phthaldialdehyde (OPA) in the presence of *N*-acetylcysteine to highly yield fluorescent derivatives. The OPA derivatives were separated on a reversed-phase C_{18} column and detected at 340 nm excitation wavelength and 455 nm emission wavelength. Gizzerosine did not appear on the HPLC chromatogram

can be further improved by proportionately reducing the volumes of the H_2O, acid solution, and base solution used for the extraction of gizzerosine, or by adding a greater volume of sample at the expense of water in the HPLC vial. Using gizzerosine standard, the rates of its recovery from water, ethanol, and 6 M HCl hydrolysates were 97.8 ± 0.73%, 74.2 ± 1.5%, and 95.6 ± 1.2% (mean ± SEM, n = 5), respectively. Gizzerosine was stable in deionized-distilled water, ethanol, 1.5 M $HClO_4$, and acid hydrolysates at room temperatures (20–25 °C) for at least 5 days.

When a known amount of gizzerosine standard was added to foodstuffs, the precision (agreement between replicate measurements) of the HPLC method, as evaluated by the relative deviation (mean of absolute deviation/mean of replicate measurements × 100%; Wu and Meininger 2008), was 1.7–2.0%. The accuracy (the nearness of an experimental value to the true value) of the HPLC method, as determined with the addition of a known amount of gizzerosine standard to foodstuffs and expressed as the relative errors [(measurement value – true value)/true value × 100%; Wu and Meininger 2008)], was 1.8–2.3% for the above biological samples (n = 5).

11.3.4 Analysis of Gizzerosine in Foodstuffs

Using water and acid hydrolysis extraction methods, gizzerosine was not detected in seven U.S. poultry meal samples, three fishmeal samples, three U.S. soybean meal samples, and three U.S. corn grain samples. The detection limit for gizzerosine in all these foodstuffs was at least 0.21 mg/kg feedstuff, as indicated previously.

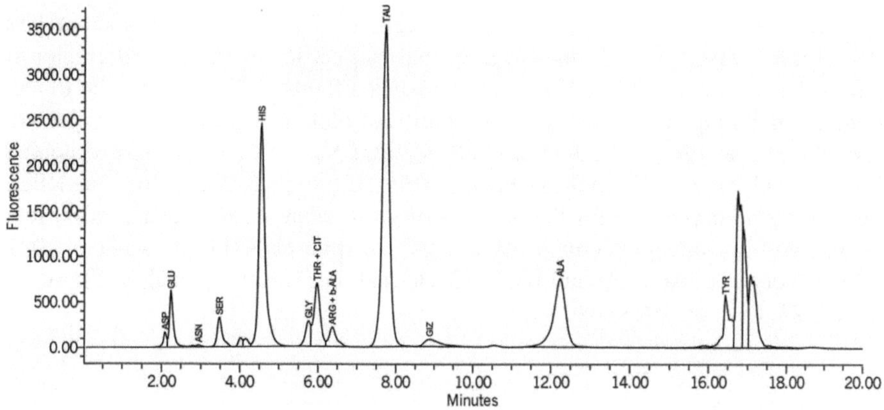

Fig. 11.5 An HPLC chromatogram for the analysis of amino acids in gizzerosine-containing fishmeal. A fishmeal sample (0.1 g) was mixed with 1 ml of 1.5 M $HClO_4$ and 1 ml of 1000 nmol/ml gizzerosine standard. After vortexing for 1 min, 0.5 ml of 2 M K_2CO_3 was added to the mixture for neutralization. The solution was centrifuged at 600 × *g* for 10 min, and the supernatant fluid was used for the analysis of gizzerosine and natural amino acids involving the pre-column derivatization with *o*-phthaldialdehyde (OPA) in the presence of *N*-acetylcysteine to yield highly fluorescent derivatives. The OPA derivatives were separated on a reversed-phase C_{18} column and detected at 340 nm excitation wavelength and 455 nm emission wavelength

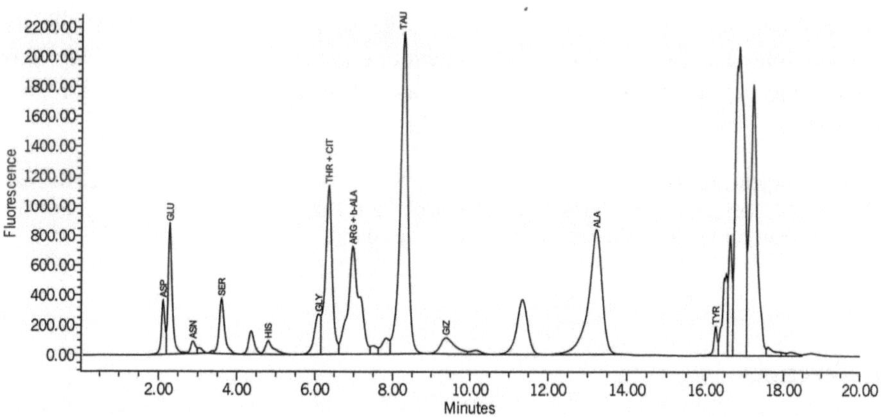

Fig. 11.6 An HPLC chromatogram for the analysis of amino acids in gizzerosine-containing chicken meat and bone meal. A sample (0.1 g) of chicken meat and bone meal was mixed with 1 ml of 1.5 M $HClO_4$ and 1 ml of 1000 nmol/ml gizzerosine standard. After vortexing for 1 min, 0.5 ml of 2 M K_2CO_3 was added to the mixture for neutralization. The solution was centrifuged at 600 × *g* for 10 min, and the supernatant fluid was used for the analysis of gizzerosine and natural amino acids involving the pre-column derivatization with *o*-phthaldialdehyde (OPA) in the presence of *N*-acetylcysteine to yield highly fluorescent derivatives. The OPA derivatives were separated on a reversed-phase C_{18} column and detected at 340 nm excitation wavelength and 455 nm emission wavelength

The detection limit for gizzerosine in foodstuffs (at least 0.21 ppm) is about 10% of its lowest dose (i.e., 2.2 ppm in diet) that can cause severe gizzard erosion in chicks (Okazaki et al. 1983). Thus, these analyzed foodstuffs would be safe to feed animals, including cats and dogs. Biosecurity of diets is essential for the growth, development, health, and production of all animals (Wu 2018). Because physiological AAs are well separated from gizzerosine (Figs. 11.3, 11.5, and 11.6), our HPLC method is also applicable for the analysis of gizzerosine in the plasma, milk, fetal fluids, gastrointestinal contents, cells, and tissues of animals (Furukawa et al. 2021; Gilbreath et al. 2020a, b; He and Wu 2022; He et al. 2023; Herring et al. 2022; Rezaei et al. 2022; Li et al. 2023a, b).

11.4 Conclusion

We developed an improved method for the analysis of gizzerosine by HPLC involving pre-column derivatization with OPA and fluorescence detection. There was no deed for the cleanup of food samples or derivatization products. The time for the assay was reduced from 49 to 20 min per sample through the use of *N*-acetyl-cysteine instead of 2-mercaptoethanol or ethanethiol for the pre-column derivatization of gizzerosine with OPA. Gizzerosine was well separated from all natural AAs present in physiological fluids (including plasma) and proteins. This simple, rapid, sensitive, specific, reproducible, accurate, and automated technique is very useful for the analysis of gizzerosine in foodstuffs (including fishmeal and meat-bone meal) and biological samples.

Acknowledgements This work was supported by North American Renderers Association and Texas A&M AgriLife Research (H-8200). We thank Prof. Hiromasa Kiyota (Tohoku University, Sendai, Japan) for kindly providing gizzerosine for our study.

References

Aldrich G (2018) Fish meal: supply, preservation biggest challenges for dog and cat diets. Petfood Industry. https://www.petfoodindustry.com/ingredient-issues/proteins/article/15464537/fish-meal-supply-preservation-biggest-challenges-for-dog-and-cat-diets. Accessed Oct 19 2023

Becker MI, Carrasco I, Beltran C, Torres M, Jaureguiberry B, Ioannes AED (1998) Development of monoclonal antibodies to gizzerosine, a toxic component present in fish meal. Hybridoma 17:373–381

Dai ZL, Wu ZL, Jia SC, Wu G (2014a) Analysis of amino acid composition in proteins of animal tissues and foods as pre-column *o*-phthaldialdehyde derivatives by HPLC with fluorescence detection. J Chromatogr B 964:116–127

Dai ZL, Wu ZL, Wang JJ, Wang XQ, Jia SC, Bazer FW, Wu G (2014b) Analysis of polyamines in biological samples by HPLC involving pre-column derivatization with *o*-phthalaldehyde and *N*-acetyl-L-cysteine. Amino Acids 46:1557–1564

Furukawa K, He WL, Bailey CA, Bazer FW, Toyomizu M, Wu G (2021) Polyamine synthesis from arginine and proline in tissues of developing chickens. Amino Acids 53:1739–1748

Funaba M, Tanak T, Kaneko M, Iriki T, Hatano Y, Abe M (2001) Fish meal vs. corn gluten meal as a protein source for dry cat food. J Vet Med Sci 63:1355–1357

Gilbreath KR, Nawaratna GI, Wickersham TA, Satterfield MC, Bazer FW, Wu G (2020a) Metabolic studies reveal that ruminal microbes of adult steers do not degrade rumen-protected or unprotected L-citrulline. J Anim Sci 98:skz370

Gilbreath KR, Bazer FW, Satterfield MC, Cleere JJ, Wu G (2020b) Ruminal microbes of adult sheep do not degrade extracellular L-citrulline. J Anim Sci 98:skaa164

Gjevre AG, Kaldhusdal M, Eriksen GS (2013) Gizzard erosion and ulceration syndrome in chickens and turkeys: a review of causal or predisposing factors. Avian Pathol 42:297–303

He WL, Wu G (2022) Oxidation of amino acids, glucose, and fatty acids as metabolic fuels in enterocytes of developing pigs. Amino Acids 54:1025–1039

He WL, Posey EA, Steele CC, Savell JW, Bazer FW, Wu G (2023) Dietary glycine supplementation enhances postweaning growth and meat quality of pigs with intrauterine growth restriction. J Anim Sci 101:skadskad354

Herring CM, Bazer FW, Johnson GA, Seo H, Hu SD, Elmetwally M, He WL, Long DB, Wu G (2022) Dietary supplementation with 0.4% L-arginine between days 14 and 30 of gestation enhances NO and polyamine syntheses and water transport in porcine placentae. J Anim Sci Biotechnol 13:134

Hou YQ, Jia SC, Nawaratna G, Hu SD, Dahanayaka S, Bazer FW, Wu G (2015) Analysis of L-homoarginine in biological samples by HPLC involving pre-column derivatization with *o*-phthalaldehyde and *N*-acetyl-L-cysteine. Amino Acids 47:2005–2014

Hou YQ, Li XL, Dai ZL, Wu ZL, Bazer FW, Wu G (2018) Analysis of glutathione in biological samples by HPLC involving pre-column derivatization with o-phthalaldehyde. Methods Mol Biol 1694:105–115

Hu SD, He WL, Bazer FW, Johnson GA, Wu G (2023) Synthesis of glycine from 4-hydroxyproline in tissues of neonatal pigs. Exp Biol Med 248:1206–1220

Ito Y, Noguchi T, Naito H (1985) Fluorometric determination of gizzerosine, a histamine H_2-receptor agonist discovered in feedstuffs, employing high-performance liquid chromatography. Anal Biochem 151:28–31

Li P, Wu G (2020) Composition of amino acids and related nitrogenous nutrients in feedstuffs for animal diets. Amino Acids 52:523–542

Li P, Wu G (2022) Functional molecules of intestinal mucosal products in animal nutrition and health. Adv Exp Med Biol 1354:263–277

Li P, Wu G (2023a) Amino acid nutrition and metabolism in domestic cats and dogs. J Anim Sci Biotechnol 14:19

Li P, Wu G (2023b) Composition of nutrients in rendered animal-sourced feedstuffs. In: Phillips CJC (ed) encyclopedia of animal nutrition, 2nd edn. CABI, Wallingford, pp 576–579

Li P, He WL, Wu G (2021) Composition of amino acids in foodstuffs for humans and animals. Adv Exp Med Biol 1332:189–210

Li XL, Bazer FW, Johnson GA, Burghardt RC, Wu G (2023a) Dietary supplementation with L-citrulline improves placental angiogenesis and embryonic survival in gilts. Exp Biol Med 248:702–711

Li XY, He WL, Wu G (2023b) Dietary glycine supplementation enhances the growth performance of hybrid striped bass (*Morone saxatilis* ♀× *Morone chrysops* ♂) fed soybean meal-based diets. J Anim Sci 101:skad345

Manosalva H, Ioannes AED, Becker MI (2004) Development of monoclonal antibodies bearing the internal image of the gizzerosine epitope and application in a competitive ELISA for fish meal. Hybrid Hybridomics 23:45–54

Masumura T, Sugahara M, Noguchi T, Mori K, Naito H (1985) The effect of gizzerosine, a recently discovered compound in overheated fish meal, on the gastric acid secretion in chicken. Poult Sci 64:356–361

Murakita H, Gotoh T (1990) Phthalaldehyde post-column derivatization for the determination of gizzerosine in fish meal by high performance liquid chromatography. J Chromatogr A 515:527–530
Ohta Y, Ohashi H, Enomoto S, Machida Y (1988) Simple measurement of gizzerosine in fish meals by high-performance liquid chromatography. Agric Biol Chem 52:2817–2821
Okazaki T, Noguchi T, Igarashi K, Sakagami Y, Seto H, Mori K, Naito H, Masumura T, Sugahara M (1983) Gizzerosine, a new toxic substance in fish meal, causes severe gizzard erosion in chicks. Agric Biol Chem 47:2949–2952
Rezaei R, San Gabriel A, Wu G (2022) Dietary supplementation with monosodium glutamate enhances milk production by lactating sows and the growth of suckling piglets. Amino Acids 54:1055–1068
Rosselot G, Lopez-Lastra M, McMurtry JP (1996) Determination of gizzerosine activity in fish meal with a homologous radioimmunoassay. Poult Sci 75:873–880
Shimasaki Y, Kiyota H, Sato M, Kuwahara S (2005) Facile synthesis of (S)-gizzerosine, a potent inducer of gizzard erosion in chicks, using successive zinc-mediated and palladium-catalyzed coupling reactions. Synthesis 3191–3192
Sugahara M, Hattori T, Nakajima T (1992) Effect of heating hours and temperature for the production of gizzerosine in fishmeal. Anim Sci Technol (jpn) 63:1028–1031
Tao Z, Sato M, Wu K, Kiyota H, Yamaguchi T, Nakano T (2012) A simple, rapid method for gizzerosine analysis in fish meal by paper electrophoresis. Fish Sci 78:923–926
Tao ZH, Hu QX, Xu XJ, Kiyota H, Chen ZX, Xie SY, Qiao N (2019) Analytical method to evaluate gizzerosine in fishmeal after diazonium derivatization using high-performance liquid chromatography. Food Anal Methods 12:331–337
Torres T, Manosalva H, Carrasco I, Ioannes AED, Becker MI (1999) Procedure for radiolabeling gizzerosine and basis for a radioimmunoassay. J Agric Food Chem 47:4231–4236
Uhlig S, Kaldhusdal M, Gjevre A-G (2011) LC–MS–MS determination of gizzerosine in poultry feed. Chromatographia 74:133–138
Wagener WWD, Koch KR, Wessels JPH, Post BJ (1991) An investigation into the occurrence of the toxic amino acid gizzerosine in fish meal. J Sci Food Agric 54:147–152
Wu G (2018) Principles of animal nutrition. CRC Press, Boca Raton, Florida
Wu G (2022) Amino acids: biochemistry and nutrition, 2nd edn. CRC Press, Boca Raton, Florida
Wu G, Meininger CJ (2008) Analysis of citrulline, arginine, and methylarginines using high-performance liquid chromatography. Methods Enzymol 440:177–189
Xiao M-W, Bai X-L, Xu P-L, Zhao Y, Yang L, Liu Y-M, Liao X (2017) Rapid determination of gizzerosine in fish meals using microchip capillary electrophoresis with laser-induced fluorescence detection. Food Addit Contam A 34:760–765

Index

G. Wu (ed.), *Nutrition and Metabolism of Dogs and Cats*, Advances in Experimental Medicine and Biology 1446, https://doi.org/10.1007/978-3-031-54192-6

MIX
Papier aus verantwortungsvollen Quellen
Paper from responsible sources
FSC® C105338

If you have any concerns about our products,
you can contact us on
ProductSafety@springernature.com

In case Publisher is established outside the EU,
the EU authorized representative is:
Springer Nature Customer Service Center GmbH
Europaplatz 3, 69115 Heidelberg, Germany

Printed by Libri Plureos GmbH
in Hamburg, Germany